William Alexander Talbot

Systematic List of the Trees, Shrubs and Woody-Climbers of the

Bombay Presidency

William Alexander Talbot

Systematic List of the Trees, Shrubs and Woody-Climbers of the Bombay Presidency

ISBN/EAN: 9783337225919

Printed in Europe, USA, Canada, Australia, Japan

Cover: Foto ©berggeist007 / pixelio.de

More available books at **www.hansebooks.com**

OF THE

TREES, SHRUBS AND WOODY-CLIMBERS

OF THE

BOMBAY PRESIDENCY

BY

W. A. TALBOT, F.L.S.,

DEPUTY CONSERVATOR OF FORESTS.

———:⊃:———

Bombay:
PRINTED AT THE GOVERNMENT CENTRAL PRESS.

1894.

PREFACE.

THE following list contains the botanical names and principal synonyms of the trees, shrubs and woody climbers, indigenous in the Bombay Presidency and Sind. Vernacular names have as far as possible been added. A synopsis of the natural orders and keys to the genera and species will assist Forest and other officers to refer to the various authors quoted, for detailed information. The sequence of the genera and species in the Flora of British India of Hooker has been followed throughout this list. The Fl. Br. I. is now nearly complete, as far as regards the trees and woody plants of India; only the palms and bamboos remain to be done. Speci-mens of most of the species, mentioned in this list, have been sent to the herbarium of the Botanic Gardens of Calcutta; whilst many of the critical species have been referred to Kew for deter-mination. The Konkan is the moist region of varying breadth, lying along the coast from Bombay southwards. The adjacent hills or ghāts, with a heavy rainfall, are referred to as the Konkan ghāts. The Deccan includes Khándesh and the other dry districts of the presidency. Sind and North Kánara are referred to separately by name. North Kánara is a very important forest region of three thousand square miles of more or less continuous jungle. A num-ber of plants not hitherto found in this Presidency, also several new species, are included in this list, which it is hoped may form the framework of a future Forest Flora of the Bombay Presidency.

LIST

OF THE

PRINCIPAL BOOKS AND PAPERS REFERRED TO.

BALFOUR. Timber Trees of India, Madras, 1870.

BEDDOME. Flora Sylvatica of the Madras Presidency, 1869-73.

BENTHAM AND HOOKER. Genera Plantarum.

BIRDWOOD. Catalogue of the Flora of Mahabeleshwnr and Matheran. Bo. Nat. Hist. Soc. Jour. Vol. 2.

BRANDIS. Forest Flora of North-West and Central India.

DALZELL AND GIBSON. Bombay Flora, 1861.

DE CANDOLLE. Monographic Phanerogamarum, Vols. 1-4-5.

———————Prodromus Systematis Naturalis Regni Vegetabilis, 1824-73.

DYMOCK. Marathi Names of Plants, Bo. Nat. Hist. Soc. Jour., Vol. 2.

GAMBLE. A Manual of Indian Timbers, 1881.

GRAHAM. Catalogue of the Plants growing in Bombay and its vicinity, 1839.

GRIFFITH. Palms of British India, 1850.

HOOKER. Flora of British India, Vols. 1 to 6.

KURZ. Forest Flora of British Burmah, 1877.

MUNRO. Monograph of the Bambusaceæ. Trans. Linn. Soc., 1868.

ROXBURGH. Flora Indica, 1832, reprint, 1874.

WIGHT. Icones Plantarum Indiæ Orientalis, 1840-1853.

———Prodromus Floræ Peninsulæ Indiæ Orientalis, 1834.

KEY TO NATURAL ORDERS.

CLASS I. DICOTYLEDONS.

Pith surrounded with concentric layers of wood and bark. Leaves net veined. Flowers often 5-merous. Embryo with 2 cotyledons; radicle forming a tap-root.

DIVISION 1. ANGIOSPERMS.

Ovules in a closed vessel and fertilized through a stigma. Embryo with 2 cotyledons.

SUB-CLASS I. POLYPETALÆ.

Calyx and corolla present. Petals several, distinct, rarely wanting.

A. Thalamifloræ.—Torus small, rarely disciform. Ovary superior. Stamens hypogynous.

1. *Ranunculaceæ.*—Herbs or climbing shrubs. Leaves alternate or opposite. Sepals deciduous. Stamens indefinite. Seeds without an arillus.

2. *Dilleniaceæ.*—Trees or shrubs. Leaves alternate, simple; lateral veins prominent. Sepals persistent. Carpels usually distinct. Seeds with an arillus.

3. *Magnoliaceæ.*—Shrubs or trees. Leaves alternate, simple; sepals and petals trimerous, imbricated, deciduous. Stamens indefinite. Carpels distinct. Albumen uniform.

4. *Anonaceæ.*—Trees or shrubs, often scandent. Leaves alternate, simple. Stipules 0. Sepals 3, petals 6, in 2 series of 3 each. Stamens indefinite. Carpels distinct in fruit (united in *Anona*). Albumen ruminate.

5. *Menispermaceæ.*—Climbing or twining shrubs. Leaves alternate. Stipules 0. Flowers small, unisexual, deciduous, usually trimerous Stamens definite, free, or connate, opposite the petals. Embryo usually curved.

6. *Capparideæ.*—Trees or shrubs often scandent or herbs. Leaves alternate. Petals 4. Stamens indefinite. Ovary stalked, seeds on parietal placentas. Albumen 0.

7. *Resedaceæ.*—Herbs or shrubs. Leaves entire or pinnatisect. Flowers bracteate. Disc usually conspicuous (0 in *Oligomeris*). Stamens numerous. Fruit a capsule or berry. Seeds numerous.

8. *Bixineæ.*—Trees or shrubs. Leaves simple, alternate. Sepals 4 or 5. Stamens indefinite, free or connate. Seeds few or many, on parietal placentas. Albumen fleshy.

9. *Pittosporeæ.*—Trees or shrubs. Leaves simple, alternate exstipulate. Sepals unequal. Stamens 5. Embryo minute. Seeds albuminous.

10. *Tamariscineæ.*—Small trees or shrubs. Leaves alternate, scalelike. Flowers regular, bisexual; sepals and petals each 4, 6. Stamens as many as petals or twice as many. Ovary 1-celled. Seeds tufted.

11. *Guttiferæ.*—Resinous trees, juice often coloured. Leaves coriaceous, opposite, without stipules. Flowers unisexual or polygamous. Calyx imbricate. Stamens indefinite, free or connate. Ovary syncarpous.

12. *Ternstrœmiaceæ.*—Trees or shrubs. Leaves alternate. Flowers regular, bisexual. Stamens indefinite, free or connate at base. Ovary 3-5-celled. Embryo oily.

13. *Dipterocarpeæ.*—Trees (1 genus climbing shrubs). Leaves alternate, entire, often with large stipules. Flowers hermaphrodite. Calyx-segments usually enlarged in fruit. Cotyledons thick. Albumen 0.

14. *Malvaceæ.*—Herbs, shrubs or trees. Leaves alternate, stipulate, simple, often palmi-nerved. Stamens numerous, monadelphous; anthers 1-celled. Ovary syncarpous with axile placentation. Fruit a dehiscent capsule.

15. *Sterculiaceæ.*—Trees, shrubs or herbs. Leaves alternate, simple, or digitate. Stamens monadelphous or free; anthers 2-celled.

16. *Tiliaceæ.*—Trees, shrubs or herbs. Leaves alternate, simple with deciduous stipules. Stamens indefinite, usually free; anthers 2-celled.

B. Disciflorae.—Torus usually disciform, free or adnate to the calyx or ovary or to both, rarely absent. Ovary superior or more or less immersed in the disc.

17. *Lineæ.*—Herbs or shrubs, rarely trees. Leaves simple, alternate, stipulate. Stamens definite, more or less connate below. Disc glandular or absent. Styles free or connate. Albumen fleshy.

18. *Malpighiaceæ.*—Climbing shrubs. Leaves opposite, entire. Flowers bisexual, pentamerous. Stamens 10. Ovary 3-lobed. Fruit winged.

19. *Geraniaceæ.*—Herbs or shrubs, rarely trees. Leaves simple or compound. Stamens definite. Ovary angular or lobed.

20. *Rutaceæ.*—Trees or shrubs. Leaves simple or compound, glandular-dotted. Disc annular thick. Stamens as many or twice as many as petals, rarely indefinite.

21. *Simarubeæ.*—Trees or shrubs with bitter bark. Leaves alternate, usually compound. Flowers small. Filaments generally hairy. Ovary lobed.

22. *Ochnaceæ.*—Shrubs or trees. Leaves alternate, simple, shining. Flowers bisexual, regular. Anthers linear. Ovary deeply lobed. Fruit of 3-10 distinct 1-seeded drupes.

23. *Burseraceæ.*—Trees or shrubs with resinous or balsamic juice. Leaves alternate, compound. Stamens 8-10. Ovary entire. Seeds 1 or few.

24. *Meliaceæ*—Trees or shrubs. Leaves usually compound, alternate. Flowers panicled. Stamens 8-10, monadelphous (except in *Cedrela* and *Chloroxylon*). Ovary entire.

25. *Chailletiaceæ.*—Trees or shrubs. Leaves simple, alternate. Stipules present. Petals 2-cleft. Ovary entire.

26. *Olacineæ.*—Trees or shrubs sometimes climbing. Leaves simple, alternate. No stipules. Petals free or connate usually valvate. Fruit 1-seeded. Seeds albuminous.

27. *Ilicineæ.*—Trees or shrubs with alternate simple leaves. Petals free, imbricate. Ovary 3-6-celled. Seeds albuminous.

28. *Celastraceæ.*—Trees or shrubs. Leaves simple, opposite or alternate. Stamens 5 alternating with the petals or sometimes only 3. Ovary more or less immersed in the disc, entire or lobed.

29. *Rhamnaceæ.*—Trees or shrubs. Leaves simple, alternate or opposite. Stamens opposite the petals and equal in number. Ovary entire, often inferior.

30. *Ampelideæ.*—Shrubs or herbs with jointed stems, often scandent. Leaves alternate, simple or compound. Calyx small. Petals valvate. Stamens as many as petals and opposite to them.

31. *Sapindaceæ.*—Usually trees. Leaves alternate, compound or simple. Flowers polygamous, often irregular. Stamens free, often anisomerous.

32. *Sabiaceæ.*—Trees or shrubs. Leaves alternate, simple or compound. Stamens often unequal in size and opposite the petals.

33. *Anacardiaceæ.*—Trees often resinous. Leaves alternate, simple or compound. Stipules 0. Flowers small. Ovary 1-celled (in *Spondias* 2-5-celled). Cells 1-ovuled. Fruit a drupe.

C. **Calciflorae.**—Calyx gamosepalous. Stamens usually perigynous or epigynous, definite or indefinite. Ovary superior or inferior.

34. *Moringeæ.*—Trees. Leaves alternate, bi-tri-pinnate. Flowers bisexual. Petals unequal. Stamens 10, 5 fertile opposite to the petals alternating with 5 sterile. Capsule 3-valved, pod-like.

35. *Connaraceæ.*—Trees or shrubs. Leaves 1-3-foliate or pinnate. Flowers regular. Stamens definite. Carpels hirsute, 1-5, free.

36. *Leguminosæ.*—Trees, shrubs or herbs. Leaves usually alternate, compound. Flowers irregular (regular in *Mimoseæ*). Carpel 1, free.

37. *Rosaceæ.*—Trees, shrubs or herbs. Leaves alternate, simple or compound. Flowers regular. Ovary of 1 or more free or combining carpels, often adherent to the calyx-tube.

38.   *Rhizophoracew.*—Trees or shrubs. Leaves opposite, coriaceous. Stipules deciduous. Flowers regular. Petals usually fringed. Stamens twice as many as petals or more. Ovary usually inferior.

39.   *Combretaceæ.*—Trees or shrubs. Leaves simple, alternate or opposite. No stipules. Ovary inferior 1-celled. Fruit winged or angled, 1-seeded. Albumen 0.

40.   *Myrtaceæ.*—Trees or shrubs. Leaves usually opposite, translucent, glandular-dotted, entire. Stamens indefinite. Ovary inferior, with axile placentation.

41.   *Melastomaceæ.*—Herbs or shrubs. Leaves opposite, entire, without stipules. Petals twisted in bud. Stamens 10 or fewer, perigynous. Anthers opening at the apex with pores or slits. Ovary inferior. Seeds exalbuminous.

42.   *Lythraceæ.*—Trees, shrubs or herbs. Leaves simple, opposite or alternate. No stipules. Calyx free, lobes valvate in bud. Stamens definite or indefinite; anthers dehiscing longitudinally. Fruit a capsule. Seeds exalbuminous.

43.   *Samydaceæ.*—Trees or shrubs. Leaves alternate, distichous. Stipules deciduous. Flowers regular. Stamens definite or indefinite. Ovary free. Fruit a 1-celled capsule. Albumen fleshy.

44.   *Datiscaceæ.*—Trees rarely herbs. Leaves alternate. Stipules 0. Flowers unisexual, petals often absent. Ovary inferior often open at the apex. Placentas and styles as many as calyx-lobes. Seeds numerous, minute.

45.   *Cacteæ.*—Shrubs or herbs, with succulent stems and minute leaves. Flowers large bisexual. Stamens indefinite. Fruit fleshy.

46.   *Araliaceæ.*—Trees, shrubs or herbs. Leaves alternate, simple or compound. Flowers umbellate or capitate, petals caducous. Ovary inferior.

47.   *Cornaceæ.*—Trees or shrubs. Leaves entire, usually opposite. Stipules 0. Flowers regular. Ovary inferior. Fruit succulent, 2-celled. Albumen fleshy.

Sub-Class II. GAMOPETALÆ.

Calyx and and corolla present.   Petals united (gamopetalous).

A.   Ovary inferior (perigynous *gamopetalæ*).

48.   *Caprifoliaceæ.*—Shrubs or herbs, often climbing, rarely trees. Leaves opposite, simple or pinnate. Flowers regular or irregular. Fruit a berry or drupe.

49.   *Rubiaceæ.*—Trees, shrubs or herbs. Leaves opposite. Stipules free or connate. Flowers usually regular. Stamens as many as corolla-lobes, alternating with them. Ovary 2 or more celled. Cells 2 or many ovuled. Albumen fleshy or horny.

50.   *Compositæ.*—Herbs, shrubs, rarely trees. Leaves alternate or opposite. Flowers in involucrate heads. Anthers united in a tube round

the style. Calyx 0 or reduced to a pappus of hairs, scales or bristles. Ovary 1-celled, 1-ovuled.

51. *Goodenovieæ.*—Herbs or shrubs, rarely trees. Leaves alternate. Stipules 0. Stamens 5, free or connate in a ring. Ovary 1-2-celled. Seeds albuminous.

B. Ovary free (hypogynous *gamopetalæ*).

52. *Myrsinæ.*—Trees or shrubs. Leaves alternate. Stipules 0. Stamens as many as lobes of the corolla. Fruit fleshy or hard, usually indehiscent. 1 or few seeded.

53. *Sapotaceæ.*—Trees or shrubs, usually with milky juice. Leaves simple, alternate. Stipules 0. Stamens as many as corolla-lobes or in 2-3-series. Ovary 2 or more celled. Fruit 1-few seeded.

54. *Ebenaceæ*—Trees or shrubs. Leaves alternate, entire. No stipules. Flowers often diœcious. Anthers introrse. Fruit a berry. Seeds albuminous.

55. *Styraceæ.*—Trees or shrubs. Leaves alternate, simple. Stamens often indefinite. Ovary more or less inferior. Fruit 1-seeded. Albumen fleshy.

56. *Oleaceæ.*—Trees or shrubs, often scandent. Leaves opposite, entire or pinnate. Stipules 0. Stamens 2. Ovary 2-celled. Seeds with or without albumen.

57. *Salvadoraceæ.*—Trees or shrubs. Leaves opposite, entire. Stipules minute. Stamens 4. Ovary 2-celled. Fruit a 1-seeded berry.

58. *Apocynaceæ.*—Trees, shrubs or herbs, with often milky juice. Leaves opposite or whorled. Stamens 5, free. Fruit of 1-2-follicles, drupes or berries. Seeds often plumose.

59. *Asclepiadeæ.*—Shrubs or herbs, often climbing and with milky juice. Leaves opposite, entire. Stipules 0. Anthers connate round the stigma. Pollen in 1-2-waxy masses in each anther-cell. Fruit of 2 follicles. Seeds usually winged or plumose. Albumen copious.

60. *Loganiaceæ.*—Trees, shrubs or herbs. Leaves opposite, simple. Flowers regular, 4-5-merous. Ovary free, 2-celled. Albumen copious.

61. *Boragineæ.*—Trees or shrubs. Leaves alternate. Stipules 0. Inflorescence cymose. Flowers 4-5-merous. Ovary 2-4-celled, often 2-4-lobed. Fruit a drupe or dry, and separating into 2-4-nuts.

62. *Convolvulaceæ.*—Herbs or shrubs, often climbing. Leaves alternate. Stipules 0. Flowers large, regular, bisexual, pentamerous. Fruit capsular or succulent and indehiscent. Seeds few. Albumen 0 or scanty.

63. *Solanaceæ.*—Herbs or shrubs, rarely small trees. Leaves alternate. Stipules 0. Flowers regular, pentamerous. Fruit a berry or capsule. Seeds albuminous.

64. *Gesneraceæ.*—Herbs or shrubs. Leaves opposite or alternate. Flowers hermaphrodite, irregular. Fruit a capsule or berry. Albumen 0 or scanty.

65. *Bignoniaceæ.*—Trees, rarely erect or climbing shrubs. Leaves opposite, often compound, exstipulate. Flowers irregular. Stamens 4, in pairs or 2 only. Ovary 2-celled; ovules many. Fruit a capsule; seeds winged.

66. *Acanthaceæ.*—Herbs or shrubs. Leaves opposite, simple. Stipules 0. Flowers irregular. Fruit a capsule. Seeds usually supported on hooks of the placenta. Seeds often elastically hairy when wetted.

67. *Verbenaceæ.*—Trees, shrubs or herbs. Leaves opposite. Flowers irregular. Calyx gamosepalous, often enlarged in fruit. Ovary 4-celled. Cells 1-ovuled.

68. *Labiatæ.*—Herbs or shrubs. Leaves opposite. Flowers irregular. Stamens fewer than corolla-lobes; 2 or 4 in pairs. Ovary deeply lobed. Style gynobasic. Albumen 0 or scanty.

Sub-Class III. MONOCHLAMYDEÆ.

Perianth simple or 0 (double in some *Euphorbiaceæ*).

69. *Nyctaginaceæ.*—Herbs, shrubs or trees. Leaves usually opposite. Stipules 0. Flowers regular, hermaphrodite. Base of perianth persistent and enclosing the ovary and fruit. Seeds albuminous. Embryo curved.

70. *Polygonaceæ.*—Herbs or shrubs. Leaves simple, alternate with sheathing stipules. Flowers small, hermaphrodite or unisexual. Ovary superior. Ovule 1 erect. Seeds albuminous. Embryo straight or slightly curved.

71. *Myristicaceæ.*—Trees or shrubs. Leaves alternate, entire. Flowers small, diœcious. Stamens united in a column. Albumen ruminate. Fruit fleshy, 2-valved. Seed arillate.

72. *Laurineæ.*—Trees or shrubs. (*Cassytha* a parasite). Leaves simple, alternate or opposite. Flowers hermaphrodite or unisexual. Anther-cells opening by recurved valves. Ovary 1-celled. Ovule 1, pendulous. Albumen 0.

73. *Thymelaceæ.*—Trees or shrubs with tenacious bark. Leaves alternate or opposite. Flowers hermaphrodite. Stamens definite. Ovary free, 1-celled, 1-ovuled. Ovule pendulous. Albumen fleshy or 0.

74. *Elæagnaceæ.*—Trees or shrubs, often scandent and covered more or less with silvery scales. Leaves alternate or opposite entire. Flowers hermaphrodite. Base of perianth-tube persistent round the free 1-celled ovary. Ovule 1, erect.

75. *Loranthaceæ.*—Parasitic evergreen shrubs. Leaves coriaceous, entire, usually opposite; sometimes wanting. Flowers hermaphrodite. Ovary inferior. Ovule solitary, erect. Albumen green, fleshy.

76. *Santalaceæ.*—Trees, shrubs or herbs. Leaves entire, alternate or opposite. No stipules. Stamens opposite perianth-lobes. Ovary inferior, 1-celled. Fruit 1-seeded. Albumen fleshy.

77. *Euphorbiaceæ.*—Trees, shrubs or herbs with milky juice. Leaves alternate or opposite. Flowers unisexual. Perianth various, rarely double. Ovary free, usually 3-celled. Cells with 1-2 suspended ovules in each. Fruit a capsule or drupe. Seeds oily.

78. *Urticaceæ.*—Herbs, shrubs or trees. Leaves simple, usually alternate. Stipules present. Flowers unisexual. Stamens opposite the perianth-segments. Ovary free, usually 1-celled. Fruit 1-seeded.

79. *Casuarineæ.*—Trees with jointed branchlets. Leaves reduced to toothed sheaths at the nodes. Flowers monœcious, in spikes. Male flowers monandrous. Female flower without a perianth. Fruit a globose head of bracts, valvately opening.

80. *Salicineæ.*—Trees or shrubs. Leaves alternate. Flowers diœcious, in lateral catkins. Ovary free, 1-celled; ovules indefinite. Seeds numerous, minute, with a tuft of hairs.

Division 2. GYMNOSPERMS.

Ovules not enclosed in an ovary and fertilised by direct contact with the pollen. Cotyledons often more than 2.

81. *Gnetaceæ.*—Climbing shrubs with jointed stems. Leaves opposite, rarely 0. Flowers unisexual, enclosed in sheathing bracts.

Class II. MONOCOTYLEDONS.

Pith, wood and bark not distinct. Bundles of fibres scattered in the cellular tissue of the stem. Bark firmly adherent on the outside. Embryo with 1 cotyledon. Root fibrous.

82. *Palmæ.*—Trees or climbers. Leaves pinnately or palmately divided. Perianth 6-leaved. Ovary free of 3 distinct or united carpels.

83. *Gramineæ.*—Herbs, rarely shrubs or trees. Leaves alternate, simple on long split sheaths. Flowers usually bisexual in the axils of the glumes. Embryo at the base of the farinaceous albumen.

SYSTEMATIC LIST

OF THE

TREES, SHRUBS AND WOODY-CLIMBERS

OF THE

BOMBAY PRESIDENCY.

ORDER 1. RANUNCULACEÆ.

HERBS or climbing shrubs. Leaves alternate or opposite. Flowers regular, bisexual. Sepals 4 or more, deciduous, often petaloid. Petals many or 0. Stamens many. Carpels free. Fruit a head of many 1-seeded akenes. Seeds albuminous; embryo minute.

Petals 0... ... ... CLEMATIS.
Petals many, linear ... ... NARAVELIA.

CLEMATIS, L.

Climbing shrubs. Leaves opposite, usually compound. Inflorescence axillary or terminal. Sepals 4-8, petaloid, valvate. Petals 0. Stamens many. Carpels each with 1 ovule. Fruit a head of akenes usually with long feathery styles.

Filaments glabrous.
Leaves simple or trifoliate. Connective of anthers not
produced.
Leaves large, entire or remotely serrate ... ... C. smilacifolia.
Leaves small, entire, 1-3-toothed or lobed ... ... C. triloba.
Leaves pinnate, bipinnate or 2-ternate ... ... C. Gouriana.
Connective of anthers produced more or less. Leaves
simply pinnate, leaflets reticulate ... ... C. hedysarifolia.
Filaments hairy. Leaves pinnate; leaflets densely
villous ... ... C. Wightiana.

C. smilacifolia, Wall. in Asiat. Researches, XIII. 414; Fl. Br. I. 1.3; Dalz. & Gibs. Bomb. Fl. 1. Konkan; North Kánara in evergreen forests, not common. Fl. R.S. Fr. Dec.

C. triloba, Heyne in Roth. Nov. Sp. 251; Fl. Br. I. 1. 3. Vern. morbel or ranjae; Dalz. & Gibs. Bomb. Fl. 1. Mával; West Konkan. Dalz. Fl. Sept. Introduced and common in hedges about Poona.

C. Gouriana, Roxb. Fl. Ind. 2. 670; Fl. Br. I. 1. 4; Wight Ic. 933; Dalz. & Gibs. Bomb. Fl. 1. Indian traveller's joy. Moriel, M. Throughout the presidency in dry forests and along rivers and nálás. Fl. Oct. Fr. Dec.

C. hedysarifolia, D. C. Syst. 1. 148; Fl. Br. I. 1. 4. Konkan; Belgaum; North Kánara, along banks of nálás and rivers. Fl. Oct. Nov. Fr. Dec

C. **Wightiana,** Wall. Cat. 4674 ; Fl. Br. I. 1. 5 ; Wight Ic. t. 935 ; Dalz. & Gibs. Bomb. Fl. 1. Hills of the Deccan ; Konkan, Mahábaleshvar. Fl. Jan, Fr. H.S.

NARAVELIA, DC.

Climbing shrubs. Peduncles 1-flowered axillary. Leaves trifoliate ; terminal leaflet transformed into a tendril. Sepals 4-5. Petals 6-12, linear, longer than calyx. Akenes with long bearded styles.

N. **zeylanica,** D. C. Syst. 1. 167 ; Fl. Br. I. 1. 7 ; Dalz. & Gibs. Bomb. Fl. 1 ; *Atragene zeylanica,* Linn. Throughout the presidency, common in the moist forests of North Kánara. Fl. Oct. Fr. Nov. Dec.

ORDER 2. DILLENIACEÆ.

Shrubs, herbs or trees with large penniveined leaves, distinctly alternate Sepals imbricate persistent. Petals 5-4. Stamens indefinite. Seeds with or without an aril.

DILLENIA, Linn.

Trees. Leaves serrate. Flowers solitary or fascicled, yellow or white. Sepals 5. Petals 4-5 deciduous. Stamens numerous in many series, anthers linear dehiscing longitudinally or by terminal pores. Carpels 5-20, styles as many as ovaries. Fruit globose enclosed in the thickened calyx. Seeds exarillate.

> Leaves persistent. Flowers large white   ...   ... *D. indica.*
> Flowers much smaller yellow. Leaves deciduous   ... *D. pentagyna.*

D. **indica,** Linn. ; Fl. Br. I. 1. 36 ; Brandis For. Fl. 1 ; *D. speciosa,* Thunb. Dalz. & Gibs. Bomb. Fl. 2. Bedd. Fl. Sylv. t. 103. *Mota karmal,* M. ; *chalta,* H. Konkan (Sávantvádi) ; Kolába (Alibág) ; Belgaum. Fl. June. Fr. Feb.

D. **pentagyna,** Roxb. Cor. Pl. 1. t. 20 ; Fl. Br. I. 1. 38 ; Brandis For. Fl. 2 ; Bedd. Fl. Sylv. t. 104 ; Dalz. & Gibs. Bomb. Fl. 2. *Karmal, karambel, kurveil,* M. ; *kanagola,* K. Throughout the presidency in deciduous forests, common. Fl. Mch., Apl. Fr. May.

ORDER 3. MAGNOLIACEÆ.

Trees or shrubs with convulute stipules. Sepals and petals trimerous, imbricated. Stamens many, free on torus. Ovaries numerous, often spirally on torus. Seeds albuminous. Embryo minute.

MICHELIA, Linn.

Trees. Leaves simple, alternate. Filaments flat. Gynophore stalked. Carpels numerous, each with 3 or more ovules. Fruit a spike of 2-valved, 1-12-seeded carpels. Seeds with a fleshy testa.

M. **Champaca,** Linn. ; Fl. Br. I. 1. 42 ; Brandis For. Fl. 3. *Kud champa,* M. ; *kola sampige,* K. ; Vern. *champa.* An evergreen tree commonly cultivated near temples ; throughout the presidency. Flowers at various seasons. Ripe fruit in the cold season.

Order 4. ANONACEÆ.

Trees or shrubs. Leaves alternate, entire, exstipulate. Flowers bisexual, rarely unisexual. Sepals and petals trimerous. Stamens indefinite, the connective often produced. Ovaries numerous, rarely solitary, free (in *Anona* connate); styles short or stigmas sessile; ovules 1 or more in each cell. Carpels 1 or more, sessile or stalked, 1 or more seeded. Seeds large. Albumen ruminate.

Stamens many. Anther cells concealed by the over-
lapping or produced connectives.
Petals 2 seriate, imbricate in bud ... ... UVARIA.
Petals valvate in bud, inner subsimilar or 0.
Ovaries many ; peduncles hooked ... ... ARTABOTRYS.
Ovules 2-6, 1-seriate on ventral suture ... ... UNONA.
Ovules 1-2, basal. ... ... POLYALTHIA.
Petals 2-seriate, valvate in bud, inner dissimilar
conniving and arching over stamens ... ... GONIOTHALAMUS.
Petals valvate in bed, inner similar but smaller ... ANONA.
Anther cells not concealed by the connectives.
Petals valvate, inner largest. Ovules definite ... MILIUSA.
Petals valvate. Ovules indefinite ... ... SACCOPETALUM.
Petals valvate, inner shortest. Ovules 2-4... ... OROPHEA.
Petals imbricate, subequal. Ovules 2-8 ... ... BOCAGEA.

UVARIA, L.

Scandent shrubs. Leaves simple, stellately tomentose or glabrous. Flowers terminal or leaf-opposed ; cymose fascicled or solitary,
yellow, purple or brown. Sepals 3-valvate. Petals 6. Stamens
indefinite. Connective foliaceous or truncate-dilated. Ovaries indefinite, linear oblong, ovules many, 2-seriate. Ripe carpels many, few
or 1-seeded.

U. **Narum,** Wall. Cat. 6473; Fl. Br. I. 1. 50; *U. lurida,* D. & G.
Bomb. Fl. 3. *Naram panal,* M. Konkan ; North Kánara, in evergreen
forests. Fl. Nov. Fr. ripe Dec.

ARTABOTRYS, R. Brown.

Scandent shrubs. Leaves shining evergreen. Flowers solitary
or fascicled on woody hooked peduncles. Sepals 3-valvate. Petals 6,
2-seriate. Stamens numerous, connective truncate-dilated beyond
the anther cells. Torus plano-convex. Ovaries few or many, ovules
2 in each erect. Fruit a berry.

Flowers and fruit glabrous ... ... *A. odoratissimus.*
Flowers and fruit tomentose ... ... *A. zeylanicus.*

A. **odoratissimus,** R. Br. in Bot. Reg. 423 ; Fl. Br. I. 1. 54; Kz.
For. Fl. Burm. 1. 31. Cultivated, but not indigenous, in the Bombay
Presidency.

A **zeylanicus,** H.f. & T. Fl. Br. I. 1. 54; Bedd. Ic. Pl. Ind. Or. t. 48.
Common in the evergreen forests of North Kánara, towards the south ;
abundant in the forests rear the Gairsoppah Falls. Fl. Nov. Fruit
remains two years on stem.

UNONA, L.

Trees or shrubs often climbing. Flowers usually solitary, axillary or terminal. Sepals 3-valvate. Petals 6, nearly equal, or the inner 3 absent. Torus flat or slightly concave. Stamens numerous, closely packed, 4-angled; top of connective truncate or globose. Ovaries numerous, ovules 2 or more, 1-seriate. Ripe carpels stalked, often moniliform.

> Carpels not constricted between the seeds.
> Small tree. Flowers subsessile, villous, corolla whitish...　*U. pannosa.*
> Carpels moniliform.
> Erect or climbing shrub. Flowers yellow, petals ½ in.
> broad ...　...　...　...　...　...　...　*U. discolor.*
> Climbing shrub. Petals ¼ in. across, very long ...　...　*U. Lawii.*

U. pannosa, Dalz. in Hook. Kew Jour. Bot. III. 207; Fl. Br. I. 1. 58: Bedd. Ic. Pl. Ind. Or. t. 52; *U. farinosa,* Dalz. & Gibs. Bomb. Fl. 3. Konkan, Párwár ghát and Tulawari. D. & G. Fl. Mch. Oct.

U. discolor, Vahl. Symb. II. 63, t. 36; Fl. Br. I. 1. 59; Dalz. & Gibs. Bomb. Fl. 3; *U. Dunalii,* H. f. & T. Dalz. & Gibs. Bomb. Fl. 3. Konkan (Sávantvádi); North Kánara, in evergreen forests. Fl. Aug.

U. Lawii, H. f. & T. Fl. Ind. 132; Fl. Br. I. 1. 59; Bedd. Ic. Pl. Ind. Or. t. 73. North Kánara and Konkan, in evergreen forests. Fl. H. and R. seasons. Fr. H.S.

POLYALTHIA, Blume.

Trees or shrubs. Flowers solitary or fascicled, axillary or leaf opposed. Sepals 3, usually valvate. Petals 6, 2-seriate, flat or the inner vaulted. Torus convex. Ovaries indefinite, ovules 1-2 basal and erect. Berries stalked, globose or oblong, 1-seeded.

> Petals linear. Flowers solitary　...　...　...　*P. coffeoides.*
> Petals linear. Peduncles few or many flowered　...　*P. fragrans.*
> Petals ovate-oblong, green. Peduncles 1-3 flowered...　*P. cerasoides.*
> Petals oval, red-brown. Flowers solitary ...　...　*P. suberosa.*

P. coffeoides, Benth. & H. f. l. c. Fl. Br. I. 1. 62. North Kánara in evergreen forests. This was named at Calcutta from fruiting specimens; it may possibly be identical with *P. fragrans.* Fr. April.

P. fragrans, Benth. & H. f. l. c. Fl. Br. I. 1. 63; Bedd. Ic. Pl. Ind. Or. t. 54; *Guatteria fragrans,* Dalz. & Gibs. Bomb. Fl. 4. *Gauri,* K. Konkan; on the southern ghâts of North Kánara, common in evergreen forests. Fl. Nov. Jr. Apl. May.

P. cerasoides, Benth. & H. f. l. c. Fl. Br. I. 1. 63; Bedd. Fl. Sylv. t. 1; Brandis For. Fl. 5; *Guatteria cerasoides,* W. & A. Prodr. 10; Dalz. & Gibs. Bomb. Fl. 3. *Hoom,* M.; *rubbina,* K. Throughout the ghâts of the presidency in deciduous forests, chiefly; "in the evergreen forests of the Sátára district." Brandis. Fl. Mch. Fr. R.S.

P. suberosa, Benth. & Hook. f. Fl. Br. I. 1. 65; *Guatteria suberosa,* DC. A small tree with corky bark. Peduncles usually solitary. Petals ½ in. red-brown, silky. "*Concans,*" Graham. Fl. Fr. "throughout the year." Roxb.

GONIOTHALAMUS, Blume.

Small trees or shrubs. Flowers solitary or fascicled. Sepals 3, valvate. Petals 6, outer thick flat, inner smaller shortly clawed. Stamens many, connective produced into an oblong process. Ovaries many; ovules 2 superposed in each. Berries 1-seeded.

G. **cardiopetalus**, H. f. & T. Fl. Ind. 107 ; Fl. Br. I. 1. 75; Bedd. Fl. Sylv. 8. North Kánara ghâts in evergreen forests, common on the Ankola ghâts. Fl. Meb. Fr. R. & C.S.; ovoid, sessile.

ANONA, L.

Trees or shrubs. Flowers solitary or fascicled. Sepals 3, valvate. Petals 3 or 6. Stamens indefinite, on a hemispherical torus, top of connective ovoid. Ovaries many, subconnate; ovules 1 erect. Carpels united into a large fleshy fruit ; seeds numerous, imbedded in pulp.

Fruit tubercled. Flowers solitary ... ... *A. squamosa.*
Fruit smooth. Flowers 2-3 together ... ... *A. reticulata.*
Fruit very large, muricate ... ... *A. muricata.*

A. **squamosa**, Linn. Fl. Br. I. 1. 78; Bedd. Fl. Sylv. 9; Brandis For. Fl. 6. Sweet sop. Custard apple. *Sita phal*, H. Naturalized throughout the presidency ; usually cultivated. Fl. June July.

A. **reticulata**, Linn. Fl. Br. I. 1. 78 ; Bedd. Fl. Sylv. 9. Bullock's heart. *Rámphal*, H. Cultivated throughout the presidency.

A. **muricata**, D.C. Syst. 467 ; Bedd. Fl. Sylv. 9 ; Dalz. & Gibs. Bomb. Fl. Suppl. 2. Sour sop. Cultivated near Bombay.

MILIUSA, Lesch.

Trees or shrubs. Flowers solitary, fascicled or cymose. Sepals 3, small, valvate. Petals 6, 2-seriate ; outer smaller like the sepals. Torus elongated, cylindric. Stamens definite or indefinite ; connective hardly apiculate. Ovaries indefinite, ovules 1-2, rarely 3-4. Ripe carpels globose or oblong 1, many seeded.

M. **indica**, Lesch. in A. DC. Mem. Soc. Genev. V. 36 ; Fl. Br. I. 1. 86. Bedd. Ic. Ind. Or. t. 85. In evergreen forests from the Konkan to Mysore. common in North Kánara in the ghât forests. Fl. throughout the year. Fr. ripe near Kárwár in August.

SACCOPETALUM, Bennett.

Trees. Flowers axillary, solitary or fascicled. Sepals 3, small, valvate. Petals 6, valvate, in 2 series. Stamens indefinite ; connective produced into a conspicuous appendage. Ovaries indefinite, ovules 6 or more. Ripe carpels subglobose.

S. **tomentosum**, H. f. & T. Fl. Ind. 152. Fl. Br. I. 1. 88 ; Bedd. Fl Sylv. 10; Brandis For. Fl. 7 ; Dalz. & Gibs. Bomb. Fl. 4. *Hoom*, H. ; *wumb*, K. Ghâts of North Kánara and Konkan. Fl. Apl. Fr. June.

OROPHEA, Blume.

Trees or shrubs. Flowers small, axillary solitary, fascicled or cymose. Inner petals clawed, cohering into a mitriform cap. Stamens
6-12, fleshy. Ovaries 3-15; ovules 4. Ripe carpels often globose.

O. zeylanica, H.f. & T. Fl. Ind. 111; Fl. 1.Br. 1. 90; Bedd. Fl. Sylv.
11. "A shrub or very small tree found on the North Kánara ghâts."
Bedd.

BOCAGEA, St. Hilaire.

Trees. Flowers small, terminal, axillary or fascicled on woody
tubercles, 1-2-sexual. Sepals ovate, imbricate. Petals 6 in 2
series, nearly equal, usually concave. Stamens 6-21, imbricate in
2 or more series, thick, fleshy, connective produced. Ovaries 3-6;
ovules 1 or 2-8 on the ventral suture. Ripe carpels globose stalked.

B. Dalzellii, H. f. & T. Fl. Br. I. 1. 92; *Sagerœa laurina,* Dalz. &
Gibs. Bomb. Fl. 2; *Sagerœa Dalzellii,* Bedd. Ic. Pl. Ind. Or. t. 42.
Sagcri, har-kinjal, undie, M. Konkan; North Kánara, in evergreen
forests. Fl. Oct. Nov.

ORDER 5. MENISPERMACEÆ.

Climbing shrubs. Leaves alternate. Flowers very small, diœc-
cious. Stamens definite, often monadelphous. Ovaries 3 or 1, ovules
solitary, usually amphitropous. Albumen even, ruminate or 0.

Ovaries 3. Seed oblong or subglobose. Cotyledons
   foliaceous.
Filaments free ... ... TINOSPORA.
Filaments connate ... ... ANAMIRTA.
Seed horse-shoe-shaped; cotyledons linear; albumen
   copious.
Petals 6, minute. Ovaries 3-12 ... ... TILIACORA.
Petals 6. Ovaries 3-6 ... ... COCCULUS.
Ovary 1. Albumen scanty. Endocarp muricate.
Sepals 6-10, free. Petals of ♀ 3-5, free ... ... STEPHANIA.
Sepals 4, free. Petals of ♀ 1... ... ... CISSAMPELOS.
Sepals 4-8, connate. Petals of ♀ 1 ... ... CYCLEA.

TINOSPORA, Miers.

Climbing shrubs. Flowers in racemes or panicles. Sepals 6 in 2
series, inner larger membranous. Petals 6 smaller. Male fl. Stamens 6; filaments thickened at top. Female fl. Staminodes 6, clavate. Ovaries 3. Drupes 1-3, dorsally convex, ventrally concave.
Putamen tubercled, dorsally keeled, intruding. Albumen ruminate.

Leaves woolly beneath. Flowers green ... ... *T. malabarica.*
Leaves glabrous. Flowers yellow ... ... *T. cordifolia.*

T. malabarica, Miers Contrib. III. 32; Fl. Br. I. 1. 97; Dalz. &
Gibs. Bomb. Fl. 5. Moist forests of the Konkan and North Kánara.

T. cordifolia, Miers Contrib. III. 31; Fl. Br. I. 1. 97; Brandis For.
Fl. 8; Dalz. & Gibs. Bomb. Fl. 5. *Gulaveli, gulbel, gulwail, giroli,*
M.; *gulo,* Vern. Dalz. Konkan and North Kánara forests, common.
Fl. H.S. Fr. R.S.

ANAMIRTA, Coleb.

A climbing shrub. Flowers panicled. Sepals 6, a little unequal. Petals 0. Male fl. Anthers sessile on a column. Female fl. Staminodes 9, clavate, 1-seriate. Ovaries 3 on a short gynophore. Stigma reflexed. Drupes stalked ; putamen woody. Seed globose embracing the hollow intruded endocarp. Albumen almost ruminate.

A. Cocculus, W. & A. Prodr. I. 446 ; Fl. Br. I. 1. 98 ; Brandis For. Fl. 8 ; Dalz. & Gibs. Bomb. Fl. 4 ; Vern. *kakaphula, kakmari.* Konkan ; North Kánara. " Berries size and colour of a black cherry." Roxb.

TILIACORA, Coleb.

A climbing shrub. Flowers in axillary panicles. Sepals 6, 2-seriate, outer much smaller. Petals 6, minute, cuneate. Male fl. Stamers 6, free. Female fl. Carpels 9-12 : styles short subulate. Drupes stalked ; putamen sulcate. Albumen oily, ruminate.

T. racemosa, Coleb. in Trans. Linn. Soc. XIII. 67 ; Fl. Br. I. 1. 99. Konkan, not mentioned by Dalz. & Gibs. in the Bombay Flora ; not observed in North Kánara. Fl. throughout the year. Brandis.

COCCULUS, DC.

Climbers. Sepals 6, inner 3 larger. Petals 6, shorter than the sepals, entire or 2-cleft. Male fl. ; stamens 6, free, embraced by the petals. Female fl. Staminodes 6 or 0. Ovaries 3-6. Drupes laterally compressed, style scar nearly basal. Endocarp tuberculate, horse-shoe-shaped.

Leaves more or less glabrous, suborbicular, long petioled.
Flowers in large panicles. Drupe 1 in. in diam. ... *C. macrocarpus.*
Leaves ovate, villous. Flowers in short panicles ... *C. villosus.*
Leaves oblong or trapezoid, glabrate. Male flowers
fascicled. Female flowers solitary. Drupe 1/12 to 1/6 in. *C. Leæba.*

C. macrocarpus, W. & A. Prodr. 13 ; Fl. Br. I. 1. 101 ; Dalz. & Gibs. Bomb. Fl. 5. *Vatoli, vatyel,* M. Konkan ; North Kánara along banks of rivers and nálás and in moist forests locally abundant. Fl. Feb. Mch. Fr. May.

C. villosus, DC. Prod. I. 98 ; Fl. Br. I. 1. 101 ; Dalz. & Gibs. Bomb. Fl. 5 ; Brandis For. Fl. 9. *Kursan, zamir,* Sind ; *vasanvel,* M. ; *vasandi,* Sans. Common in hedges throughout the presidency. Fl. Feb. Mch.

C. Leæba, DC. Prod. I. 99 ; Fl. Br. I. 1. 102 ; Brandis For. Fl. 9. Deccan ; Sind, in dry and arid regions. Fl. throughout the year.

STEPHANIA, Lour.

Climbing shrubs with peltate leaves. Flowers in axillary, cymose umbels. Male fl. Sepals 6-10. Petals 3-5, fleshy. Anthers 6, connate, at top of staminal column, transversely dehiscent. Female fl. Sepals 3-5. Petals of the male. Ovary 1 ; style 3-6-divided. Endocarp compressed, horse-shoe-shaped, dorsally tubercled ; sides perforated.

S. hernandifolia, Walp. Rep. 1. 96; Fl. Br. I. 1. 103; Dalz. & Gibs. Bomb. Fl. 6. Throughout the Konkan and North Kánara, in moist situations. Fl. Aug. Fr. Oct.

CISSAMPELOS, Linn.

Climbers. Leaves peltate. Male fl. : cymose. Sepals 4. Petals 4, connate, forming a 4-lobed cup. Stamens monadelphous, anthers united into a peltate disc, bursting transversely. Female fl. : racemed in the axils of leafy bracts. Perigonium of 2, 2-nerved sepals adnate to the bracts. Ovary 1 ; style short, 3-toothed. Endocarp horse-shoe-shaped, dorsally tubercled, sides excavated.

C. Pareira, Linn. Fl. Br. I. 1. 103 ; Dalz. & Gibs. Bomb. Fl. 5. Brandis For. Fl. 10. *Paharvel, paharmul.* M. Throughout the presidency ; common in hedges. Fl. Mch. Oct.

CYCLEA, Arnott.

Climbers. Leaves usually peltate. Flowers in axillary panicles. Male fl. : sepals 4-8, connate into an inflated calyx. Petals 4-8, connate. Anthers 4-6, connate into a peltate disc, crowning the staminal column, bursting transversely. Female fl. Sepal 1. Petal 1. Ovary 1 ; style short, 3-5-lobed, lobes radiating. Drupe ovoid ; endocarp horse-shoe-shaped, dorsally tubercled, sides 2-locellate.

> Calyx subglobose, 6-8-lobed. Corolla urceolate ... *C. Burmanni.*
> Calyx campanulate, 4-lobed. Corolla a 4-lobed cup ... *C. peltata.*

C. Burmanni, Miers Contrib. III. 239, t. 121 ; Fl. Br. I. 1. 104 ; Dalz. & Gibs. Bomb. Fl. 6. *Pakur,* Vern. Hilly parts of the Konkan and gháts, Dalz. Fl. Jany.

C. peltata, H. f. & T. Fl. Ind. 201 ; Fl. Br. I. 1. 104, Dalz. & Gibs. Bomb. Fl. 6. *Paryel,* M. Konkan and North Kánara, in moist forests, from the sea-level upwards. Fl. R.S. Fr. Jany.

Order 6. CAPPARIDEÆ

Herbs, shrubs or trees, erect or climbing. Leaves alternate Sepals 4, free or connate. Petals 4, rarely 2 or 0. Stamens indefinite, 8, 6, or 4, at the base of or on a long or short gynophore. Disc 0 or tumid or lining the calyx-tube. Ovary sessile or stalked, 1-celled. Fruit capsular or berried. Seeds angled or reniform, exalbuminous.

> Climbing, unarmed shrubs.
> Fruit moniliform, a long fleshy berry .. ... Mærua.
> Fruit cylindric, dehiscent ... ... Cadaba.
> Climbing, armed shrubs. Fruit globose ... ... Capparis.
> An unarmed tree. Leaves digitately 3-5-foliate ... Cratæva.

MÆRUA, Forsk.

Climbing shrubs. Leaves simple. Flowers corymbose. Calyx-tube lined by a disc ; lobes 4, valvate. Petals 4. Stamens numerous inserted high on gynophore, filaments exserted. Ovary long stalked, 1-celled, ovules many on 2-4 parietal placentas. Berry fleshy, moniliform, 1 or more seeded.

M. arenaria, H. f. & T. Fl. Br. I. 1. 171 ; *Niebhuria oblongifolia,* D.C.; Dalz & Gibs. Bomb. Fl. 8. Dhárwár, Gujarát and throughont the Deccan, in hedges; absent from the Konkan and North Kánara. Fl. and Fr. C.S.

CRATÆVA, L.

Trees. Leaves trifoliate. Flowers large, yellow, in terminal corymbs. Sepals 4, cohering below with the convex disc. Petals 4, long-clawed. Stamens indefinite, filaments filiform, free. Ovary on a long gynophore with 2 placentas bearing numerous ovules Fruit a large berry.

C. religiosa, Forst.; DC. Prod. 1. 243; Fl. Br. I. 1.172 ; Brandis. For. Fl. 16. *C. Roxburghii,* Br. *C. Nurvala.* Ham. Dalz. & Gibs. Bomb. Fl. 8. Bedd. Fl. Sylv. 14. *Bitusi,* K. ; *nirrala, kumla, karwan,* M. ; *varvanna,* H. Throughout the presidency, near the banks of rivers and nálás, in moist shady places, also planted near Mussulman tombs. Fl. Apl., May. Fr. R.S.

CADABA, Forsk.

Shrubs. Leaves simple or trifoliate. Flowers solitary, corymbose or racemed. Sepals 4, unequal. Petals 4-2-0, clawed, hypogynous. Disc spathulate with a tubular claw. Stamens 4-6, unilateral ; filaments filiform, exserted, spreading. Ovary-long stalked, 1-celled ; ovules many. Fruit fleshy, cylindric, indehiscent or tardily dehiscent.

Straggling shrubs. Stamens 4. Fruit dehiscent ... *C. indica*
...... Stamens 5. Fruit indehiscent... *C. farinosa.*
A small rigid tree. Stamens 5... ... ... *C. heterotricha.*

C. indica, Limk. ; DC. Prod. 1. 244; Fl. Br. I. 1. 172 ; Dalz. & Gibs. Bomb. Fl. 9. Throughout the dry districts of the presidency, on old walls and in dry situations. Fl. Jan., Feb.

C. farinosa, Forsk.; DC. Prod. 1. 244; Fl. Br. I. 1. 173. Sind.

C. heterotricha, Stocks in Hook. Ic. Pl. t. 839 ; Fl. Br. I. 1. 173. Sind ; on rocks near Cape Monze, Stocks.

CAPPARIS, Linn.

Trees or shrubs, erect or climbing, usually armed with twin stipulary thorns. Leaves simple or 0. Flowers usually large. Sepals 4. Petals 4. Stamens indefinite, filaments filiform. Ovary on gynophore, ovules many. Fruit fleshy, many seeded. Cotyledons convolute. Flowers solitary axillary or 2-3-fascicled.

Flowers, large solitary, white, filaments purple ... *C. spinosa.*
Flowers with lower petals yellowish brown .. ... *C. zeylanica.*
Flowers blue, lower petals with yellow basal spots ... *C. Heyneana.*

Flowers solitary green. Fruit ribbed, scarlet	...	*C. divaricata.*
Flowers umbelled racemed or panicled.		
Tree or shrub, erect, glabrous. Flowers red..	...	*C. aphylla.*
Climber, branches glabrous. Flowers white 5 inches across		*C. Moonii.*
Climber, branches pubescent. Flowers white 2 inches across		*C. Roxburghii.*
Erect pubescent tree. Flowers white	...	*C. grandis.*
Flowers in simple umbels.		
Spines straight acicular	...	*C. longispina.*
Spines recurved.		
Fruit pisiform, black	...	*C. sepiaria.*
Fruit size of a cherry, several seeded	...	*C. pedunculosa.*
Flowers seriate in lines on the branches.		
Brown tomentose climber. Petals staineus purple.		*C. horrida,*
Glabrous climber. Petals pubescent, white ...	...	*C. tenera.*

C. spinosa, Linn. Fl. Br. 1. 1. 173. Brandis For. Fl. 14 ; *C. Murrayana,* Grah. Cat. Bomb. Pl. 9 ; Dalz. & Gibs. Bomb. Fl. 9. Caper plant. *Kalvari,* Sindhi. At Mahábaleshvar, and in most nálás and rivers, but along the gháts as far north as Málsej ; Dalzell. Sind, Gujarát, on dry rocks and stony hills. Fl. Apl., June. Fr. Nov.

C. zeylanica, Linn. Fl. Br. 1. 1. 174 ; *C. brevispina.* DC. ; W. & A. Prod. 24 ; Dalz. & Gibs. Bomb. Fl 9. *Wagutty,* M. North Kanara ; Dhárwár ; Western Deccan and Konkan ; in hedges and along banks of nálás, also in deciduous forests. Fruit bright scarlet 1½ in. ovoid. Fl. Feb. Fr. R.S.

C. Heyneana, Wall. Cat. 6985 ; Fl. Br. I. 1. 174 ; *C. formosa,* Dalz. & Gibs. Bomb. Fl. 9. *Chayruka,* H. Common on the North Kanara and South Konkan gháts, in evergreen forests. Fl. April, May. Fr. R S.

C. divaricata, Lamk. ; DC. Prod. 1. 252 ; Fl. Br. I. 1. 174 ; *C. stylosa,* D. C. ; Wall. Cat. 6. 980 ; Dalz. & Gibs. Bomb. Fl. 10 ; Bedd. Fl. Sylv. 13. Common all over the Deccan both on stony ground and on the black soil in "bábul" forests. Fl. Feb., Mch. Fr. ripe Aug. Leaves often linear.

C. aphylla, Roth. ; DC. Prod. 1. 246. Brandis For. Fl. 14 ; Bedd. Fl. Sylv. 13. Dalz. & Gibs. Bomb. Fl. 9 ; Fl. Br I. 1. 174. *Shipri gidda,* K. ; *kiral,* Sind ; *ker,* Guj. ; *kera,* M. Sind ; Gujarát ; Deccan and generally throughout the driest parts of the presidency. Fl. Mch., Apl. Fr. Sept., Oct.

C. Moonii, Wight I. 11. 35 ; Fl. Br. I. 1. 175. Along the North Kánara and Konkan gháts. in evergreen forests. Fl. Dec., Feb. Fr. R.S. Flowers large, an ornamental shrub.

C. Roxburghii, DC. Prod. 1. 247 ; Dalz. & Gibs. Bomb. Fl. 9 ; Fl. Br. I. 1. 175. *Poorvi,* M. On the gháts, Dalz. ; North Kánara ; Kumta and Ankola gháts, in evergreen forests. This is probably not distinct from *C. Moonii,* Wgt.

C. grandis, Linn. f. ; DC. Prod. 1. 248 ; Fl. Br. I. 1, 176 ; Dalz. & Gibs. *Kavotel.* M. ; *torate,* K. : *puchowuda,* Vern. Sparingly found on the gháts and in the Deccan. Dalz. In the forests of the Dhárwár district bordering on North Kánara, fairly common. Fl. Oct. Fr. C.S.

C. pedunculosa, Wall. Cat. 6. 993 ; Fl. Br. I. 1. 176 ; Dalz. & Gibs. Bomb. Fl. 9. *Kolisna*, M. Konkan gháts; at Mahábaleshvar and in the thickest jungles generally, Dalz.

C. longispina, Hook. f. & T. Fl. Br. I. 1. 176. Along the North Kánara gháts in open situations near evergreen forests. Fl. Feb., Mch. Fr. R.S.

C. sepiaria, Linn. ; DC. Prod. 1. 247 ; Brandis For. Fl. 15 ; Dalz. & Gibs. Bomb. Fl. 10. *Kanthar*, Guj. Throughout the dry parts of the presidency ; very common in hedges and open situations in deciduous forests. Fl. Feb., May. Fr. R.S.

C. horrida, Linn. f. ; DC. Prod. 1. 246 ; Fl. Br. I. 1. 173 ; Dalz. & Gibs. Bomb. Fl. 10. Brandis For. Fl. 15. *Ardandu*, H. ; *wag, gowindi*, M. Throughout the presidency and Sind ; common in hedges and in open situations from the sea-level upwards. Fl. Nov., Apl. Fr. R.S.,

C. tenera, Dalz. in Hook. Kew Jour. Bot. II. 41; Dalz. & Gibs. Bomb. Fl. 9. Fl. Br. I. 1. 179. On the southern gháts of North Kanara, in evergreen forests. Fl. April, May. Fr. R.S.

ORDER 7. RESEDACEÆ.

Herbs or shrubs. Leaves alternate. Stipules 0 or minute. Flowers spiked or racemose, often polygamous. Calyx 4-7-divided : sepals imbricate in bud. Petals 2-7, entire or lobed, open in bud. Disc often unilateral, conspicuous. Stamens numerous, inserted on the disc. Ovary 1-celled of 2, 6 connate carpels, often lobed at the top. Ovules many on 2-6 placentas. Fruit a capsule or berry. Seeds many.

OCHRADENUS, Delile.

Branched shrubs. Leaves small linear. Flowers minute. Calyx 5-partite. Petals 0. Stamens 10-20. Ovary ovoid, 3-beaked ; ovules on 3 placentas. Berry many-seeded ; or sometimes seeds few large.

Ochradenus baccatus, Del. Fl. Æg. 15, t. 31, f. I. Fl. Br. I. 1. 182. Sind.

ORDER 8. BIXINEÆ.

Trees or shrubs. Leaves alternate ; stipules minute or 0. Flowers regular, 1-2-sexual. Stamens usually hypogynous ; anthers bursting by slits or pores. Disc thick, often glandular. Ovary 1 or few-celled. Ovules parietal. Fruit dry or fleshy. Seeds albuminous.

Flowers bisexual.	Petals broad. Capsule 3, 5-celled	...	...	COCHLOSPERMUM.
	,, ,, Capsule 2-valved...	...	...	BIXA.
	Petals small ...	...	...	SCOLOPIA.
Flowers diœcious	Petals, 0 ...	...	...	FLACOURTIA.
	Petals with a scale...	...	...	HYDNOCARPUS.

COCHLOSPERMUM, Kunth.

Trees or shrubs. Leaves digitately lobed. Flowers large, yellow,
bisexual. Sepals 5, deciduous. Petals 5, contorted in bud. Stamens
indefinite, inserted on a disc without glands. Ovary with numerous ovules on 3-5 parietal placentas. Capsule 3-5-valved. Seeds
cochleate ; testa hard, woolly ; embryo curved.

C. gossypium, DC. Prod. 1. 527 ; Fl. Br. I. 1. 190 ; Brandis For.
Fl. 17 ; Bedd. Fl. Sylv. 14. *Gunglay, gulgul,* M. ; *ganeri,* Bhil. In the
dry deciduous forests of the Deccan ; common on the Khándesh Sátpudás.
Fl. Feb., Apl. Fr. June, July.

BIXA, L.

A shrub or small tree. Leaves simple ; stipules minute. Flowers
in terminal panicles, bisexual. Sepals 5, deciduous. Petals 5,
contorted in bud. Anthers opening by 2 terminal pores. Ovary
1-celled; ovules many, on 2 parietal placentas. Capsule loculicidally
2-valved ; placentas on the middle of each valve. Seeds many ;
testa pulpy, red.

B. Orellana, Linn. Fl. Br. I. 1. 190 ; Brandis For. Fl. 17 ; Dalz. &
Gibs. Bomb. Fl. Suppl. 5. *Arnotto ; kisri, sendri,* Vern. Cultivated
throughout the presidency; indigenous in America. Fl. Sept. Fr. Jany.

SCOLOPIA, Schreber.

Spinous trees. Leaves alternate, entire. Flowers in axillary
racemes, small. Sepals 4-6. Petals 4-6, imbricate in bud. Stamens
many, connective produced. Ovary 1-celled ; ovules few, on 3-4
parietal placentas. Berry 2-4-seeded. Seeds with long funicles.

Scolopia crenata, Clos. l. c. 250 ; Fl. Br. I. 1. 191 ; Bedd. Fl. Sylv.
15 ; *Phoberus crenatus,* W. & A. Dalz. & Gibs. Bomb. Fl. 11. North
Kánara ghats ; ghats south of Rám ghát, Dalz. Common in the forests
near Nilkund, North Kánara. Fl. Mch. Fr. ripe Aug.

FLACOURTIA, Commers.

Trees or shrubs, usually spinous. Leaves toothed or crenate.
Flowers diœcious. Sepals 4-5, small, imbricate. Petals 0. Stamens
numerous, anthers short, versatile. Ovary on a glandular disc. Styles
2 or more, stigmas 2-notched or 2-lobed. Fruit a few-seeded berry.
Seeds obovoid, testa coriaceous ; cotyledons orbicular.

Trees. Fruit large.
Racemes pubescent. Leaves hairy beneath   ...   *F. montana.*
Racemes and leaves quite glabrous.
Fruit size of a small plum. Trees or shrubs   ...   *F. Cataphracta.*
Fruit smaller than those of the above species
(in the variety *sapida,* size of a pea). Stig-
mas 5-11   ...   .   *F. Ramontchi.*
Shrub. Stigmas 3-4   ...   .   *F. sepiaria.*

F. montana, Grah. Cat. Bomb. Pl. 10 ; Fl. Br. I. 1. 192 ; Dalz. &
Gibs. Bomb. Fl. 10. *Han sampige,* K. ; *attak, champer,* M. Konkan,
North Kánara, in evergreen forests. Fruit edible. Fl. C. S. Fr. April.

F. Cataphracta, Roxb. in Willd. Sp. Pl IV. 830; Fl. Br. I. 1. 193 ;
Dalz. & Gibs. Bomb. Fl. 10. *Jaggum,* Vern. This species given on
the authority of the Bombay Flora is said to be found in the Warri
country on the banks of rivers.

F. Ramontchi, L. Herit. Stirp. 59, t. 30, 31; Fl. Br. I. 1. 193 ;
Brandis For. Fl. 18 ; Bedd. Fl. Sylv. 16; Dalz. & Gibs. Bomb. Fl. 10.
Paker, kaker, bhekal, M. ; *Kanju,* H. Throughout the presidency ; some-
times cultivated. Fl. Nov., Mch. Fr. R.S. Aug.

F. sepiaria, Roxb. Cor. Pl. I. 48, t. 68; Fl. Br. I. 1. 194 ; Dalz. &
Gibs. Bomb. Fl. 11 ; Brandis For. Fl. 18. Bedd. Fl. Sylv. 16. *Tambat,*
Vern. Common in the Deccan towards the ghâts. Fl. C.S. Fr. R.S.

HYDNOCARPUS, Gærtn.

Trees. Leaves alternate, entire or serrate. Flowers solitary or
in small racemes, diœcious. Sepals 4-5, imbricate. Petals 5-9, each
with a basal scale or the scales cohering in a cup. Male fl.: stamens 5-8. Female fl. : staminodes 5 or more. Ovary 1-celled ; stigmas
3-6, dilated lobed ; ovules many on 3-6 parietal placentas. Fruit
globose with a hard rind, many-seeded. Albumen oily.

H. Wightiana, Blume, Rumph. IV. 22 ; Fl. Br. I. 1. 196 ; Dalz. &
Gibs. Bomb. Fl. 11 ; Bedd. Fl. Sylv. 16. *Kastel, kantel,* M. ; *toratti,* K
South Konkan ; North Kánara, in evergreen forests. Fl. Apl. Fr. R.S.

Order 9. PITTOSPOREÆ.

Trees. Leaves entire, alternate, exstipulate. Flowers hermaphrodite. Sepals 5. Petals 5, imbricate. Stamens 5, opposite sepals.
Ovary 1-celled. Ovules many, parietal or axile. Capsule woody,
2-valved dehiscent. Seeds in pulp.

PITTOSPORUM, Banks.

Trees. Leaves quite entire, exstipulate. Petals connate at base.
Capsule 1-celled, 2-valved. Placenta in middle of each valve. Seeds
smooth, in pulp.

Glabrous ... ... *P. floribundum.*
Tomentose ... ... *P. dasycaulon.*

P. floribundum, W. & A. Prod. 154 ; Fl. Fr. I. 1. 199 ; Dalz. & Gibs.
Bomb. Fl. 44 ; Bedd. Fl. Sylv. 17 ; Brandis For. Fl. 19. *Yekkadi,* M.
Konkan and North Kánara, in the ghát forests. Fl. Apl., Sept. Fr. C.S.

P. dasycaulon, Miquel in Herb. Hohenack. 775 ; Bedd. Fl. Sylv.
236. Fl. Br. I. 1. 199. *Gapsundi,* M. Konkan ; North Kánara ; Belgaum ; common in the evergreen ghát jungles. Fl. Nov., Dec.
Fr. Feb., Mch.

Order 10. TAMARASCINEÆ.

Shrubs or small trees. Leaves scale-like, imbricating. Flowers
racemose or paniculate, regular, bisexual. Sepals 5. Petals 5,
imbricate. Stamens 5-10, free or connate, inserted on a lobed disc.
Ovary syncarpous ; ovules numerous. Seeds crested, winged or
covered with down. Albumen 0.

TAMARIX, Linn.

Flowers white or red. Sepals and petals free. Ovary 1-celled ; ovules at bottom of the ovary. Seeds smooth, with a long coma.

Stamens 5.
Flowers in large panicles ... ... *T. gallica.*
Flowers in close cylindrical spikes ... .. *T. dioica.*
Flowers in interrupted spikes.
Stamens 10 ... ... *T. articulata.*
Leaves not punctate, sheathing amplexicaul. ... ... *T. ericoides.*
Leaves punctate, closely sheathing. ... ... *T. stricta.*

T. gallica, Linn. Fl. Br. I. 1. 248 ; Brandis For. Fl. 20 ; Bedd. Fl. Sylv. 20. *Lei, lai, jhaw,* Sind. Along the banks of rivers and near the sea-coast of Sind. Fl. July, Aug. Fr. Dec., Mch.

T. dioica, Roxb. Hort. Beng. 22 ; Fl. Br. I. 1. 249 ; Brandis For. Fl. 21 ; Bedd. Fl. Sylv. 20. *Gaz, lán, jau,* Sind. Forms extensive forests along the Indus. Fl. May, July. Fr. C.S.

T. articulata, Vahl. Symb. II. 48, t. 32 ; Fl. Br. I. 1. 249 ; Brandis For. Fl. 22 ; Bedd. Fl Sylv. 20. *Asreli,* Sind. Upper and Middle Sind eastwards to the Jumna.

T. stricta, Boiss. Fl. Or. 1. 778 ; Fl. Br. I. 1. 249. Sind. Stocks. Fl. May, July. Fr. Sept.

T. ericoides, Rottl. in Nov. Act. Nat. Cur. Berol. IV. 214, t. 4. Fl. Br. I. 1. 249 ; *Trichaurus ericoides,* W. & A. ; Dalz. & Gibs. Bomb. Fl. 14. *Jao, sarub, sarata,* M. Common in the beds of the Konkan and Deccan rivers ; in the Kálánaddi of North Kánara near Sulgeri. Flowers during the cold weather Nov. to Jan. Fr. Feb.

Order 11. GUTTIFERÆ.

Trees or shrubs with resinous juice. Leaves simple, opposite, coriaceous. Flowers unisexual or polygamous. Sepals imbricate. Ovary 1 to many-celled. Fruit baccate. Seeds large. Albumen 0.

Embryo a solid tigellus with minute cotyledons or 0.
Calyx of four or five sepals ... ... GARCINIA.
Calyx closed in bud bursting into 2 valves .. ... OCHROCARPUS.
Embryo of 2 fleshy cotyledons with a small radicle.
Ovary 1-celled, 1-ovuled ... ... CALOPHYLLUM.
Ovary 2-celled, 4-ovuled ... ... MESUA.

GARCINIA, Linn.

Trees with yellow juice. Leaves evergreen. Flowers axillary or terminal, often fascicled, polygamous. Sepals 4, 5 decussate. Petals 4, 5, imbricate. Stamens of male flower many. Ovary of female or hermaphrodite flower 2 to 12-celled. Ovules 1 in each cell, attached to the inner wall. Fruit a berry.

Flowers tetramerous.
Fruit globose, 4 to 8-celled, not grooved, pulpy. ... *G. indica.*
Fruit ovoid, 6 to 8-grooved. ... ... *G. Cambogia.*
Fruit globose, 4-celled, not pulpy. ... ... *G. Morella.*
Flowers generally pentamerous.
Fruit large, yellow. ... ... *G. Xanthochymus.*
Fruit smaller, green. ... ... *G. ovalifolius.*

G. indica, Chois. in DC. Prod. 1. 561; Fl. Br. I. 1. 261; *G. purpurea,* Roxb. Dalz. & Gibs. Bomb. Fl. 31; Bedd. Fl. Sylv. 21. *Murgal,* K.; *bhairnd, ratamba,* M.; *kokum,* H. Wild mangosteen of the English and brindon of the Portuguese. North Kánara and South Konkan, in evergreen forests; often planted. Fl. Dec., Feb. Fr. ripe Apl., May.

G. Cambogia, Desrouss. in Lamk. Encycl. III. 701; Fl. Br. I. 1. 261; Bedd. Fl. Sylv. t. 85. *Oopayi mara,* K On the southern ghâts of North Kánara, in evergreen forests. Fl. cold and hot seasons. Fr. ripe R.S.

G. Morella, Desrouss. in Lamk. Encycl. III. 701; Fl. Br. I. 1. 264; *G. pictoria,* Roxb. Bedd. Fl. Sylv. t. 87. Bedd. Fl. Sylv. t. 86. *Arsina gurgi, hardala,* K. In the Siddápur táluka evergreen forests of North Kánara. Fl. Nov. Fr. ripe Feb., Mch. *G. pictoria,* Roxb., is probably distinct from *G. Morella,* Desrouss. The staminodes and stigmas of the flowers are different. It is common in the North Kánara evergreen forests.

G. Xanthochymus, H. f. Fl. Br. I. 1. 269; *Xanthochymus pictorius,* Roxb. Bedd. Fl. Sylv. t. 88; Dalz. & Gibs. Bom. Fl. 31. *Janagi; deavkai,* K. Abundant in the evergreen forests of North Kánara. Fl. Apl., May. Fr. ripe C.S.

G. ovalifolia, Hook. f. Fl. Br. I. 1. 269; *Xanthochymus ovalifolius.* Roxb. Bedd. Fl. Sylv. 21; Dalz. & Gibs. Bomb. Fl. 31. *Haldi,* M.; *tawir,* Vern. *Var. 3, macrantha, is common on the Gairsoppah ghát.* Fl. Jany. Evergreen forests of North Kánara and Konkan ghâts. Fl. H.S. Fr. C.S.

OCHROCARPUS, Thouars.

Evergreen trees. Leaves opposite or 3-verticellate. Calyx opening into 2 or 3 valves. Petals 4-7. Stamens indefinite. Berry 1-4-seeded.

O. longifolius, Benth. & H. f. Gen. Plant. 1. 980; Bedd. Fl. Sylv. t. 89; Fl. Br. I. 1. 270; *Calysaccion longifolium,* Wgt. Ic. t. 1999; Dalz. & Gibs. Bom. Fl. 32, *Wundy, punay, surungi, suragi,* K.; *gardundy,* Vern. North Kánara and Konkan evergreen ghát forests. Fl. Mch. Fr. R.S.

CALOPHYLLUM, Linn.

Evergreen trees. Leaves opposite, shining, with numerous parallel veins at right angles to midrib. Sepals and petals 4-12, imbricate. Stamens numerous. Ovules solitary, erect. Fruit a drupe.

Petals 4.　Tree quite glabrous...　　...　　...　　...*C. inophyllum.*
　　　"　　Young parts tomentose　　...　　...　　...*C. tomentosum.*
Petals 3-0 ...　　...　　...　　...　　...　　...　　...*C. Wightianum.*

C. inophyllum, Linn. Fl. Br. I. 1. 273; Bedd. Fl. Sylv. 22; Dalz. & Gibs. Bomb. Fl. 31. *Undi,* M.; *vuma, hona,* K. Common along the banks of rivers near the coast; also commonly planted or indigenous along the coast from Bombay southwards, Fl. Dec. Fr. H.S.

C. tomentosum, Wgt. Ic. t. 110; Fl. Br. I. 1. 274; Bedd. Fl. Sylv. 22; *C. angustifolium*, Dalz. & Gibs. Bomb. Fl. 32; of Roxb. (?). *C. elatum*, Bedd. Fl. Sylv. l. c. 22. Poon spar tree. *Nagari*, M.; *surhoni*, K. In many of the ghát evergreen forests of North Kánara. Common on the Nilkund and Gairsoppab gháts; attains a great size. Fl. H.S. Fr. July. The prominent close set veins on both surfaces of the leaves are characteristic.

C. Wightianum, Wall. Cat. 4847; Fl. Br. I. 1. 274; Bedd. Fl. Sylv. t. 90; *C. spurium*, Choisy; Dalz. & Gibs. Bomb. Fl. 32. *Bobbi*, M.; *Irai*, K. Very common along the banks of North Kánara rivers; it has a very characteristic yellow bark. Fl. Dec. Fr. Mch.

MESUA, L,

Trees. Leaves opposite, rigidly coriaceous, veins inconspicuous, very numerous. Flowers large, polygamous, of hermaphrodite axillary solitary. Sepals and petals 4 each, imbricate. Stamens numerous, free or connate at the base. Ovary 2-celled; style long, stigma peltate. Ovules 2 in each cell, erect. Fruit woody, 1-4-seeded; seeds with a thin fragile testa.

M. ferrea, Linn. Fl. B. I. 1. 277; Dalz. & Gibs, Bomb. Fl. 31. *M. coromandelina*, Bedd. Fl. Sylv. t. 64. *Nagchampa*, M.; *nagasampige*, K. Sparingly throughout the evergreens of North Kánara and South Konkan. Fl. Mch. Fr. May. Very ornamental when in full bloom.

ORDER 12. TERNSTRŒMIACEÆ.

Shrubs or trees. Leaves simple, alternate, usually coriaceous. Flowers regular, hermaphrodite or rarely unisexual. Sepals 5, rarely 4-7, free or slightly connate. Petals 5, rarely 4-9. Stamens numerous, free or shortly connate at the base, usually adnate to the base of the deciduous corolla. Ovary free, sessile 3-5-celled; styles as many, free or connate; ovules 2 or many in each cell. Fruit baccate or capsular. Seeds few or numerous, placentas axile. Albumen scanty or 0.

> Fruit a berry; albumen fleshy. Anthers basifixed ... ...EURYA.
> Fruit a capsule; albumen 0. Anthers versatile ... ...GORDONIA.

EURYA, Thunb.

Shrubs. Leaves crenate-serrate. Flowers diœcious in axillary fascicles, bracteoles persistent. Sepals 5. Petals 5 united at the base. Stamens 15 or less. Ovary usually 3-celled. Styles 3, free or united; ovules many in the inner angle of each cell. Fruit a berry. Albumen fleshy.

E. japonica, Thunb. Fl. Jap. 191, t. 25; Bedd. Fl. Sylv. t. 92; Fl. Br. I. 1. 284. Konkan, Stocks. Fl. May, Sept.

GORDONIA, Ellis.

Evergreen trees. Leaves usually crenate. Flowers large, white, fragrant, solitary axillary or collected at the ends of the branches,

2-4 bracteolate. Sepals usually 5, unequal, graduating from the bracts to the petals. Petals free, the innermost larger. Stamens 5-adelphous or all connate, adnate to the petals. Ovary 3-5 celled : stigma stout, spreading ; ovules 4-8 in each cell. Capsule oblong, woody angled, dehiscent. Seeds flat-winged, albumen 0.

G. obtusa, Wall. Cat. 1459 ; Fl. Br. I. 1. 291 ; Bedd. Fl. Sylv. t. 83. Konkan gháts.

ORDER 13.   DIPTEROCARPEÆ.

Resinous trees or climbing shrubs. Leaves alternate, simple. Stipules usually small. Flowers regular, bisexual, sweet-scented. Calyx 5-lobed ; 2 or more segments accrescent in fruit. Petals 5. Stamens pentamerous. Connective often aristate. Ovary 1-3-celled. Ovules 2 in each cell, pendulous, laterally affixed or erect. Fruit winged. Seeds 1-2.

> Calyx much enlarged in fruit.
> Calyx in fruit with a distinct tube, wings 2, erect ...DIPTEROCARPUS.
> Wings 5, unequal, spreading ...   . ANCISTROCLADUS.
> Calyx in fruit with an obscure tube.
> Fruit 3-5-winged   ...   .SHOREA.
> Fruit 2-winged   ...   .HOPEA.
> Calyx not winged in fruit   ...   .VATERIA.

DIPTEROCARPUS, Gaertn.

Large trees, stellately pubescent. Leaves coriaceous simple, stipules large valvate enclosing the terminal bud. Flowers large, racemed. Calyx-tube free. Petals pubescent externally. Stamens numerous. Ovary 3-celled ; ovules 2 in each cell. Fruit 1-seeded. enclosed in the calyx-tube, free ; wings 2, erect.

D. turbinatus, Gaertn. f. Fruct. III. 51, t. 188 ; Fl. Br. I. 1. 295; *D. indicus*, Bedd. Fl. Sylv. t. 94. Woodoil tree. *Challane*, K. On the southern gháts of North Kánara, in evergreen forests ; common on the Gairsoppah ghát. Fl. Dec., Jan. Fr. ripe May. Certainly indigenous.

ANCISTROCLADUS, Wall.

Climbing shrubs with circinately hooked branches. Leaves coriaceous, entire. Flowers in terminal or lateral panicles. Calyx-tube short ; lobes unequally enlarged. Stamens 5-10. Ovary 1-celled, inferior ; styles 3, articulated to rounded disc ; ovule solitary. Fruit crowned with the accrescent calyx wings.

A. Heyneanus, Wall. Cat. 7262 ; Fl. Br. I. 1. 229 ; Dalz. & Gibs. Bomb. Fl. 34. *Kardor, kurdul*, M. On the gháts of the Konkan and North Kánara, in evergreen forests. Fl. Mch. Fr. April.

SHOREA, Roxb.

Resinous trees. Leaves entire. Flowers bracteate, in lax cymose panicles. Calyx-tube short, adnate to the torus, all seg-

ments enlarged in fruit. Stamens indefinite, connective subulate,
produced. Ovary 3-celled, ovules 2 in each cell. Fruit enclosed
in the winged calyx. Seed 1, cotyledons fleshy.

S. Talura, Roxb. Fl. Ind. II. 618; Fl. Br. I. 1. 304; *S. laccifera,*
Heyne; Bedd. Fl. Sylv. 26. In the evergreen forests of the Sirsi táluka
of North Kánara, rare. Fl. Jan. Fr. April.

HOPEA, Roxb.

Resinous trees. Leaves entire, firm. Flowers in lax panicles of
unilateral racemes. Calyx-tube very short, segments imbricate.
Petals connate, deciduous. Stamens 15, connective long cuspidate.
Ovary 3-celled, cells 2-ovuled. Nut 1-seeded, 2-winged.

H. Wightiana, Wall. Cat. 6295; Fl. Br. I. 1. 309; Bedd. Fl. Sylv.
t. 96. *Haiga,* K.; *kavsi,* M. Common along the banks of the North
Kánara ghât rivers and nálás. Fl. Mch., June. Fr. R.S.

VATERIA, L.

Resinous trees. Leaves entire, firm. Flowers usually paniculate. Calyx with a short tube adnate to the torus, divisions scarcely
accrescent in fruit. Stamens 15. Ovary 3-celled, cells 2-ovuled.
Fruit ovoid 1-seeded, seated on the reflexed calyx. Cotyledons
thick unequal, radicle superior.

V. indica, L. Fl. Br. I. 1. 313; *V. malabarica,* Blume; Bedd. Fl.
Sylv. t. 84. Piney varnish tree; Indian copal tree. *Dhupada,* K. On
the southern ghâts of North Kánara, in evergreen forests; commonly
planted along road sides in the North Kánara district. Fl. Mch., Apl.
Fr. R.S.

ORDER 14. MALVACEÆ

Soft wooded shrubs or trees. Leaves simple, rarely compound,
alternate, often palminerved. Bracteoles 3 or more. Sepals 5-
valvate. Petals 5 twisted-imbricate. Stamens numerous monadelphous. Anthers 1-celled. Ovary 2, many celled. Fruit of dry
cocci, dehiscent or indehiscent. Albumen scanty, often mucila-
ginous or 0.

Staminal column tubular, entire.
Stigmas spreading. Ovary 10-celled   ...   .DECASCHISTIA.
Ovary 2-5-celled.   Bracteoles 5 or more   ...   ...HIBISCUS.
Stigmas coherent.   Bracteoles 3-5   ...   . THESPESIA.
Bracteoles 3, large cordate   ...   .GOSSYPIUM.
Staminal column tubular at base only.   Filaments
    5-delphous or free.
Leaves simple   ...   .KYDIA.
Leaves digitate   ...   .ADANSONIA.
Filaments with single anthers   ...   .BOMBAX.
Filaments with 2-3 anthers   ...   .ERIODENDRON.

DECASCHISTIA, W. & A

Shrubs. Leaves entire or lobed. Flowers large. Bracteoles 10.
Styles 10, connate below. Capsule loculicidally 10-valved, hispid.

D. **trilobata**, Wgt. Ic. t. 88 ; Fl Br. I. 1.332 ; Dalz. & Gibs. Bomb. Fl. 21. On the Konkan and the Supa ghâts of North Kánara, in open situations. Fl. Oct., Nov. Fr. C.S.

HIBISCUS, Medik.

Herbs, shrubs or trees. Leaves stipulate. Inflorescence axillary. Bracteoles 5 or more. Calyx 5-divided, valvate, sometimes spathaceous. Styles 5, connate below. Fruit a capsule, loculicidally 5-valved.

> Climbing prickly shrub. Bracteoles forked at apex ...*H. furcatus.*
> Bracteoles more than 5 free or connate at base. Capsule
>   small hispid. Seeds glabrescent ... ...*H. collinus.*
> Tree. Bracteoles connate at base. Capsule spuriously
>   10-celled. Seeds pilose ... ...*H. tiliaceous.*

H. **furcatus**, Roxb. Fl. Ind. III. 204 ; Fl. Br. I. 1. 335 ; Dalz. & Gibs. Bomb. Fl. 19. Common in many of the North Kánara and Konkan moist forests. Fl. Jany., Feb. Fr. H.S.

H. **collinus**, Roxb. Fl. Ind. III. 198 ; Fl. Br. I. 1. 338. Konkan Fl. Br. I. Fl. R.S.

H. **tiliaceous**, Linn. Fl. Br. I. 1. 343 ; Bedd. Fl. Sylv. 29 ; *Paritium tiliaceum*, W. & A. Dalz. & Gibs. Bom. Fl. 17. Konkan and North Kánara, along the sea-coast ; also along the banks of tidal rivers, near the sea. Fl. Jany. Fr. Mch., Apl. " The flowers yellow in the morning turn red in the afternoon." Bedd.

THESPESIA, Corr.

Trees or shrubs. Leaves entire or lobed. Inflorescence axillary. Bracteoles 5-8, deciduous. Calyx truncate, 5-toothed or 5-partite. Corolla large, yellow. Staminal tube 5-toothed at apex. Ovary 4-5-celled ; style club-shaped, 5-furrowed ; ovules few in each cell. Capsule dehiscent or indehiscent. Seeds glabrous or tomentose.

> A shrub. Leaves lobed. Seeds glabrescent ... ... *T. Lampas.*
> A tree. Leaves entire. Seeds pilose or powdery .. *T. populnea.*

T. **Lampas**, Dalz. & Gibs. Bomb. Fl. 19 ; Fl. 8 Br. I. 1. 345. *Ranbhendy*, M. In the Konkan and North Kánara forests from the sea-level upwards. Very common. Fl. Aug., Oct. Fr. Nov., Dec.

T. **populnea**, Corr. in Ann. Mus. IX. 290 ; Fl. Br. I. 1. 345 ; Dalz. & Gibs. Bom. Fl. 150 ; Bedd. Fl. Sylv. t. 63. *Bhendy*, tulip or Portia tree. Along the sea-shores of the Konkan and North Kánara ; often cultivated as a road-side tree. Fl. C.S. Fr. R.S.

GOSSYPIUM, L.

Herbs, shrubs or low trees. Leaves palmately lobed. Flower large yellow with a purple centre or all purplish. Bracteoles 3, large leafy, cordate, black glandular dotted like the calyx. Calyx cup-shaped, truncate or 5-toothed. Petals convolute or spreading.

Ovary 5-celled; style clavate 5-grooved with 5 stigmas; ovules many in each cell. Capsule 3-5-valved. Seeds clothed with woolly hairs; cotyledons leafy, black-dotted.

G. Stocksii, Mast. Fl. Br. I. 1. 346. Limestone rocks on the coast of Sind; also near Karáchi. Fl. Dec., Jany. Fr. Feb., Mch· *G. herbaceum*, Linn., *G. barbadense*, Linn., aud *G. aiboreum*, Linn., are cultivated species.

KYDIA, Roxb.

Trees. Leaves palminerved, usually lobed. Flowers unisexual. Calyx 5-lobed, persistent, surrounded at the base with the 4-6 leafy bracteoles, which are accrescent in fruit. Petals 5, longer than calyx attached to the base of the staminal column. Stamens monadelphous; the tube divided into 5 divisions, each bearing 3 anthers, which are imperfect in the female flower. Ovary 2-3-celled; style 3-cleft; stigmas 3, peltate, imperfect in the male flower; ovules 2 in each cell. Capsule subglobose, 3-valved. Seeds reniform, furrowed. Young leaves, branches, inflorescence, capsule, &c., covered with a grey stellate tomentum.

K calycina, Roxb. Fl. Ind. III. 188; Fl. Br. I. 1. 348; Dalz. & Gibs. Bomb. Fl. 24; Bedd. Fl. Sylv. t. 3; Brandis For. Fl. 29. *Warung, iliya*, M.; *bellaka*, K. Common in deciduous forests on the North Kánara and Konkan ghâts. Fl. July, Nov. Fr. Jany· Yields a strong fibre.

ADANSONIA, L.

Tree with a very thick trunk at the base. Leaves digitate, deciduous. Flowers axillary, large, pendulous, long peduncled. Calyx 5-cleft, 2-bracteolate. Petals 5. Staminal tube cylindrical. dividing above into numerous filaments. Ovary ovoid, silky-tomentose; style filiform, bent after flowering. Fruit a pendulous capsule with mealy pulp.

A. digitata, Linn. Fl. Br. I. 1. 348; Brandis For. Fl. 30; Dalz. & Gibs. Bomb. Fl. Suppl. 9. *Gonik chentz*, M.; *goruk amla*, H. Baobab, monkey bread tree. Cultivated in many parts of the presidency. Fl. May, June.

BOMBAX, L.

Trees. Leaves digitate, deciduous. Flowers axillary or subterminal. Calyx cup-shaped, truncate or 5-7-lobed. Petals large, scarlet, obovate. Stamens pentadelphous and divided into numerous filaments. Ovary 5-celled, stigmas 5; ovules many in each cell. Capsule 5-valved, woolly within. Seeds woolly.

Leaflets 5-7. Stamens 75   ...   . *B. malabaricum.*
Leaflets 7-9. Calyx prickly. Stamens about 600 ... *B. insigne.*

B. malabaricum, DC. Prod. 1. 479; Fl. Br. I. 1. 349; Brandis For. Fl.31; Bedd. Fl. Sylv. t. 32; *Salmalia malabarica*, Schott. Dalz. & Gibs.

Bomb. Fl. 22. The cotton tree. *Burla, sauri,* K. ; *sayar,* M. Throughout the presidency in deciduous forests. Fl. Feb., Mch. Fr. Apl., May. Ovary green, glabrous.

B. insigne, Wall. Pl. Asiat. Rar. 1. 71, t. 79, 80 ; Fl. Br. I. 1. 349. North Kánara, in deciduous forests, common on the Kyga ghát. Fl. Feb. Mch. Fr. Apl. Ovary red, tomentose.

ERIODENDRON, DC.

Trees. Leaves digitate. Flowers tufted at the ends of the branches, or axillary. Staminal bundles 5, each bearing 2-3 sinuous or linear anthers. Ovary 5-celled ; style cylindrical dilated, stigma obscurely 5-lobed ; ovules many in each cell. Capsule 5-valved, silky within. Seeds with silky hairs.

E. anfractuosum, DC. Prod. 1. 479 ; Fl. Br. I. 1. 350 ; Dalz. & Gibs. Bomb. Fl. 22 ; Bedd. Fl. Sylv. 30. *Shameula, kutsavar,* Vern. Khándesh, Dalz. Fl. H.S.

Order 15. STERCULIACEÆ.

Trees or shrubs. Leaves stipulate, alternate, simple or digitate. Flowers regular, uni or bisexual. Petals 5 or 0. Stamens monadelphous indefinite or definite, or free and definite, with or without alternating staminodes. Ovary free, 2-5-celled, rarely 1 carpel. Fruit dry or fleshy dehiscent or indehiscent.

Flowers unisexual or polygamous. Petals 0.
Andrœcium columnar or sessile.

Anthers numerous. Fruit dehiscent	... STERCULIA.
Anthers 5. Fruit indehiscent	... HERITIERA.
Flowers hermaphrodite. Petals deciduous.	
Staminodes present. Fruit a capsule.	
Capsule membranous	... KLEINHOVIA.
Capsule woody, not tubercled.	
Seeds not winged	... HELICTERES.
Seeds winged	... PTEROSPERMUM.
Capsule tubercled, like a mulberry in appearance	... GUAZUMA.
No staminodes. Capsule loculicidal	... ERIOLÆNA.

STERCULIA, Linn.

Trees. Leaves simple, palmately lobed or digitate. Flowers polygamous. Calyx tubular, often coloured. Petals 0. Ovary of 4-5, sessile or stalked carpels. Ovules 2 or more in each. Follicles woody or membranous. Seeds albuminous, rarely winged.

Follicles woody. Seeds not winged	
Leaves digitate	... *S. fœtida*
Leaves palmately lobed.	
Carpels bristly...	... *S. urens.*
Carpels rusty villous	... *S. villosa.*
Leaves simple 1-nerved	... *S. guttata.*
Follicles membranous	... *S. colorata.*
Seeds winged	... *S. alata.*

S. fœtida, Linn.; DC. Prod. 1. 483 ; Fl. Br. I. 1. 354 ; Bedd. Fl. Sylv. 31. *Jungle badam,* H. Konkan. Fl. Apl., May. Fr. C.S.

S. urens, Roxb. Fl. Ind. III. 145 ; Fl. Br. I. 1. 355 ; Dalz. & Gibs. Bomb. Fl. 231 ; Bedd. Fl. Sylv. 32; Brandis For. Fl. 33. *Karai, kandol, saldawar,* M. ; *kurda, kalauri,* Vern. *Kud* in the Sátpudás of Khándesh. In the Deccan districts, in dry deciduous forests ; often associated with *Boswellia* in the Khándesh Sátpudás ; near the coast on rocky soil in the Konkan and North Kánara. Fl. Jany. Fr. Apl. May.

S. villosa, Roxb. Fl. Ind. III. 153 ; Fl. Br. I. 1. 355 ; Brandis For. Fl. 32; Dalz. & Gibs. Bomb. Fl. 22. *Sarda,* M. ; *savaya,* K. In deciduous forests from Gujarát southwards to North Kánara. Fl. Dec., Jany. Fr. H.S., R.S.

S. guttata, Roxb. Fl. Ind. III. 148 ; Fl. Br. I. 1. 355 ; Bedd. Fl. Sylv. t. 105 ; Dalz. & Gibs. Bomb. Fl. 23. *Kookur, goldar, koketi,* M.; *happu savaga,* K· Throughout the forests of the Konkan and North Kánara usually in or near moist evergreens. Fl. Jany. Fr. ripe R.S.

S. colorata, Roxb. Fl. Ind. III. 146. ; Fl. Br. I. 1. 359 ; Bedd. Fl. Sylv. 32 ; Dalz. & Gibs. Bomb. Fl 23. *Kowsey,* M. *;'haikoi, khavas, khanshi,* Vern. Throughout the presidency in deciduous forests, nowhere abundant. Fl. Mch., May. Fr. May, June.

S. alata, Roxb. Fl. Ind. III. 152; Fl. Br. I. 1. 360 ; *S. Haynii,* Bedd. Fl. Sylv. t. 230· In the evergreen forests of North Kánara at the foot of the Burboli ghát, apparently rare· The winged seeds are characteristic· Ripe fruit C.S.

HERITIERA, Ait.

Trees. Leaves simple, silvery scaly beneath. Calyx 4-7 cleft. Petals 0. Staminal column slender bearing a single ring of 5 anthers. Ovary carpels 5, 1 ovule in each. Fruit woody, keeled or winged on back.

H. littoralis, Dryand. in DC. Prod. 484; Fl. Br. I. 1. 363 ; Dalz. & Gibs. Bomb. Fl. 22 ; Bedd. Fl. Sylv. 33. Along the banks of the Kálánadi and near the coast of North Kánara at Kárwár, but nowhere abundant. Fl. R.S· Fr. Jany ; remains long on tree·

KLEINHOVIA, Linn.

Tree. Leaves entire, palminerved. Inflorescence terminal. Sepals deciduous. Petals 5 unequal. Staminal column dilated into a 5-fid cup. Stamens 15. Ovary 5-celled. Capsule membranous inflated, loculicidally 5-valved. Seeds ; tubercled.

K. Hospita, Linn. ; DC. Prod. 1. 488 ; Fl. Br. I. 1. 364 ; Dalz. & Gibs. Bomb. Fl. 23 ; Bedd. Fl. Sylv· 33. Southern Konkan, Nimmo (a doubtful native, Dalz.)

HELICTERES, Linn.

Trees or shrubs. Leaves simple. Calyx tubular, 5-fid. Petals 5, clawed. Staminal column adnate to gynophore. Anthers 5-10. Ovary 5-lobed, 5-celled. Carpels dehiscent, spirally twisted.

H. Isora, Isora, Linn.; Roxb. Fl. Ind. III. 143; Fl. Br. I. 1. 365; Dalz. & Gibs. Bomb. Fl. 22; Bedd. Fl. Sylv. 33; Brandis For. Fl. 34. *Kevani, muradsing,* M.; *kavargi,* K.; *murrori-ka-jhar,* H. Throughout the forests of the presidency; abundant in many of the North Kánara ghát forests. Yields a strong fibre. Fl. Aug., Dec. Fr, ripe Mch.

PTEROSPERMUM, Schreb.

Trees or shrubs, scaly or stellate tomentose. Leaves leathery, oblique, simple or lobed. Bracteoles entire or laciniate, persistent or caducous. Sepals and petals 5 each, deciduous. Staminal column short, adnate to the gynophore. Filamented anthers in 3s between each pair of 5 ligulate staminodes. Connective apiculate, Capsule loculicidally 5-valved. Seeds winged. Albumen thin or 0.

Bracteoles linear entire, caducous	...	... ...	*P. suberifolium.*
Bracteoles lacinate or palmately lobed...	...	...	*P. acerifolium.*
Bracteoles pinnatisect, segments linear	...	...	*P. reticulatum.*
Bracteoles deeply gashed	..	... ...	*P. Heyneanum.*

P. suberifolium, Lam. Ill. t. 570, f. II.; Fl. Br. I. 1. 367; Dalz. & Gibs. Bomb. Fl. 24; Bedd. Fl. Sylv. 34. *Muchcunda,* Vern. Southern gháts of North Kánara, in evergreen forests, also in the Konkan forests. Fl. C.S.

P. acerifolium, Willd.; D.C. Prod. 1. 500; Bedd. Fl. Sylv. 35; Brandis For. Fl. 35. On the southern gháts of North Kánara, in evergreen forests, abundant on the Devimone ghát. Fl. C.S. Fruit remains long on tree.

P. reticulatum, W. & A. Prod. 1. 69; Fl. Br. I. 1. 369; Bedd. Fl. Sylv. 34. In the evergreen forests of North Kánara near the Falls of Gairsoppah, rare. Flowers during the cold weather, the ripe fruit remains long on the tree.

P. Heyneanum, Wall. Cat. 1. 169; Fl. Br. I. 1. 369; Bedd. Fl. Sylv. 34. *P. Lawianum,* Nimmo. Dalz. & Gibs. Bomb. Fl. 24. Dhárwár and southern gháts, Dalz. Fl. C.S.

ERIOLÆNA, DC.

Trees. Leaves simple or lobed, cordate, downy or stellately tomentose beneath. Flowers on axillary few flowered peduncles. Bracteoles 3-5, often laciniate. Calyx deeply 5-cleft. Petals flat with a broad tomentose claw. Stamens numerous, all fertile, monadelphous, in many rows, the outer gradually shorter. Style 1, stigma 10-lobed. Capsule woody, 5-10-celled, loculicidal. Seeds numerous in each cell, winged above.

Bracteoles pinnatisect.
Peduncles 3-flowered.   Style pilose   ...   ... *E. Stocksii*.
Peduncles many flowered.
Style pubescent   ...   . *E. Hookeriana*.
Style glabrous   ...   . *E. Candollei*.
Bracteoles entire   ...   . *E. quinquelocularis*.

E. Stocksii, H. f. & T. ms.   Fl. Br. I. 1. 370.   Konkan, Stocks.

E. Hookeriana, W. & A. Prod. 70; Fl. Br. I. 1. 370; Bedd. Fl· Sylv. 35; Brandis For. Fl. 36.   Konkan, and in the dry forests of the Satpudás, Fl. Mch., Apl.   Fr. C.S.

E. Candollei, Wall. Pl. As. Rar. 1. 51, t. 64; Cat. 1175; Fl. Br. I. 1. 370; Dalz. & Gibs. Bomb. Fl 24. *Hadang*, K. Throughout the deciduous forests of North Kánara; Belgaum and Konkan. Wood much used in the construction of country carts. Fl. H.S.   Fr. C.S.

E. quinquelocularis, Wgt. Ic. t. 882; Fl. Br. I. 1. 371; Bedd. Fl. Sylv. 35.   In the forests of the Konkan and Belgaum gháts.   Fl. July.

GUAZUMA, Plum.

A tree. Leaves simple, tomentose. Flowers in axillary cymes. Sepals 5, at first spathaceous. Petals 5, concave at the base, prolonged at apex into 2 narrow processes. Stamens 10 connate into a column, tubular below, above of 5 fertile 3-antheriferous filaments, opposite the petals and 5 lanceolate staminodes opposite the sepals: anther lobes divergent. Ovary sessile, 5-lobed, 5-celled; ovules numerous in each cell. Capsule woody tubercled.

G. tomentosa, Kunth, DC. Prod. 1. 485; Fl. Br. I. 1. 375; Bedd. Fl. Sylv. t. 107. Dalz. & Gibs. Bomb. Fl. Suppl. 10. Cultivated in gardens throughout the presidency. Fl. July, Aug.   Fr. C.S.

ORDER 16. TILIACEÆ.

Mostly trees or shrubs. Leaves alternate, simple or lobed. Stipules deciduous. Flowers regular, hermaphrodite or unisexual. Sepals 3-5, valvate. Petals free, equal in number to sepals. Stamens numerous, rarely definite, springing from a dilated torus. Ovary free, 2-10-celled, placentation axile. Fruit fleshy or dry, 2-10 celled. Seeds albuminous or exalbuminous.

Anthers opening by slits.
Fruit not prickly ...   . GREWIA.
Fruit prickly   ...   . ERINOCARPUS.
Anthers opening by a terminal pore   ...   ... ELÆOCARPUS.

GREWIA, Linn.

Trees or shrubs. Leaves entire, often 3-nerved. Flowers axillary or terminal. Sepals 5 distinct. Petals clawed, glandular at base, sometimes wanting. Stamens numerous on raised torus. Ovary 2-4-celled; stigma shortly lobed. Fruit a drupe, fleshy or fibrous, entire or 2-4-lobed; stones 1 or more seeded. Cotyledons flat; albumen fleshy or 0.

Inflorescence terminal or extra axillary.
Scandent. Leaves asperous, 3-nerved. Drupe glabrescent
of 4 separate flattened lobes ... ... *G. umbellifera.*
Erect shrub. Leaves scabrous. Drupe 4-lobed, bristly ... *G. columnaris.*
Scandent. Leaves glabrescent. Drupe globose scarcely
lobed, yellow, pilose ... ... *G. orientalis.*
Scandent. Leaves thinly stellate hairy beneath. Drupe
1-celled, 1-seeded ... ... *G. umbellata.*
Scandent. Leaves scabrous. Drupe deeply 4-lobed, lobes
1-seeded, reddish, glabrous ... ... *G. heterotricha.*
Diffuse shrub. Leaves small, glabrous. Drupe small,
2-lobed, orange-red ... ... *G. populifolia.*
Inflorescence axillary.
Leaves hoary beneath.
Shrub. Leaves pubescent on both sides, 5-nerved. Drupe
slightly 2-lobed, grey... ... ... *G. orbiculata.*
Tree. Leaves often lobed, 3-5-nerved, stipules falcate
auricled. Drupe 1-4-lobed, black ... ... *G. tiliæfolia.*
Tree. Leaves roundish, tomentose, 5-nerved; stipules
subacute. Drupes subturbinate, 1-2-celled... ... *G. asiatica.*
Leaves not hoary beneath.
Shrub. Leaves scabrous. Drupe globose, long pedicellate,
rind crustaceous, hairy, 1-4-lobed size of pea, stones
pitted ... .- ... ... *G. pilosa.*
Shrub. Leaves 5-nerved, rugose above, villous beneath.
Drupe globose, rind crustaceous, 1-4-lobed, size of
cherry, red, pilose .. .. *G. villosa.*
Drupe fleshy didymous.
Tree. Leaves glabrous lanceolate, serrate. Drupe small,
2-lobed, black ... ... *G. lævigata.*
Drupes fleshy, 1-4-lobed.
Tree. Leaves glabrescent. Drupe deeply 4-lobed, pur
plish ... ... *G. Ritchiei.*
Tree or shrub. Leaves scabrous. Drupes yellow, sub-
globose, obscurely 4-lobed, size of a large pea ... *G. abutilifolia.*
Shrub. Leaves glabrescent, or pilose above, densely
tomentose beneath. Drupe glabrescent, size of a small
cherry, 4-lobed ... ...- *G. hirsuta.*
Shrub. Leaves velvetty beneath. Flowers polygamous.
Drupe 4-lobed, hairy, brownish, ½-inch .. ... *G. polygama.*
Inflorescence terminal, panicled. Leaves glabrescent.
Drupe entire, size of pea, glabrous ... ... *G. Microcos.*

G. umbellifera, Bedd. Fl. Sylv. 37; Fl. Br. I. 1. 393. This is a large scandent shrub, the stem of which is covered with blunt thorns. Albumen fleshy, white, copious. Common along the North Kánara ghâts, in evergreen forests. Flowers during the rainy season. Fruit ripe in February.

G. columnaris, Sm.; DC. Prod. 1. 510; Fl. Br. I. 1. 383; Dalz. & Gibs. Bomb. Fl. 26. North Kánara, in evergreen forests; mentioned in Dalz. & Gibs. Bomb. Fl., but no locality is given. It is also found on the Bababuden hills of Mysore. Fl. H.S. Fr. Oct.

G. oriontalis, Linn.; DC. Prod. 1. 50; Fl. Br. I. 1. 384; Dalz. & Gibs. Bomb. Fl. 26. Southern ghâts, Dalz.; also on the Bababuden hills at 3,000 ft. elevation. Fl. June. Fr. Sept., Oct.

G. umbellata, Roxb. Fl. Ind. II. 591; Fl. Br. I. 1. 385. Konkan.

G. heterotricha, Mast Fl. Br. I. 1. 385. A lofty climber, common in the evergreen forests of North Kánara from Ainshi southwards, not a tree. Fl. C.S. Fr. H.S.

G. populifolia, Vahl. DC. Prod. 1. 511; Fl. Br. I. 1. 385; Brandis For. Fl. 38. *Gingo*, Vern.; *gungo* in Sind. Throughout the dry districts of the presidency and Sind. Fl. Fr. Sept., Oct. in the Dhárwár district.

G. salvifolia, Heyne in Roth. Nov. Sp. 239; Fl. Br. I. 1. 386. Dry forests near Bádámi, Bijápur district. Fl. June. Fr. Sept., Oct. Perhaps not distinct from *G. excelsa*, Vahl.

G. orbiculata, Rottl. in Nov. Act. Nat. Cur. Berol. 1803, 205; Fl. Br. I. 1. 386. Konkan.

G. tiliæfolia, Vahl. Symb. 1. 35; Fl. Br. I. 1. 386; Dalz. & Gibs. Bomb. Fl. 26; Bedd. Fl. Sylv. 37; Brandis For. Fl. 41. *Dadsal, butale,* K.; *dhamani,* M. Throughout the deciduous forests of the presidency, common in North Kánara, where it attains a large size. Wood elastic, valuable for carriage shafts. Flowers during May. Fruit ripe during rainy season.

G. asiatica, L.; W. & A. Prod. 1. 79; Fl. Br. I. 1. 386; Dalz. & Gibs. Bomb. Fl. 26; Brandis For. Fl. 40; Bedd. Fl. Sylv. 37. *Dhamin phalsa,* H. Poona district, also cultivated. Fl. Feb., Mch. Fr. May, June.

G. pilosa, Lam. Dict. III. 43; Fl. Br. I. 1. 388; Dalz. & Gibs. Bomb. Fl. 26; Brandis For. Fl. 39. Throughout the presidency, in deciduous forests. Fl. May, July Fr. Sept., Oct.

G. villosa, Willd. in Nov. Act. Nat. Cur. Berol. 1803, 205. Fl. Br. I. 1. 388; Dalz. & Gibs. Bomb. Fl. 25; Brandis For. Fl. 39. Throughout the presidency, in the dry districts. Fl. June, Sept.

G. lævigata, Vahl. Symb. 1. 34; Fl. Br. I. 1. 389. Brandis For. Fl. 42; Bedd. Fl. Sylv. 37. North Kánara and Konkan near village sites and in deciduous forests, not common. Fl. R.S. Fr. R.S., C.S.

G. Ritchiei, Mast. Fl. Br. I. 1. 389. Konkan gháts.

G. abutilifolia, Juss. in Ann. Mus. IV. 92; Fl. Br. I. 1. 390; Dalz. & Gibs. Bomb. Fl. 26. Bedd. Fl. Sylv. 37. Throughout the presidency, common in North Kánara along nálás and in moist shady places. Flowers in June. Fruit on deflexed villous pedicils ripe, August April.

G. hirsuta, Vahl. Symb. 1. 34; Fl. Br. I. 1. 391. Dhárwár district, in deciduous open forest, common. Fl. Aug. Fr. C.S.

G. polygama, Roxb. Fl. Ind. II. 588; Fl. Br. I. 1. 391; Dalz. & Gibs. Bomb. Fl. 26; Brandis For. Fl. 42. *Gowli,* Vern. Along the gháts from Bombay southwards. Fl. July., Aug. Fr. Nov., Dec.

G. Microcos, L.; DC. Prod. 1. 510; Fl. Br. I. 1. 392; Dalz. & Gibs. Bomb. Fl. 26. *Shirul, asólin,* M. Common in evergreen forests throughout the Konkan and North Kánara. Fl. Fr. R.S.

ERINOCARPUS, Nimmo.

A tree. Leaves cordate, toothed, 5-7-nerved. Flowers large, yellow, in terminal panicles. Sepals 5. Petals 5, clawed. Stamens many on a raised torus, free, or united at the base. Ovary 3-celled, cells 2-ovuled, style filiform, stigma minute. Fruit woody, triangular, spiny, 1-celled, indehiscent. Seed pendulous, albumen fleshy.

E. **Nimmoanus**, Grah. Cat. Bomb. Pl. 21. Fl. Br. I. 1. 394 ; Dalz. & Gibs. Bomb. Fl. 27 ; Bedd. Fl. Sylv. t. 110. *Chor, choura, chira, haladi*, M. ; *adwi bhendy*, K. Throughout the presidency, common in North Kánara and Konkan deciduous forests. Fl. Sept , Oct. Fr. C.S.

ELÆOCARPUS, Linn.

Trees. Leaves simple, alternate. Flowers in axillary racemes, generally bisexual. Sepals 5. Petals 5, fringed or lobed, inserted round the base of the thick glandular disc. Stamens numerous, inserted on the disc between the glands; anthers linear, opening by a terminal pore. Ovary sessile, 2-5-celled; style columnar. Fruit a drupe with a bony stone, 1-5-celled. Seeds pendulous, 1 in each cell, albumen fleshy.

Drupe 5-celled, 5-seeded ... ... *E. Ganitrus.*
Drupe 1-seeded.
Petiole glandular at apex. Anthers tipped with a few
hairs. Drupe oblong, slightly falcate ... ... *E. serratus.*
Petiole glandular. Drupe ovoid, straight. Anthers with-
out hairs ... ... *E. oblongus.*
Anthers with a long awn.
Stamens 70. Awns erect ... ... *E. tuberculatus.*
Stamens 50. Awns erect ... ... *E. aristatus.*
Stamens 20. Awns a length reflexed ... ... *E. Munroii.*

E. **Ganitrus**, Roxb. Fl. Ind. 11. 592 ; Fl. Br. I. 1. 400 ; Dalz. & Gibs. Bomb. Fl. 27 ; Brandis For. Fl. 43 ; Bedd. Fl. Sylv. 38. *Rudrack,* H. On the higher ghats of the Konkan, Dalz. Fl. C.S.

E. **sorratus**, L.; W. & A. Prod. 1. 82 ; Fl. Br. I. 1. 401 ; Bedd. Fl. Sylv. 38. Brandis For. Fl. 43. On the southern ghats of North Kánara, in evergreen forests, common near the Falls of Gairsoppah. Fl. C.S. Fr. Apl.

E. **oblongus**, Gaertn.; W. & A. Prod. 82 ; Fl. Br. I. 1. 403 ; Dalz. & Gibs. Bomb. Fl. 27 ; Bedd. Fl. Sylv. 38. *Khas,* M. On the Konkan and North Kánara ghats, in evergreen forests. The North Kánara tree has the petiole and nerves underneath more or less covered with reddish tomentum. Fl. Mch., Apl. Fr. R.S.

E. **tuberculatus**, Roxb. Fl. Ind. II. 594 ; Fl. Br. I. 1. 404 ; Bedd. Fl. Sylv. t. CXIII. *Monocera tuberculata*, W. & A. Dalz. & Gibs. Bomb. Fl. 27. Southern ghats of the presidency from Belgaum southwards in evergreen forests. Fl. Jany. Fr. R.S.

E. **aristatus**, Roxb. Fl. Ind. II. 559 ; Fl. Br. I. 1. 405. Konkan and North Kánara ghats, in evergreen forests. Fl. Apl. Fr. August.

E. **Munroii**, Mast. Fl. Br. I. 1. 407. *E. glandulifera*, Hook. f. Bedd. Fl. Sylv. 38. Konkan, Stocks. Fl. Sept., Oct. Fr. C.S.

ORDER 17. LINEÆ.

Herbs or shrubs. Leaves alternate, simple, entire. Inflorescence various. Flowers regular, bisexual. Sepals 5, free or connate below, imbricate. Petals 5, usually caducuous, often contorted. Stamens 4-5 with as many staminodes, or 8-10, filaments united at the base into a ring. Glands 5, entire or 2-lobed, usually adnate to the staminal ring or obsolete. Ovary entire 3-5-celled; styles 3-5, free or more or less connate, stigmas terminal. Ovules 1-2 in each cell, inserted in the inner angle, pendulous. Fruit usually splitting into 3-5-cocci, rarely a drupe. Seeds 1-2 in each cell, testa sometimes winged albumen fleshy or 0.

HUGONIA, L.

Climbing, tomentose shrubs, with woody tendrils. Leaves alternate, serrate, stipulate. Inflorescence of yellow flowers, lower peduncles converted into spiral hooks. Sepals 5. Petals 5. Stamens 10 with glandular swellings on the basal ring between the filaments. Ovary 5-celled; styles 5, filiform; ovules 2, collateral in each cell. Drupe globose. Seeds compressed, albuminous.

H. **Mystax**, Linn. Fl. Br. I. 1. 413; Dalz. & Gibs. Bomb. Fl. 17-From the Konkan southwards near the sea-coast atVengurla, not observed in North Kánara. Fl. Fr. Aug.

ORDER 18. MALPIGHIACEÆ.

Climbing shrubs. Leaves opposite, entire. Calyx often with sessile glands. Petals 5, usually clawed. Stamens 10. Ovary 3-celled. Cells each 1 seeded. Fruit of 1 or more winged samaras.

Styles 1-2. Calycine gland 1, large, adnate to pedicil   ... HIPTAGE.
Styles 3. Calycine glands 0. ...   . ASPIDOPTERYS.

HIPTAGE, Gærtn.

Climbing shrubs. Leaves opposite. Stipules 0. Flowers white, fifth petal coloured. Calyx 5-partite; glands large, adnate to pedicil. Fruit of 1-3, 2-3-winged samaras.

H. **Madablota**, Gærtn. Fruct. II. 169. t. 116; Fl. Br. I. 1. 418; Dalz. & Gibs. Bomb. Fl. 33; Brandis For. Fl. 44. *Haladwail*, M. *Bokhi, utimukta,* Vern. Throughout the presidency near water-courses and moist places. Fl. Fr. Jan. Feb.

ASPIDOPTERYS, A. Juss.

Climbing shrubs. Leaves opposite, eglandular. Flowers small, yellow or white. Calyx short, eglandular. Petals 5, not clawed. Ovary 3-lobed; sides winged. Fruit of 1-3 samaras.

Ovary hairy. Samara oblong.   ...   .— *A. Roxburghiana.*
Ovary glabrate. Samaras suborbicular, very membranous. *A. canarensis.*
White tomentose. Samaras orbicular, not membranous... *A. cordata.*

A. Roxburghiana, A. Juss. in Archiv. Mus. Hist. Nat. III. 511 ; Fl. Br. I. 1. 420; Dalz. & Gibs. Bomb. Fl. 33. Throughout the ghát forests of the Konkan ; not seen in North Kánara.

A. canarensis, Dalz. in Hook. Kew Journ. Bot. III. 37 ; Fl. Br. I. 1. 420. In the evergreen forests of the Supa gháts of North Kanara, rare. Fl. Mch. Fr. May.

A. cordata, A. Juss. in Archiv, Mus. Nat. Hist. III. 513 ; Dalz. & Gibs. Bomb. Fl. 34. Fl. Br. I. I. 421. Common throughout the Konkan and North Kánara, in moist forests. Fl. Sept, Oct. Fr. C.S.

ORDER 19. GERANIACEÆ.

Herbs rarely trees. Leaves opposite or alternate. Flowers hermaphrodite. Sepals 5, rarely 4 or more. Petals equal sepals or 0. Torus with 5 glands alternating with petals. Stamens as many or double or treble the sepals. Ovary of 3-5 carpels. Fruit capsular or berried.

AVERRHOA, Linn.

Trees. Leaves imparipinnate. Flowers small regular. Sepals 5. Petals 5. Stamens 10 or 5 and 5 staminodes. Ovary 5-lobed. Ovules many.

Leaflets glabrous and glaucous beneath...   ...   ... *A. Carambola.*
Leaflets pubescent beneath   ...   . *A. Bilimbi.*

A. Carambola, Linn.; DC. Prod. I. 689 ; Fl. Br. I. 1. 439 ; Brandis For. Fl. 45; Bedd. Fl. Sylv. 39 ; Dalz. & Gibs. Bomb. Fl. Suppl. 16. *Camaranga,* Vern. Cultivated throughout the presidency ; run wild in some of the forests near villages of North Kánara. Fl. H. and R.S.

A. Bilimbi, Linn.; DC. Prod. 1. 689; Fl. Br. I. 1. 439 ; Brandis For. Fl. 46; Bedd. Fl. Sylv. t. 117; Dalz. & Gibs. Bomb. Fl. Suppl. 16. *Bilimbi,* Vern. Cultivated in the Konkan and North Kánara. Fl. H.S. Fr. R.S.

ORDER 20. RUTACEÆ.

Trees or shrubs. Leaves dotted with pellucid glands, opposite or alternate, simple or compound, exstipulate. Flowers in axillary cymes or panicles, regular, bisexual. Sepals and petals 4-5 each. Stamens 4-5 or 8-10. Anthers introrse. Ovary of 4-5 carpels. Styles as many, free or united. Ovules 2 or more in each cell. Fruit various. Albumen fleshy or 0.

Flowers polygamous.
Ovary lobed, styles free.
Leaves opposite, trifoliate, unarmed   ...   . EVODIA.
Leaves alternate, trifoliate or pinnate, armed ..   ... ZANTHOXYLUM.
Ovary entire. Style 1.
Leaves 1-foliate   ...   . ACRONYCHIA.
Leaves trifoliate   ...   . TODDALIA.
Flowers hermaphrodite.

Ovules solitary or 2 in each cell.
Style very short, persistent   .   ... GLYCOSMIS.
Style articulate, deciduous.
Unarmed, leaves compound. Filaments linear   ... MURRAYA.
Unarmed. Filaments dilated   ...   . CLAUSENA.
Armed. Calyx 3-lobed   ...   . TRIPHASIA.
Armed. Calyx 4-5 lobed ...   . LIMONIA.
Armed. Calyx cupular   ...   . LUVUNGA.
Leaves 1-foliate. Disc elongate ...   . PARAMIGNYA.
Leaves 1-foliate. Disc cupular ...   . ATALANTIA.
Ovules many in each cell.
Leaves 1-foliate   ...   . CITRUS.
Leaves pinnate   ...   . FERONIA.
Leaves trifoliate   ...   . ÆGLE.

EVODIA, Forst.

Trees. Leaves glabrous. Flowers small in axillary cymes. Sepals 4-5. Petals 4-5. Stamens 4-5 at base of disc. Ovary pubescent, 4-lobed, 4-celled. Cells 2-ovuled. Stigma 4-lobed. Fruit capsular. Testa crustaceous. Albumen fleshy.

E. Roxburghiana, Benth. Fl. Hongk. 59; Fl. Br. I. 1. 487; *E. triphylla,* Bedd. Fl. Sylv.; Anal. Gen. XLI: t. 6. f. 2; *Xanthoxylon triphyllum,* Dalz. & Gibs. Bomb. Fl. 45. Throughout the ghâts of the Konkan and North Kánara in moist situations, common in the deciduous forests near Yellápur, North Kánara. Flowers during the rainy season. Fruit ripe Sept., Jan.

ZANTHOXYLUM, Linn.

Trees or scandent shrubs. Leaves with the leaflets entire or crenate. Flowers small, in axillary or terminal cymes. Calyx 3-8-fid. Petals 3-5. Stamens 3-5. Ovary of 1-5 oblique carpels. Fruit of 1-5 globose 1-seeded dehiscent carpels. Albumen fleshy.

Shrub, leaves trifoliate. Cymes axillary   ...   ... *Z. ovalifolium.*
Tree. Leaflets 8-20-pairs. Cymes terminal   ...   ... *Z. Rhetsa.*

Z. ovalifolium, Wgt. Ill. 169; Fl. Br. I. 1. 492; Bedd. Fl. Sylv. Anal. Gen. 42. Western ghâts of North Kánara probably, but not yet noted by any one from the Bombay Presidency.

Z. Rhetsa, DC. Prod. 1. 728; Fl. Br. I. 1. 495; Dalz. & Gibs. Bomb. Fl. 45; Bedd. Fl. Sylv. Anal. Gen. 41. *Tirphul, tisul, cochli,* M.; *jummina,* K.; *pepuli,* H. Throughout the Konkan and North Kánara in moist evergreen forests, very common in the forests near Yellápur, also on the seacoast near Kárwár. This tree is distinguished by its prickly stem like that of *Bombax malabaricum.* Flowers from July till November. Fruit ripe during cold weather.

TODDALIA, Juss.

Rambling prickly shrubs or trees. Leaves alternate, leaflets sessile. Flowers small. Calyx 2-5-lobed. Petals 2-5. Stamens 2-4-5 or 8 with the alternate imperfect. Ovary entire or 4-parted in the male flowers, in the female flowers 2-7-celled. Ovules 2 in each cell. Fruit gland-dotted, 2-7-celled. Albumen fleshy.

Prickly shrubs, leaves trifoliate.... ... ... *T. aculeata.*
Unarmed trees. Leaves with 6-10 leaflets ... ... *T. bilocularis.*

T. aculeata, Pers.; DC. Prod. II. 83; Fl. Br. I. 1. 497; Dalz. & Gibs. Bomb. Fl. 46. Bedd. Fl. Sylv. 42. Common throughout the western ghâts of the Bombay Presidency in moist evergreen forests. Flowers August, December. Fruit ripe during the rainy season.

T. bilocularis, W. & A. Prod. 149; Fl. Br. I. 1. 497; *Dipetalum biloculare,* Dalz. & Gibs. Bomb. Fl. 46. Evergreen forests of North Kánara and the Konkan, Dalz.

ACRONYCHIA, Forst.

Trees. Leaves subopposite. Flowers axillary cymose. Calyx 4-lobed. Petals 4, spreading, revolute. Stamens 8, alternate longer. Ovary tomentose, in hollow tip of disc. Fruit a 3-5-celled drupe. Seeds often exserted, testa black. Albumen copious.

A. laurifolia, Bl. Bijd. 245; Fl. Br. I. 1. 498; *A. pedunculata,* Bedd. Fl. Sylv. 42; *Cyminosma pedunculata,* DC.; Dalz. & Gibs. Bomb. Fl. Suppl. 17; *Clausena simplicifolia,* Dalz. & Gibs. Bomb. Fl. 30. In moist forests on the North Kánara ghâts; common on the hills near Kárwár at about 1,000 feet elevation. Flowers during August. Fruit ripe in January.

GLYCOSMIS, Correa.

Shrubs or trees. Leaves 1-5-foliate, leaflets polymorphous, alternate. Flowers small. Calyx 4-5-partite. Petals 4-5. Stamens 8-10. Anthers often with an apical or dorsal gland. Ovary 2-5-celled. Berry white fleshy 1-3-seeded.

G. pentaphylla, Corr. in Ann. Mus. VI. 384; Fl. Br. I. 1. 499; Dalz. & Gibs. Bomb. Fl. 29; Brandis For. Fl. 49; Bedd. Fl. Sylv. 43. *Kirmira,* Vern. Very common throughout the moist evergreen forests of the Konkan and North Kánara. Fl. throughout the year.

MURRAYA, Linn.

Shrubs or small trees. Leaves pinnate. Flowers solitary or in terminal corymbs or axillary cymes. Calyx 5-divided. Petals 5, free. Stamens 10, inserted around disc. Filaments alternate shorter. Ovary 2-5-celled. Berry 1-2-celled, 1-2-seeded. Seeds with a woolly or glabrous testa.

Leaves 3-8-foliate. Corymbs few flowered ... ... *M. exotica.*
Leaves 10-20-foliate. Corymbs many flowered ... *M. Kœnigii.*

M. exotica, Linn.; Fl. Br. I. 1. 502; Bedd. Fl. Sylv. 44; Brandis For. Fl. 48; *M. paniculata,* Jack. Dalz. & Gibs. Bomb. Fl. 29. *Kunti,* Vern. In the evergreen forests of the Konkan and North Kánara ghâts, often cultivated in gardens.

M. Kœnigii, Spreng. Syst. Veg. II. 315; Fl. Br. I. 1. 502. Bedd. Fl. Sylv. 44; Brandis For. Fl. 48; *Bergera Kœnigii,* Dalz. & Gibs.

Bomb. Fl. 29. *Karhi-nimb, kudia nim,* M. ; *karhepah, gandla,* Vern.
Common in moist evergreen forests of the Konkan and North Kánara ;
forms a large proportion of the undergrowth in some of the high timber
deciduous forests of the Supa sub-division of North Kánara. Flowers
during the hot season, Fruit ripens during the rainy season.

CLAUSENA, Burm.

Shrubs or trees. Leaves imparipinnate. Flowers small. Calyx
4-5 divided. Petals 4-5 membranous. Stamens 8-10 inserted round
disc, alternate shorter. Ovary stipitate 4-5-celled ; cells 2-ovuled.
Berry small.

Inflorescence in terminal panicles   ...   . *C. indica.*
Inflorescence in axillary racemes   ...   . *C. Willdenovii.*

C. indica, Oliv. in Jour. Linn. Soc. V. Suppl. II. 36 ; Fl. Br. I. 1.
505 ; Bedd. Fl. Sylv. 45 ; *Piptostylis indica,* Dalz. & Gibs. Bomb. Fl. 29.
Common in evergreen forests on the ghâts of North Kánara from Ainshi
southwards. Párwár ghát, Dalz. Flowers April, May. Fruit ripens
during the rainy season.

C. Willdenovii, W. & A. Prod. 96 ; Fl. Br. I. 1. 506 ; Bedd. Fl. Sylv.
44 ; Dalz. & Gibs. Bomb. Fl. 30 ; *Cookia dulcis,* Bedd. in Madras Jour.
1861. On the ghâts of North Kánara and Belgaum near the Goa frontier,
in evergreen forests. Flowers during February. Fruit ripens in June.

TRIPHASIA, Lour.

A spiny shrub. Leaves alternate, trifoliate. Leaflets crenate.
Flowers solitary or in 3-flowered cymes. Calyx 3-divided. Petals
3. Stamens 6, filaments dilated at the base. Ovary 3-celled ; cells
1-ovuled. Berry small, 1-3-seeded. Seeds immersed in mucilage.

T. trifoliata, DC. Prod. 1. 536 ; Fl. Br. I. 1. 507. Dalz. & Gibs.
Bomb. Fl. Suppl. 12. China *limbu.* Cultivated in gardens throughout
the presidency.

LIMONIA, Linn.

Shrubs or small trees. Leaves imparipinnate, petiole winged.
Calyx 4-5 divided. Petals 4-5. Stamens 8-10 inserted around an
annular disc. Ovary oblong, 4-5-celled. Ovules 1-2 in each cell.
Berry globose. Seeds imbedded in mucilage.

L. acidissima, Linn. ; Fl. Br. I. 1. 507 : Dalz. & Gibs. Bomb. Fl. 29 ;
Bedd. For. Fl. 45. Brandis For. Fl. 47. In the dry deciduous forests
of the Belgaum district near Gokák, Dalz. Fl. April and May.

LUVUNGA, Hamilt.

Climbing spinous shrubs. Leaves trifoliate. Flowers axillary.
Calyx cupular. Petals 4-5. Stamens 8-10, around a cupular disc.
Ovary 2-4-celled, ovules 2 in each cell. Berry large, ellipsoid ; rind
thick. Seeds large ovoid.

L. oleuthorandra, Dalz. in Hook. Kew Journ. Bot. II. 258. Fl. Br.
I, 1. 509 ; Dalz. & Gibs. Bomb. Fl. 30. In the moist evergreen forests
of the Konkan and North Kánara.

PARAMIGYNA, Wgt.

Armed climbing shrubs. Leaves 1-foliate. Flowers large, axil
lary. Calyx cupular. Petals 4-5. Stamens 8-10. Ovary 3-5.
celled. Berry ovoid, contracted at base.

P. monophylla, Wgt. Ill. 1. 109; Fl. Br. I. 1. 510; Dalz. & Gibs.
Bomb. Fl. 30. *Kurwa wagutti*, M.; *ranyced*, Vern. Very common
throughout the evergreen forests of the Konkan and North Kánara.

ATALANTIA, Corr.

Spinous shrubs or trees. Leaves simple, coriaceous. Flowers
axillary or terminal. Calyx 5-lobed or irregularly split. Petals
3-5. Stamens 6-8, rarely 15-20; filaments usually connate. Ovary
2-5-celled. Berry globose, rind leathery.

Ovary sessile. Calyx irregularly lobed, split to base
on 1 side.... ... ... *A. monophylla.*
Calyx regularly or sub-regularly 4-5-lobed. Racemes
long ... ... *A. racemosa.*
Racemes short ... ... *A. ceylanica.*
Ovary stipitate ... ... *A. missionis.*

A. monophylla, Correa DC. Prod. 1. 535; Fl. Br. I. 1. 511; Dalz. &
Gibs. Bomb. Fl. 28. *Ran* or *makur limbu*, M. Throughout the moist
forests of the Konkan, Belgaum and North Kánara gháts. Flowers
during Nov. Fruit ripe in April.

A. racemosa, W. & H. Prod. 91; Fl. Br. I. 1. 512; Bedd. Fl. Sylv. 46;
Sclerostylis atalantioides, Wgt.; Dalz. & Gibs. Bomb. Fl. 29. Common
in the evergreen forests of the Konkan and North Kánara. Flowers
in November. Fruit ripe in April.

A. coylanica, Oliv. in Jour. Linn. Soc. V. Suppl. II. 25; Fl. Br. I. 1.
12 ; Bedd. Fl. Sylv. 46. Phoondah ghát in the Konkan, Ritchie.

A. missionis, Oliv. in Jour. Linn. Soc. V. Suppl. 25; Fl. Br. 1. 1.
513; Bedd. Fl. Sylv. 46. North Kánara in evergreen forests near
Kárwár. Flowers in the cold season. Fruit ripe Aug.

CITRUS, Linn.

Evergreen spinous shrubs or trees. Leaves simple, coriaceous.
Calyx 4-5 cleft. Petals 4-5. Stamens 20-60, filaments more or
less connate. Ovary many celled on a large annular disc. Berry
globose, fleshy, many celled, with membranous septa filled with fusiform, distended vesicles.

All parts glabrous. Fruit vesicles concrete.
Petals coloured. Flowers often unisexual ... ... *C. medica.*
Petals white. Flowers bisexual ... ... *C. Aurantium.*
Young shoots and nerves of leaves beneath pubes-
cent. Fruit vesicles distinct ... ... *C. decumana.*

C. medica, Linn.; Fl. Br. I. 1. 514; Brandis For. Fl. 52; Dalz. &
Gibs. Bomb. Fl. Suppl. 13. Citron, lemon, sour lime, sweet lime. *Nimbu,
mitha nimbu, bijapara, mahalunga, bijori*, Vern. Cultivated throughout
the presidency and wild on the Sátpudás and western gháts.

C. Aurantium, Linn.; Fl. Br. I. 1. 515; Brandis For. Fl. 53; Dalz. & Gibs. Bomb. Fl. Suppl. 12. Sweet orange. *Narangi,* H. Cultivated in the Deccan.

C. decumana, L.; Fl. Br. I. 1. 516; Brandis For. Fl. 55 Dalz. & Gibs. Bomb. Fl. Suppl. 12. Shaddock, pumelo. Cultivated throughout the presidency. The fruit arrives at great perfection along the coast of North Kánara and Konkan.

FERONIA, Gaertn.

A spinous tree. Leaves alternate, pinnate. Flowers polygamous, in panicles or racemes. Petals usually 5, imbricate. Stamens 10-12, filaments dilated, villous at the base. Ovary 5-celled, ovules numerous in several series. Fruit globose, 1-celled. Seeds numerous; cotyledons thick, fleshy.

F. Elephantum, Correa, Roxb. Fl. Ind. II. 411; Fl. Br. I. 1. 516; Dalz. & Gibs. Bomb. Fl. 30; Brandis For. Fl. 56; Bedd. Fl. Sylv. t. 121. Elephant or wood apple. *Cawtha,* Vern. Throughout the presidency in dry situations, wild or cultivated. Flowers during the hot weather. Fruit ripe November.

ÆGLE, Corr.

Spinous trees. Leaves alternate, trifoliate. Flowers large, white, bisexual, in axillary panicles. Petals 4-5, imbricate. Stamens numerous, filaments short, subulate. Ovary ovoid, axis broad, cells 8-20, near the circumference. Fruit globose with a hard rind. Seeds many, testa woolly and mucous.

A. Marmelos, Corr.; Fl. Ind. II. 579; Dalz. & Gibs. Bomb. Fl. 31. Fl. Br. I. 1. 516; Brandis For. Fl. 57; Bedd. Fl. Sylv. t. 161. Bael tree; *baelputri,* K.; Bengal quince. Wild or cultivated throughout the presidency in dry situations. Flowers during May. Fruit ripe in November.

ORDER 21. SIMARUBEÆ.

Shrubs or trees. Leaves simple or pinnate. Flowers small, regular, diclinous. Calyx 3, 5-lobed. Petals 3-5, rarely 0, valvate or imbricate. Stamens as many or twice as many as petals, inserted at the base of disc. Ovary free, 1-6-celled, usually deeply lobed; styles 2-5; ovules usually solitary in each cell. Fruit drupaceous, capsular or samaroid, of 2-6 distinct carpels. Seed albuminous.

Ovary deeply divided.
Leaves pinnate ; fruit samaroid ... . AILANTUS.
Leaves simple ; fruit drupaceous, winged ... ... SAMADERA.
Ovary entire. Leaves 2-foliate ... ... BALANITES.

AILANTUS, Desf.

Large trees. Leaves alternate, unequally pinnate. Flowers polygamous, in large axillary panicles. Calyx 5-fid. Petals 5, valvate. Disc 10-lobed. Stamens 10, inserted at base of disc: in hermaphrodite flowers stamens 2-3. Ovary 2-5-partite; styles connate; ovules 1 in each cell. Fruit of 1-5, 1-seeded samaras ; wing large, membranous. Seed pendulous.

Leaflets toothed, glandular-hairy. Samara twisted. *A. excelsa.*
Leaflets entire, glabrous. Samara not twisted　　... *A. malabarica*

A. excelsa, Roxb. Fl. Ind. II. p. 450 ; Fl. Br. I. 1. 518 ; Dalz. & Gibs. Bomb. Fl. 46 ; Brandis For. Fl. 58 ; Bedd. Fl. Sylv. t. 122. *Maruk, varul*, Vern. Common about Broach and Baroda; Deccan, Gibs. Commonly planted in the Khándesh district.

A. malabarica, DC. Prod. II. 89 ; Fl. Br. I. 1. 518 ; Dalz. & Gibs. Bomb. Fl. 46 ; Brandis For. Fl. 58 ; Bedd. Fl. Sylv. t. 122. *Gugguldhup, muddhedup*, K. Throughout the Konkan and North Kánara in evergreen forests, from the sea-level upwards. Flowers in February and March. Fruit ripe in May.

SAMADERA, Gærtn.

Trees. Leaves simple. Flowers in axillary or terminal umbels. Calyx 3-5-partite. Petals 3-5. Disc large, conical. Stamens 8-10 ; a small scale at the base. Carpels 4-5 ; styles free at base, more or less united above. Fruit of 1-5, compressed, 1-seeded drupes ; each with a narrow wing.

S. indica, Gærtn. Frut. II. t. 156 ; Fl. Br. I. 1. 519 ; Bedd. For. Sylv. 49 ; Grah. Cat. Bom. Pl. 37. Throughout the South Konkan along river banks, not yet noted from North Kánara.

BALANITES, Delile.

Spiny shrubs or trees. Leaves 2-foliate entire. Flowers green in small axillary cymes. Calyx segments 5. Petals 5, imbricate. Disc thick conical entire. Ovary 1-celled ; ovules solitary, pendulous. Fruit a woody 1-seeded drupe. Seed pendulous.

B. Roxburghii, Planch. in Ann. Sc. Nat. Ser. 4, II. 258. Fl. Br. I. 1. 522 ; Brandis For. Fl. 59 ; Bedd. Fl. Sylv. 50 ; Grah. Cat. Bomb. Pl. 23. *Hingu, hinganbet*, Vern. Throughout the driest parts of the presidency in open situations. Flowers in April and May. Fruit ripe December.

ORDER 22. OCHNACEÆ.

Trees or shrubs. Leaves alternate, simple, coriaceous ; stipules 2. Inflorescence panicled or umbellate, bracteate. Sepals 4-5, imbricate, persistent. Petals 5, rarely 4 or 10, free, deciduous. Disc enlarged after flowering. Stamens 4-10 or indefinite, inserted on the disc. Ovary short, 1-2-celled or elongate and 1-10-celled. Fruit indehiscent, compound, each drupe or pyrene 1-4-seeded ; or capsular and 1-5-celled. Albumen fleshy or 0.

Shrub. Stamens indefinite. Fruit of 3-10 drupes ... OCHNA.
Tree. Stamens 10. Fruit of 5 or fewer drupes　... GOMPHIA.

OCHNA, L.

Shrubs. Leaves serrate, glabrous, shining. Flowers large, yellow·
Sepals 5, coloured, persistent. Petals 5-10, deciduous. Disc thick,
lobed. Stamens numerous, shorter than the petals. Ovary 3-10-
lobed; lobes 1-celled; ovules solitary in each cell. Fruit of 3-10
drupes seated on the broad disc. Seed erect, albuminous.

O. squarrosa, L. ; Fl. Br. I. 1. 523 ; Dalz· & Gibs. Bomb. Fl. Suppl.
17 ; Bedd. Fl. Sylv. 50 ; Brandis For. Fl. 60. Common in the moist forests
of the South Konkan and North Kánara on the hills near the sea ; but also
found inland at elevations up to 2,000 feet. Fl. Feb., Mch. Fr. ripe in
May, June.

GOMPHIA Schreb.

Glabrous trees or shrubs. Flowers yellow. Sepals and petals 5
each. Disc thick, lobed. Stamens 10, filaments very short. Ovary
deeply 5-6-lobed ; styles connate, stigma simple. Drupes 5 or fewer,
seated on a broad disc, 1-seeded. Seed exalbuminous.

G. angustifolia, Vahl. Symb. II. 49 ; Fl. Br. I. 1. 525 ; Grah. Cat.
Bomb. Pl. 38 ; Bedd. Fl. Sylv. 51. On the gháts of the South Konkan.

ORDER 23. BURSERACEÆ.

Balsamiferous trees or shrubs. Leaves alternate, generally compound. Flowers small, regular, hermaphrodite, polygamous. Calyx
3-6-lobed. Petals 3-6. Disc annular, usually conspicuous. Stamens equal or twice the number of petals. Ovary free, 1-5-celled.
Ovules 2 in each cell. Fruit drupaceous. Albumen 0.

Drupe dehiscent	... BOSWELLIA.
Drupe indehiscent. Calyx 5-fid, bell-shaped	... GARUGA.
Calyx 4-toothed, urceolate	... BALSAMODENDRON.
Calyx 3-fid, campanulate	... CANARIUM.

BOSWELLIA, Roxb.

Trees with white thin bark. Leaves imparipinnate, deciduous,
leaflets opposite, sessile, serrate. Flowers hermaphrodite. Calyx
5-toothed. Petals 5. Disc crenate. Stamens 10. Ovary 3-celled.
Ovules 2 in each cell, pendulous. Drupe trigonous of 3, 1-seeded
pyrenes, finally separating.

B. serrata, Roxb., ex Coleb. in Asiat. Res. IX. 379, t. 5 ; Fl. Br. I.
1. 528 ; B. thurifera, Roxb. ; Brandis For. Fl. 61 ; Bedd. Fl. Sylv. 52.
Salai, salphullie, Vern. Common in the Sátpudás of the Khándesh
district, where it forms pure forests of considerable extent. Throughout
the Deccan, to within 20 miles of the western gháts. Fl. Mch., Apl.

GARUGA, Roxb.

Trees. Leaves alternate, imparipinnate, leaflets opposite, subsessile, crenate. Disc large, lining calyx. Petals 5. Stamens 10,
equal. Drupe globose, fleshy, containing 1-5, 1-seeded pyrenes.

G. pinnata, Roxb. Cor. Pl. III. t. 208 ; Fl. Br. I. 1. 528 ; Brandis For. Fl. 62 ; Bedd. Fl. Sylv. t. 118. *Kuduk, kakad,* M. ; *halabaluji,* K. ; *kurak, kangkar,* Vern. Throughout the presidency in dry deciduous forests, often associated with *Odina Wodier.* Flowers during May. Fruit ripe in July.

BALSAMODENDRON, Kunth.

Balsamiferous trees or shrubs. Leaves compound, leaflets sessile, oblique, crenate or serrate. Flowers small, polygamous. Calyx 3-4-toothed. Petals 3-4, valvate. Stamens 6-8, alternating shorter, surrounded by disc. Drupe ovoid, 1-3 bony 1-seeded pyrenes. Seeds exalbuminous.

Leaves usually 1-foliate　　...　...　...　... *B. Mukul.*
Leaves trifoliate.
Unarmed　　...　...　...　...　...　... *B. pubescens.*
Thorny ...　...　...　...　...　...　... *B. Berryi.*

B. Mukul, Hook. ex Stocks in Hook, Kew Jour. Bot. I. 259, t. 8. Fl. Br. I. 1, 529 ; Brandis For. Fl. 64 ; Dalz. & Gibs. Bomb. Fl. Suppl. 19. *Gugal,* Vern. In the dry regions of Sind and Khándesh. Flowers in March and April.

B. pubescens, Stocks in Bomb. Trans. 1847 ; Fl. Br. I. 1. 529 ; Brandis For. Fl. 64. On the dry hills of Sind as far south as Karáchi. Flowers during March and April.

B. Berryi, Arn. in Ann. Nat. Hist. III. 86 ; Fl. Br. I. 1. 529 ; Brandis For. Fl. 65 ; Bedd. Fl. Sylv. t. 126 ; *B. gileadense,* Don. Grah. Cat. Bomb. Pl. 43. Cultivated. Bombay. Fl. Feb., Mch.

CANARIUM, Linn.

Trees. Leaves imparipinnate. Flowers hermaphrodite or polygamous. Calyx 3-lobed, valvate. Petals 3-5. Stamens 6, distinct, on margin or outside disc. Ovary 2-3-celled. Ovules 2 in each cell. Drupe trigonous with a thick, bony stone.

C. strictum, Roxb. Fl. Ind. III. 138 ; Fl. Br. I. 1. 534 ; Dalz. & Gibs, Bomb. Fl. 52 ; Bedd. Fl. Sylv. t. 128. Black dammer tree. *Raldhup,* K. ; *gugul,* Vern. In the evergreen forests of the Konkan and North Kánara, common near the Ainshi ghát of North Kánara at 2,000 feet elevation. Flowers during the hot weather. Fruit ripe next cold season, January.

ORDER 24. MELIACEÆ.

Trees or shrubs. Leaves alternate, usually pinnate. Flowers regular, paniculate, hermaphrodite or polygamo-diœcious. Calyx 3-6-lobed. Petals 3-6, free or connate at base. Stamens 5-20, outside base of disc ; filaments connate in a tube, rarely free. Ovary 2-5-celled, usually 2-ovuled. Fruit capsular, drupaceous or berried. Seeds exalbuminous.

Stamens united into a tube.
Leaves simple ...　...　...　...　...　... TURRÆA.
Leaves pinnate, *leaflets toothed.* *Seeds not winged.*
Flowers elongated. Style long　...　...　... MELIA.
Flowers globose. Style short　...　...　... CIPADLSSA.

Leaflets entire. Seeds not winged.
Flower oblong. Style long ...   . DYSOXYLUM.
Flowers globose. Style short.
Anthers included in tube. Seeds not angular.
Anthers 5   ...   . AGLAIA.
Anthers 6-10.
Berry indehiscent   ...   ... LANSIUM.
Capsule dehiscent   ...   ... AMOORA.
Anthers exserted.
Fruit a tomentose berry   ...   . WALSURA.
Fruit a glabrous capsule   ...   . HEYNEA.
Anthers included. *Seeds angular* ...   . CARAPA.
Seeds winged.
Ovary 5-celled ...   . SOYMIDA.
Ovary 3-celled ...   . CHICKRASSIA.
Stamens distinct.
Stamens 4-6   ...   .,   ... CEDRELA,
Stamens 10   ...   . CHLOROXYLON.

TURRÆA, Linn.

Trees or shrubs. Leaves entire or obtusely lobed. Flowers axillary, peduncle bracteate at the base. Calyx 4-5-fid. Petals 4-5, spathulate, elongated. Staminal tube long; anthers 10-8, within mouth of tube. Disc annular. Ovary 5 or more celled. Ovules 2 in each cell. Capsule loculicidal, valves woody, separating from the winged axis.

Glabrous ...   . *T. virens.*
Pubescent ...   . *T. villosa.*

T. virens, Linn., Fl. Br. I. 1. 541; Dalz. & Gbis. Bom. Fl. 36. Konkan gháts.

T. villosa, Benn. Pl. Jav. Rar. 182; Fl. Br. I. 1. 542; Bedd. Fl. Sylv. 64, (*Euonymus* sp.) On the Mahábaleshvar hills; in Gujarát; common in the moist forests of the Supa sub-division of North Kánara in open situations. Flowers during April and May.

MELIA, Linn.

Trees. Leaves 1-3-pinnate, leaflets toothed or entire. Flowers in axillary panicles. Calyx 5-6-lobed. Petals 5-6 free. Staminal-tube cylindrical; anthers 10-12, inserted near apex. Disc annular. Ovary 3-6-celled; style slender, stigma capitate; ovules 2, superposed. Fruit a drupe. Seeds albuminous; cotyledons foliaceous.

Leaves 1-pinnate   ...   . *M. Azadirachta.*
Leaves 2-3 pinnate.
Leaflets serrate. Flowers lilac, not mealy-pubes-
cent ...   . *M. Azedarach.*
Leaflets crenate. Flowers white mealy-pubescent. *M. dubia.*

M. Azadirachta, Linn.; Fl. Br. I. 1. 544; Bedd. Fl. Sylv. t. 13; *Melia indica*, Brandis For. Fl. 67; *Azadirachta indica*, Adr. Juss. Dalz. & Gibs. Bomb. Fl. 36. *Nim*, *nimari*, Vern. Cultivated throughout the presidency. Fl. hot season. Fr. ripe in July.

M. Azedarach, Linn.; Fl. Br. I. 1. 541; Brandis For. Fl. 68; Bedd.
Fl. Sylv. t. 14; *M. sempervirens,* Sw. Prod. 67; Dalz. & Gibs. Bomb.
Fl. Suppl. 15. Bastard cedar, bead tree, Persian lilac. *Pejri, padrai,*
M.; *mullanim,* H. Cultivated sparingly throughout the presidency.
Flowers during the hot weather. Fruit ripe during the rains.

M. dubia, Cav. Diss. (VII) 364 (1789); Fl. Br. I. 1. 545; *M. composita,*
Dalz. & Gibs. Bomb. Fl. 36; Brandis For. Fl. 69; Bedd. Fl. Sylv. t.
12. *Kariberan,* K.; *nimbarra,* Vern. In the evergreen and deciduous
forests of the North Kanara and Konkan ghâts; common on the Supa
ghâts; often planted. Flowers during March. Fruit ripe during the
rainy and cold seasons, October till January.

CIPADESSA, Blume.

Shrubs or trees. Leaves odd-pinnate. Calyx 5-toothed. Petals
5, valvate in bud. Staminal tube deeply 10-lobed, lobes linear
bifid; anthers short, subapiculate. Disc cup-shaped, adnate to the
base of the staminal tube. Ovary 5-celled; ovules 2, collateral,
pendulous in each cell. Fruit a drupe, 5-ribbed, 5-celled; cells
1-2-seeded. Seeds albuminous; embryo sub-foliaceous.

C. fruticosa, Blume Bijd. 162; Fl. Br. I, 1.545; *Mallea Rothii,* Adr. Juss.
Dalz. & Gibs Bomb. Fl. 37; Bedd. Fl. Sylv. 54. *Nal bila,* H. In the
moist forests of the ghâts; from Poona to Belgaum, also in the Konkan.
Fl. and Fr. R.S.

DYSOXYLUM, Blume.

Trees. Leaves pinnate; leaflets entire, oblique at the base,
coriaceous. Flowers in panicles. Calyx 4-5-fid, caducuous. Petals
4-5, oblong, spreading. Staminal tube cylindrical; anthers short,
6-10, included or half exserted. Disc tubular equal to or twice as
long as the ovary. Ovary 3-4-celled; ovules 2 in each cell. Capsule 1-4-celled, loculicidally dehiscent. Seeds arillate or 0, exalbuminous.

Calyx cup-shaped, subentire, ¼ the length of flower... *D. binectariferum.*
Calyx short, 4-5-divided ... ... *D. glandulosum.*

D. binectariferum, Hook. f. ex Bedd. in Trans. Linn. Soc. XXV. 212;
Fl. Br. I. 1. 546; *D. macrocarpum,* Bedd. Fl. Sylv t. 150; *Epicharis
exarillata,* Arn.; Dalz. & Gibs. Bomb. Fl. 37. *Yerindi,* Vern. In ever-
green forests along the ghâts from Khandâla to the Mysore frontier of
North Kânara, common in the large evergreen forests near the Falls of
Gairsoppah. Flowers during the rainy season. Fruit ripe in Feb.

D. glandulosum, sp. nov.—A large tree. Leaves 12 inches;
petioles angled, 4-6-inches. Leaflets 8-10-pairs, shortly petioluled,
2-9-inches long by 1½ to 3-inches broad, alternate or subopposite,
coriaceous, puberulous when young, ovate acuminate, base cuneate,
nearly glabrous, pale and strongly nerved beneath; lateral nerves
10-20-pairs with hollow glands, ciliate at the mouth, in the axils.

Flowers hermaphrodite, small, ⅛-inch, in axillary panicles, shortly pedicelled, pedicel ⅛-inch. Calyx short, deeply 4-lobed. Petals 4, broad, slightly imbricate in bud, bluntly pointed, pubescent, ¼-inch. Staminal tube urceolate, mouth crenulate, anthers 8, included. Disc entire, cup-shaped, equalling or half the ovary, ciliate on the margin, hairy within, glabrous without. Ovary 4-celled, densely white tomentose. Style equalling staminal column, stigma capitate, exserted. Fruit 2-inches, size and shape of a fig, verrucose, bright yellow when ripe, 3-4-seeded; seeds angular, with a dark-brown testa; cotyledons green.

This tree is closely allied to *D. malabaricum*, Bedd. MSS. Specimens sent to Kew were considered distinct. Specimens sent to Calcutta were doubtfully referred to *D. Beddomei*, Hiern. As it differs from both these species I have described and named it as above. It is found in the evergreen forests of North Kánara. Flowers in Feb. Fruit ripe in May.

AGLAIA, Lour.

Trees or shrubs. Leaves pinnate or trifoliate, scaly or pubescent. Flowers polygamo-diœcious, minute, paniculate. Calyx 5-lobed. Petals 5 concave, imbricated. Staminal tube urceolate or subglobose, 5-toothed at the apex or entire; anthers 5, erect, included or half exserted. Disc minute. Ovary 1-3-celled; cells 1-2-ovuled. Fruit a dry berry 1-2-celled and seeded. Seeds with a fleshy covering

A. **Roxburghiana**, Miq. Ann. Mus. Ludg-Bat. IV. 41 ; Fl. Br. I. 1. 555 ; Bedd. Fl. Sylv. t. 130. In the evergreen forests of the gháts of North Kánara from Ainshi southwards, also in the Konkan. A large tree. Fl. Nov., Dec. Fr. ripe July. There is also a variety of the above or a new species, very common in waste places and in the forests of the Konkan and North Kánara, near the coast, sometimes a small tree usually only a shrub. Panicles elongate, exceeding the leaves ; pedicils short or 0. Fruit ovoid, 2-seeded. Seeds surrounded with a thick, white, transparent, veined, glutinous, pulp. Epicarp thin leathery, lepidote, bright yellow. Fl. and Fr. R. S.

LANSIUM, Rumph.

Trees. Leaves oddpinnate. Flowers polygamo-diœcious, 5-merous in axillary panicles or racemes. Sepals and petals rounded, imbricated. Staminal tube globose, crenulated ; anthers 10, in 2 rows. Ovary globose, 3-5-celled ; cells 1-2-ovuled ; stigma 3-5-lobed. Fruit a berry, 1-5-celled ; cells 1-2-seeded. Seeds enclosed in a pulpy aril, exalbuminous.

L. **anamalyanum**, Bedd. Fl. Sylv. t. 131 ; Fl. Br. I. 1. 558. Abundant on the southern gháts of North Kánara in evergreen forests. Flowers during April and May.

AMOORA, Roxb.

Trees. Leaves unequally pinnate; leaflets oblique. Flowers axillary, paniculate or racemose. Calyx 3-5-divided. Petals 3-5, thick, concave. Staminal-tube subglobose, 6-10-crenate; anthers 6-10 included. Ovary sessile, 3-5-celled; cells 1-2-ovuled. Fruit a subglobose capsule, 3-4-celled and seeded, usually loculicidally dehiscent. Seeds with a fleshy arillus.

Leaflets usually very oblique at the base. Petals 3.
Leaflets 9-15, opposite. Panicles spicate ... ... *A. Rohituka.*
Leaflets 4-6, alternate. Male panicles dense, pyramidal, erect... ... ... *A. canarana.*
Leaflets opposite, 3-13, terminal one often hooded.
Male panicles drooping... ... ... *A. cucculata.*
Leaflets usually not oblique at the base, alternate, 3-5. Petals 4. Fruit pear-shaped ... ... *A. Lawii.*

A. Rohituka, W. & A. Prod. I. 119; Fl. Br. I. 1. 559; (Bedd. Fl. Sylv. t. 132). Brandis For. Fl. 69; *Amoora macrophylla,* Nimmo in Grah. Cat. Bomb. Pl. South Konkan and North Kánara gháts, in evergreen forests; abundant in the forests near Yellápur. Flowers during July. Fruit ripe in March.

A. canarana, Benth. & Hook. f. Gen. Plant. 1. 335; Fl. Br. I. 1. 560; throughout the evergreen forests of North Kánara from Goond southwards. Flowers during March and April.

A. cucculata, Roxb. Cor. Plant III. 54. t. 258; Fl. Br. I. 1. 560; Dalz. & Gibs. Bomb. Fl. 37; Bedd. Fl. Sylv. 55. Konkan; Párwár ghát, Dalz.

A. Lawii, Benth. & Hook. f. Gen. Plant. 1. 335; Fl. Br. I. 1. 561; Bedd. Fl. Sylv. t. 133; *Nemedra Nimmonii,* Dalz. & Gibbs. Bomb. Fl. 37. *Madrasada,* K.; *buramb,* M. Throughout the evergreen forests of the Konkan and North Kánara; very common in the Ainshi ghát forests. Flowers during December, January. Fruit ripe in June, indehiscent.

WALSURA, Roxb.

Trees. Leaves 1-5-foliate; leaflets alternate. Flowers in many flowered axillary or terminal panicles. Calyx 5-divided, imbricate. Petals 5, free, spreading, imbricate or subvalvate. Filaments 8-10, free or connate into a tube; anthers 8-10 erect, exserted. Ovary 2-3-celled immersed in the disc; style short; cells with 2 collateral ovules in each. Fruit a fleshy berry, 1-seeded. Seed enclosed in a fleshy aril.

W. piscidia, Roxb. Fl. Ind. II. 387; Fl. Br. I. 1. 564; Bedd. Fl. Sylv. 56; Dalz. & Gibs. Bomb. Fl. 37. In the evergreen forests of the Konkan and North Kánara; common in the Devimone ghát forests. Flowers in November. Fruit ripe in May.

HEYNEA, Roxb.

Trees. Leaves imparipinnate; leaflets opposite. Panicles terminal, corymbose, long peduncled; peduncles and pedicils articulate;
flowers hermaphrodite. Calyx 4-5 fid. Corolla of 4-5, oblong
petals, Staminal tube 8-10-fid, divisions linear, bidentate at apex;
anthers between the linear teeth. Ovary immersed in the disc;
stigma with a thickened ring at the base; ovules 2 in each cell.
Fruit capsular, 1-celled, 2-valved, 1-seeded. Seed enclosed in an
arillus, exalbuminous.

H. trijuga, Roxb. Fl. Ind. II. 390; Fl. Br. I. 1. 565; Dalz. & Gibs.
Bomb. Fl. 38; Brandis For. Fl. 70; Bedd. Fl. Sylv. t. 134. *Kora,* K.;
gundira, M.; *limbara,* Vern. Throughout the western ghát forests of
the presidency; common along the banks of rivers in North Kánara and
in evergreen forests. Fl. Feb., Mch. Fr. ripe in the cold season,
Oct., Feb.

CARAPA, Aubl.

Trees. Leaves 2-6-foliate; leaflets opposite, entire. Panicles
axillary, lax. Calyx 4-fid. Petals 4, reflexed. Staminal tube
globose, 8-dentate at apex, teeth bipartite; anthers 8, 2-celled, included, sessile. Disc fleshy, cup-shaped, at base of ovary and
adherent to it. Ovary 4-celled, 4-sulcate; cells 2-8-ovuled; style
short; stigma discoid. Fruit a globose capsule (3-4-in. in diameter)
6-12-seeded. Seeds large, angular; testa hard, spongy.

C. moluccensis, Lam. Encycl. Meth. 1. 621; Fl. Br. I. 1. 567; Bedd.
Fl. Sylv. t. 130. Sea-coast of the Konkan.

SOYMIDA, Adr. Juss.

A tree. Leaves paripinnate; leaflets opposite, entire. Flowers
in axillary or terminal panicles. Sepals 5, imbricate. Petals 5,
free, spreading, unguiculate. Staminal tube short; 10-lobed; lobes
bidentate; anthers sessile between the teeth, short. Disc flat.
Ovary 5-celled; cells many ovuled; style short; stigma broad, fleshy.
Capsule woody, 5-valved. Seeds numerous, winged at both ends,
albuminous.

S. febrifuga, Adr. Juss. in Mem. Mus. XIX. 251, t. 22, f. 26; Fl.
Br. I. 1. 567; Brandis For. Fl. 71; Bedd. Fl. Sylv. t. 8; Dalz. & Gibs.
Bomb. Fl. 38. Bastard cedar; Indian redwood. *Rohan, lál chundun,*
Vern.; *palara,* M. In the dry forests, on stony hills, in the Dhárwár
and Khándesh districts, also on the Konkan gháts. Fl. March. Fr. July.

CHICKRASSIA, Adr. Juss.

A large tree. Leaves paripinnate; leaflets acuminate, oblique.
Flowers in terminal panicles. Calyx 5-divided. Corolla of 5, free
contorted petals. Staminal tube cylindric, 10-crenate; anthers 10,

erect, between the crenatures. Disc 0. Ovary 3-celled ; cells multi-
ovulate ; ovules biseriate. Fruit a 3-celled, septicidally 3-valved
capsule. Seeds numerous, winged below, exalbuminous.

C. tabularis, Adr. Juss. in Mem. Mus. XIX. 251. t. 22, f. 27 ; Fl. Br.
I. 1. 568 ; Bedd. Fl. Sylv. t. 9 ; *C. Nimmonii*, Grah.; Dalz. & Gibs. Bomb.
Fl. 38. Chittagong wood tree *Lál devadari, pabba,* M. Both the
glabrous and villous varieties of this tree are common in the evergreen
forests of North Kánara and Konkan. The villous variety (*C. velutina,*
Roem) has the capsule often 4-valved. Fl. March, April ; Fr. Aug., Sep.

CEDRELA, Linn.

Immense tree. Leaves pinnate ; leaflets numerous, opposite,
oblique, entire or serrate. Panicles terminal. Calyx 5-cleft.
Petals free, 4-6, inserted on the top of the disc, alternating with
staminodes. Disc 4-6-lobed. Ovary sessile on the disc, 5-celled
cells with 8-12, biseriate ovules ; stigma discoid. Capsule 5-celled.
Seeds winged.

C. Toona, Roxb. Fl. Ind. I. 635 ; Fl. Br. I. 1. 568 ; Bedd. Fl. Sylv. t.
10 ; Dalz. & Gibs. Bomb. Fl. 38 ; Brandis For. Sylv. 72 ; *C. serrata,*
Royle. Brandis For. Sylv. 73. Toon tree. *Todu, maha nim, tuni, huruk,*
M. ; *tundu, devdári,* K. Abundant in the evergreen forests of North
Kánara and the Konkan ; immense trees of this species are common near
Yellápur. Fl. Jan. Fr. Mch., Apl. Wood fragrant.

CHLOROXYLON, DC.

A tree. Leaves abruptly pinnate ; leaflets many, unequal-sided·
Flowers small, bisexual, in terminal or axillary panicles. Calyx 5-
divided. Petals 5, clawed, imbricate. Disc 10-lobed. Stamens 10,
inserted outside base of disc ; anthers versatile. Ovary pubescent,
immersed in disc, 3-lobed and celled ; ovules 8 in each cell ; style
short, slender ; stigma capitate. Capsule 3-celled, loculicidally
3-valved. Seeds imbricate, winged ; albumen 0.

C. Swietenia, DC. Prod. I. 625 ; Fl. Br. I. 1. 569 ; Bedd. Fl. Sylv.
t. 11 ; Dalz. & Gibs. Bomb. Fl. 39 ; Brandis For. Fl. 74. *Mashwal,* K. ;
halda, billu, M. Satinwood. Common throughout the drier parts of
the Deccan and Konkan ; abundant in the dry deciduous forests of the
Dhárwár and Belgaum districts. Fl. March, April. Fr. ripe June.

ORDER 25. CHAILLETIACEÆ.

Trees or shrubs. Leaves alternate, entire. Stipules petiolar.
Flowers small, hermaphrodite or unisexual, in axillary corymbose
cymes. Sepals 5. Petals 5 free. Stamens 5. Disc of 5 glands
or scales or a glandular or lobed cup. Ovary free, 2-3-celled.
Ovules 2 in each cell, pendulous from the top. Drupe pubescent,
2-celled, compressed. Seeds solitary. Albumen 0.

Chailletia gelonioides, Hook. f. in Gen. Plant. 1. 341 ; Fl. Br. I. 1.
570 ; Bedd. Fl. Sylv. 59 ; *Moacurra gelonioides,* Roxb. Dalz. & Gibs.

Bomb. Fl. 52. This small tree or shrub is found on the gháts from the Konkan southwards in evergreen forests; it is very common in North Kánara near the Falls of Gairsoppah, where it is usually a small shrub. Fruit dehiscent, mesocarp bright red. Fl. April, June. Fr. Aug., Dec..

ORDER 26. OLACINEÆ.

Trees or shrubs. Leaves simple, usually alternate. Flowers regular, hermaphrodite or unisexual. Calyx 4-5-toothed, sometimes accrescent. Petals 4-6. Stamens equal to or twice as many as petals, all fertile or some without anthers. Disc free or adnate to ovary or to calyx. Ovary free or immersed in disc, 2-3-celled; cells 2-3-ovuled. Fruit usually an indehiscent drupe. Albumen fleshy or 0.

Stamens twice as many as petals.
Fertile stamens 10. Calyx not enlarged in fruit ... XIMENIA.
Fertile stamens 3-5. Calyx enlarged in fruit... ... OLAX.
Stamens equal and opposite to petals.
Ovary 3-5-celled ... ... STROMBOSIA.
Ovary 1-celled ... ... CANSJERA.
Stamens equal to petals, but alternate with them.
Erect shrubs or trees.
Petals glabrous ... . GOMPHANDRA.
Petals villous within ... ... MAPPIA.
Climbing shrub ... ... SARCOSTIGMA.

XIMENIA, Linn.

A shrub or low tree, often spiny. Leaves simple. Flowers bisexual or polygamous. Petals 4-5, revolute, bearded within. Ovary 3-celled at base. Ovules 1 in each cell, pendulous. Drupe ovoid with a fleshy sarcocarp, 1-seeded.

X. americana, Willd. Sp. Pl. II. 230; Fl. Br. I. 1. 574. Dhárwár and Belgaum in the driest forests. On stony ground. Scandent. Flowers during January.

OLAX, Linn.

Scandent shrubs. Prickly or not. Leaves simple. Racemes axillary. Petals 6. Stamens 3-5, staminodes 5-6, bifid. Ovary half immersed in disc, 3-celled at base, 1-celled above. Fruit more or less covered by the accrescent calyx.

Fruit ⅔ths covered by the accrescent calyx ... .. *O. scandens.*
Fruit nearly completely covered by calyx ... ... *O. Wightiana.*

O. scandens, Roxb. Fl. Ind. I. 163; Fl. Br. I. 1. 575; Brandis For. Fl. 75. In the Sátpudás near rivers and nálás. Brandis. Fl. cold season.

O. Wightiana, Wall. Cat. 6779; Fl. Br. I. 1. 575; Dalz. & Gibs. Bomb. Fl. 27; Bedd. Fl. Sylv. 60. This scandant shrub is common in the evergreen forests of the gháts of North Kánara from Ainshi southwards. Fl. Jan., Feb, Fr. ripe April, May.

STROMBOSIA, Blume.

Trees. Leaves simple. Inflorescence in short cymes. Calyx cup-shaped. Petals 5, valvate, hairy within. Stamens 5. Ovary half immersed in disc, 3-5-celled. Fruit a drupe. Seeds pendulous. Albumen fleshy.

S. ceylanica, Gardn. in Calc. Jour. Nat. Hist. VI. 350. Fl. Br. I. 1. 579; Bedd. Fl. Sylv. t. 137. *Sphærocarya leprosa*, Dalz. & Gibs. Bomb. Fl. 223. On the Konkan and North Kánara gháts, in evergreen forests. Flowers in December. Fruit ripe May. A large tree.

CANSJERA, Juss.

Climbing spiny shrubs. Leaves simple. Flowers bracteate in dense spikes. Calyx very minute. Petals 4-5. Fertile stamens 4, opposite petals; staminodes 4-5, alternate with stamens. Ovary 1-celled. Style cylindric. Ovule 1, pendulous. Fruit a drupe, stone bony.

C. Rheedii, Gmel. Syst. I. 280 ; Fl. Br. I. 1. 582 ; Bedd. Fl. Sylv. Anal. Gen. t. XXVI ; Brandis For. Fl. 75. On the Konkan and North Kánara gháts, in evergreen moist forests. Fl. Nov., Dec. Fruit ripe in May.

GOMPHANDRA, Wall.

Trees or shrubs. Leaves simple. Flowers polygamo-diœcious or hermaphrodite, in axillary cymes. Calyx 4-5-lobed. Corolla campanulate, 4-5 lobes ; lobes with a prominent rib within. Stamens 5 ; filaments thick, hollowed in front to contain the anthers, hairy at back. Disc thick, annular or 0. Ovary 1-celled. Ovules 2, collateral. Funicle dilated into an obturator. Fruit a drupe. Albumen fleshy, bipartite.

G. axillaris, Wall. Cat. 3718 ; Fl. Br. I. 1. 586 ; Bedd. Fl. Sylv. 61 ; *Platea axillaris*, Dalz. & Gibs. Bomb. Fl. 28. Common in the evergreen forests of the Konkan and North Kánara gháts. Fl. April, May. Fr. Nov.

MAPPIA, Jacq.

Trees. Leaves simple. Flowers in corymbs. Calyx 5-toothed. Petals 5. Stamens 5, alternate with the petals. Ovary superior, 1-celled. Ovules 2, pendulous. Fruit a drupe. Albumen fleshy.

M. oblonga, Miers Contrib. I. 65 ; Fl. Br. I. 1. 589 ; Dalz. & Gibs. Bomb. Fl. 28. *Gur, kalgur*, M. Common on the Konkan and North Kánara gháts in moist evergreen forests, has a very strong fœtid smell, particularly the flowers. *M. ovata*, Miers, and *M. fœtida*, Miers, are probably not distinct from *M. oblonga*, Miers. Fl. ; at different times, usually during Aug. Fr. ; ripe 2 months after flowering.

SARCOSTIGMA, W. & A.

Climbing shrubs. Leaves simple, coriaceous. Flowers diœcious, minute, in long interrupted spikes. Calyx cup-shaped. Petals 4-5, cohering in a short tube, free upwards, valvate. Males. Stamens 4-5. Females. Ovary 1-celled, ovules 2, pendulous. Funicle expanded. Drupes with a woody nut. Albumen 0.

S. Kleinii, W. & A. in Edin. New Phil. Jour. **XIV.** 299; Fl. Br. I. 1. 594; Dalz. & Gibs. Bomb. Fl. 221. In evergreen forests along the Konkan and North Kánara gháts. Chorla ghát, Dalz. Common near the Falls of Gairsoppah. Fl. Nov. Fruit ripe May.

ORDER 27. ILICINEÆ.

Trees or shrubs, usually evergreen. Leaves alternate, simple, extipulate or with minute stipules. Flowers small, in axillary cymes, fascicles or umbellules, usually diœcious. Male with imperfect ovary and female with imperfect stamens. Calyx 3-6-lobed. Petals 4-5, deciduous, imbricate in bud. Stamens as many as petals, alternating with them, hypogynous, free or adhering to the petals. Ovary free, 3-5-celled, with a short style or sessile stigma. Ovules 1-2 in each cell, pendulous. Drupe with two or more 1-seeded free, rarely connate, stones.

ILEX, L.

Calyx 4-5-lobed or partite. Corolla rotate, petals free or connate at the base. Stamens 4-5, adhering to the base of the corolla in the male, sometimes hypogynous in the female. Ovary 2-12-celled; styles 0, stigmas free or confluent on the top of the ovary. Drupe globose, with 2-16 stones.

I. malabarica, Bedd. Fl. Sylv. t. 143; Fl. Br. I. 1. 600; *I. Wightiana,* Dalz. & Gibs. Bomb. Fl. 143. On the Konkan and North Kánara gháts, in evergreen forests; along the banks of streams near Anmode, Supa sub-division of North Kanara. Fl. Feb., Mch.

ORDER 28. CELASTRACEÆ.

Shrubs or trees. Leaves simple, opposite or alternate. Flowers small, hermaphrodite or polygamous. Calyx 4-5-lobed. Petals 4-5. Stamens 3-5, alternate with petals. Ovary 3-5-celled, more or less immersed in a disc. Ovules 2 erect, 1 pendulous, or many in each cell. Fruit various. Albumen fleshy or 0.

Stamens 4-5 on or beneath margin of disc. Seeds albuminous. Leaves opposite. Fruit a capsule, dehiscent.
Ovules 2 in each cell.

Petals free. Capsule 3-5-celled	...	...	... EUONYMUS.
Petals connate. Capsule 1-celled	...	...	... MICROTROPIS.
Ovules 4 or more in each cell	...	...	... LOPHOPETALUM.
Fruit indehiscent. Ovules 2 in each cell	...		... PLEUROSTYLIA.
Leaves alternate.			
Ovary free ...	...	...	... CELASTRUS.
Ovary confluent with disc	...	...	... GYMOSPORIA.
Leaves opposite. Fruit a drupe	...	...	... ELÆODENDRON.
Stamens 3 on face of disc. Albumen 0.			
Fruit dehiscent. Seeds winged	...	...	... HIPPOCRATEA.
Fruit a berry	...	...	... SALACIA.

EUONYMUS, Linn.

Trees or shrubs. Leaves with caducous stipules. Flowers in axillary cymes. Calyx 4-5-fid. Petals 4-5. Stamens 4-5, inserted on the disc. Disc fleshy, 4-5-lobed. Ovary immersed in the disc and confluent with it, 3-5-celled, stigma 3-5-lobed, ovules 2 in each cell. Capsule loculicidally dehiscent, 3-5 celled. Seeds covered with an arillus, albuminous.

E. indicus, Heyne. in Roxb. Fl. Ind. II. 409; Fl. Br. I. 1. 608; *E. Goughii,* Wight. Dalz. & Gibs. Bomb. Fl. 47 ; Bedd. Fl. Sylv. 63. Throughout the evergreen forests of the ·Konkan and North Kánara gháts ; common on the southern gháts of North Kánara. Flowers and is in fruit from December till May. Petals orbicular, reddish, fringed.

LOPHOPETALUM, Wight.

Trees and shrubs. Leaves opposite or alternate, exstipulate. Clayx 5-lobed. Petals 5, continuous with the disc; crested or lamellate. Stamens 5, inserted on the disc. Disc thick, lobed. Ovary immersed in disc, trigonal, 3-4-celled ; ovules 4 or more in each cell in 2 series. Capsule 3-celled, opening loculicidally in 3 valves. Seeds few, winged. Albumen 0.

L. Wightianum, Arn. in Ann. Nat. Hist. III. 151 ; Fl. Br. I. 1. 615 ; Bedd. Fl. Sylv. t. 145 ; Dalz. & Gibs. Bomb. Fl. 48. *Balpale,* K. In the evergreen forests of the Konkan and North Kanára gháts. Fl. March, April. Fruit ripe at end of rainy season. A large evergreen tree with close-grained wood.

MICROTROPIS, Wall.

Trees or shrubs. Leaves opposite, extipulate. Flowers sometimes unisexual. Sepals 5. Petals 5, connate at the base into a short ring. Stamens 5, inserted on the disc or on the tube of the corolla. Disc 0 or annular. Ovary free, 2-3-celled ; stigma minutely 2-4-lobed; ovules 2 in each cell. Capsule 2-valved, 1-seeded. Seed erect, with a red testa, albumen fleshy.

Flowers sessile, clustered on the branches ...   ... *M. latifolia.*
Flowers in axillary cymes   ...   . *M. microcarpa.*

M. lati folia, Wight MSS., Fl. Br. I. 1. 613. Konkan.

M. microcarpa, Wight Ic. t. 975 ; Fl. Br. I. 1. 614. From the Konkan southwards. I have no knowledge of either of the above shrubs, which are noted as being found in the Konkan ; Fl. Br. I.

PLEUROSTYLIA, Wight.

A shrub. Leaves opposite, exstipulate. Flowers in axillary cymes. Calyx 4-lobed. Petals 5, larger than the calyx. Stamens 5, inserted below the disc ; connective dilated at the back. Disc thick, crenulate. Ovary 1-2-celled, ½ immersed in the disc; stigma peltate, on a short style; ovules 2 in each cell. Fruit indehiscent, 1-2-celled. Seeds albuminous.

P. Wightii, W. & A. Prod. 157; Fl. Br. I. 1. 617; Dalz. & Gibs. Bomb. Fl. 47; Bedd. Fl. Sylv. 66. Gháts of the Konkan, Dalz.

CELASTRUS, Linn.

Scandent shrubs. Leaves alternate; stipules minute or 0. Flowers small, usually unisexual in terminal or axillary panicles. Calyx 5-divided. Petals 5. Disc cup-shaped. Stamens 5, inserted on the margin of the disc. Ovary 2-4-celled; stigma generally lobed; ovules 2 in each cell. Capsule loculicidally dehiscent, 1-3-celled, 1-6-seeded. Seeds arillate, albumen fleshy.

C. paniculata, Willd. Sp. Pl. 1. 1125; Fl. Br. I. 1. 617; Brandis For. Fl. 82; Dalz. & Gibs. Bomb. Fl. 47; Bedd. Fl. Sylv. 66. *Oleum nigrum* plant. *Pigari,- kanguni*, M.; *kariganne*, K. Throughout the presidency, often in hedges and along river and nálá banks. Fl. March. Fr. June, Sept.

GYMNOSPORIA, W. & A.

Shrubs or small trees, often spinous. Leaves alternate, exstipulate. Flowers in axillary cymes. Calyx 4-5-divided. Petals 4-5, spreading. Stamens as many as petals, inserted beneath the disc. Disc sinuate, or lobed. Ovary 2-3-celled, immersed in the disc; ovules 2 in each cell. Fruit a capsule. Seeds arillate, albuminous.

Unarmed shrubs.
Leaves membranous, puberulous ... ... *G. puberula.*
Leaves coriaceous, glabrous ... ... *G. Rothiana.*
Spinous tree. Fruit black ... ... *G. montana.*
Spinous shrub. Fruit red ... ... *G. emarginata.*

G. puberula, Laws. Fl. Br. I. 1. 618. Bombay, Laws; also in the forests near the Ainshi ghát of North Kánara. Fl. R.S. Fr. C.S. A small shrub.

G. Rothiana, W. & A. Prod. 159; Fl. Br. I. 1. 620; Dalz. & Gibs. Bomb. Fl. 47. *Gawlin, moti yekkadi*, M. Throughout the Konkan and North Kánara evergreen forests, common in the forests near Katgal, North Kánara, at the sea-level. Fl. April. Fr. November, also at other times.

G. montana, Roxb. Fl. Ind. 1. 620 (*Celastrus*); Fl. Br. I. 1. 621; *Celastrus montana*, W. & A. Prod. 159; Bedd. Fl. Sylv. 66; Dalz. & Gibs. Bomb. Fl. 48; *C. senegalensis*, Lam. Brandis For. Fl. 81. *Hurmacha*, Vern.; *mal kanguni*, K.; *tondarshi jhad*, H.; *yekkadi, bharatti*, M. Throughout the presidency and Sind; usually in the dry deciduous forests on rocky ground. Flowers at various times.

G. emarginata, Roth. Fl. Br. I. 1. 621; Grah. Cat. Bomb. Pl. 39. *Yenkul, ingli, ikari*, M. Common on the gháts, also at Mahábaleshvar. Grah.

ELÆODENDRON, Jacq. f.

Trees or shrubs. Leaves opposite, often crenate. Flowers in axillary cymes, often polygamous. Calyx 5-cleft. Petals 5, spreading. Disc thick, angled. Stamens 5, inserted under the edge of the disc. Ovary continuous with the disc, conical, 2-5-celled; ovules 2 in each cell. Fruit an indehiscent drupe; cells usually 1-seeded. Seeds exarillate, albuminous.

E. glaucum, Pers. Synops, 1. 241; Fl. Br. I. 1. 623; Bedd. Fl. Sylv. 67; Dalz. & Gibs. Bomb. Fl. 48; Brandis For. Fl. 82. *Tamruj, bilur, luta pala, aran, burkas,* M. Throughout the presidency in deciduous forests, a small tree. On the ghâts of North Kánara and Belgaum in evergreen forests; a very large tree, common in the Goond forests of North Kánara. Fl. Feb., June. Fruit ripe in May. This tree was also in full bloom in the Dhárwár forests near Devikope in Sept., Oct.

HIPPOCRATEA, Linn.

Climbing shrubs or trees. Leaves opposite, entire or serrate; stipules small, caducous. Flowers bisexual. Calyx 5-divided. Petals 5, spreading. Stamens 3. Ovary 3-celled; surrounded by the disc. Fruit of 3 flattened carpels, distinct, 2-valved. Seeds compressed, winged below, exalbuminous.

Petals 1/6. in.   Fruit 1½ in. ...   .   *H. obtusifolia.*
Petals 1/12. in.   Fruit 3 in. long.   Cymes longer than
    the leaves   ...   .   *H. Grahami.*
Petals 1/24 in.   Cymes shorter than the leaves   ...   *H. indica.*

H. obtusifolia, Roxb. Fl. Ind. 1. 166; Fl. Br. I. 1. 623; W. & A. Prod. 104. *Daushir,* M. In the moist forests of the Konkan and North Kánara ghâts. Fl. C. S. Fruit ripe Apl., May. Cymes panicled, large.

H. Grahami, Wight Ic. t. 380; Fl. Br. I. 1. 624: Dalz. & Gibs. Bomb. Fl. 82. *Yesti,* M. Common on the ghâts of the Konkan and North Kánara, in evergreen forests. Fl. hot season. Fruit ripe Nov., Dec.

H. indica, Willd. Sp. Pl. 1. 193; Fl. Br. I. 1. 624; Dalz. & Gibs. 32; Brandis For. Fl. 83. *Kazurati, turruli,* Vern. Throughout the moist forests of the Konkan and North Kánara. Fl. C. S. Fr. R.S. Cymes much divided; flowers very small.

There is a scandent shrub in the forests near the Tinai ghât of North Kánara. (*Leaves small, ovate, obtuse, glabrous, coriaceous, crenate, 2 in. by ¾ in. Fruit 1½ in. by ½ in., obtuse, striate, 4-seeded.*) May be distinct from *H. obtusifolia,* Roxb. Fr. Sept.

SALACIA, Linn.

Climbing or straggling shrubs. Leaves opposite, exstipulate. Flowers small, usually fasciculate in the axils of the leaves; rarely terminal. Calyx 5-divided. Petals 5, imbricate. Stamens 3, recurved. Ovary conical, immersed in the disc, 3-celled ; ovules 2 or more in each cell. Fruit a berry. Seeds large angular.

Flowers 1-6, from an axil'ary tubercle.
Flowers 3-6, from each axil. Fruit globose, small ... *S. prinoides.*
Flowers 1-2, from each axil. ... ... *S. Brunoniana.*
Flowers numerous from an axillary tubercle.
Flowers fascicled. Fruit 1 in. globose, rugulose, orange-coloured ... ... *S. macrosperma.*
Flowers usually on a common peduncle, sometimes in short axillary cymes. Fruit 2-2½ in. diam. ... ... *S. oblonga.*

S. prinoides, DC. Prod. 1. 571 ; Fl. Br. I. 1. 626 ; Dalz. & Gibs. Bomb. Fl. 33 ; Bedd. Fl. Sylv. 67. *Nisul-bondi*, M. On the Konkan and North Kánara ghâts from the sea-level upwards. The scandent form is often found along the banks of rivers (Kálánadi, &c.) ; the erect shrubby form is abundant in the evergreen forests of the southern ghâts of North Kánara. Fl. at different times throughout the year.

S. Brunoniana, W. & A. Prod. 105 ; Fl. Br. I. 1. 626 ; Dalz. & Gibs. Bomb. Fl. 33. Rám ghát, Belgaum district, Dalz.

S. macrosperma, Wight Ic. t. 962 ; Fl. Br. I. 1. 628 ; Bedd. Fl. Sylv. 67. A scandent shrub on the Konkan and North Kánara ghâts ; common in the Ainshi ghát forests. The North Kánara plant has the fruit orange-coloured when ripe, which, however, turns black in dried specimens ; the calyx lobes in all my specimens are fringed with rust-coloured hairs. Fl. Jany. Fr. June.

S. oblonga, Wall. Cat. 4226 ; Fl. Br. I. 1. 628 ; Dalz. & Gibs. Bomb. Fl. 23. Bedd. Fl. Sylv. 67. In evergreen forests on the Konkan and North Kánara ghâts. The North Kánara plant, which is common in many of the forests along the ghâts from Ainshi southwards, was doubtfully referred at Kew to *S. oblonga*, Wall. It has the petals entire with spreading tips ; the corolla having the appearance of being urceolate. Fruit globose, tubercled. Fl. cold season Dec., Feb.

S. Roxburghii, Wall., is a doubtful native of the western peninsula.

Dalzell and Gibson's plant referred to in the Bomb. Fl. p. 33 is probably *S. macrosperma*, Wgt.

ORDER 29. RHAMNEÆ.

Shrubs or trees erect or scandent, prickly, spinous or unarmed. Leaves simple, alternate or opposite, often 3-5-nerved. Flowers hermaphrodite or polygamous in axillary cymes, spikate, paniculate or solitary. Calyx 4-5-cleft, lobes triangular, valvate in bud. Petals 4-5, rarely 0, inserted on the throat of the calyx-tube, cucullate or involute. Stamens 4-5 opposite the petals, often enclosed by them. Disc fleshy, filling the calyx-tube or thin and lining it. Ovary sessile, free or immersed in the disc, more or less adherent to the calyx

tube, 2-3-celled; cells 1-ovuled. Fruits various, sometimes winged.
Seed generally albuminous.

> Scandent shrubs. Fruit dry, 1-celled, 1-winged   ... VENTILAGO.
> Trees or shrubs. Fruit dry or fleshy with a 1-3-celled
> stone, not winged.
> Leaves 3-nerved ...   . ZIZYPHUS.
> Fruit dry or fleshy of 3 pyrenes or cocci. Disc lining
> the calyx-tube, or fleshy and filling it.
> Trees or shrubs. Disc thin, lining the calyx-tube   ... RHAMNUS.
> Disc fleshy, filling the calyx-tube.
> Prickly shrubs. Leaves opposite. Flowers subum-
> bellate   ...   . SCUTIA.
> Shrubs. Flowers sessile in terminal panicles. Leaves
> opposite ...   . SAGERETIA.
> Erect shrubs. Leaves alternate. Fruit size of pea, 3-
> celled, not winged...   ...   . COLUBRINA.
> Climbing shrub. Fruit 3-winged   ...   . GOUANIA.

VENTILAGO, Gærtn.

Scandent shrubs. Leaves alternate. Flowers in axillary and
terminal panicles. Calyx 5-cleft, lobes keeled within. Petals 5,
hooded, deflexed. Stamens 5, adnate to the base of the petals. Disc
5-lobed, filling the calyx-tube. Ovary immersed in the disc,
2-celled. Fruit 1-celled, 1-seeded, produced at the summit into a
long wing.

> Nut girt at the base. Ovary with a few white hairs
> at the base. Calyx glabrous within ...   . *V. madraspatana.*
> Nut girt round the middle. Ovary densely pubes-
> cent. Calyx also densely pubescent within...   ... *V. calyculata.*

V. madraspatana, Gærtn. Fruct. 1223, t. 49, f. 2; Fl. Br. I. 1. 631;
Dalz. & Gibs. Bom. Fl. 48; Brandis For. Fl. 96. *Locundie, kanwail,* M.
Throughout the moist ghát forests of the Konkan and North Kánara.
Fl. Dec., Jan. Fr. May.

V. calyculata, Tulasne in Ann. Sc. Nat. Ser. 4, VIII. 124; Fl. Br. I.
1. 631; Brandis For. Fl. 96. *Papri,* Vern. Very common in the deciduous
forests of the Konkan and Deccan; is found also in the moist forests of
North Kánara. Flowers during the rainy season. Fruit ripe February
in North Kánara. I know nothing of *V. bombaiensis,* Dalz.

ZIZYPHUS, Juss.

Trees or shrubs, sometimes climbing, armed with stipular sharp
prickles. Leaves alternate, often oblique at the base, palmately
3-5-nerved. Flowers fascicled or in axillary cymes. Calyx 5-fid,
lobes keeled within. Petals 5, hooded. Disc flat, filling the short
calyx-tube. Ovary immersed in the disc, 2-4-celled. Fruit fleshy
or dry, containing a woody or bony 1-3-seeded stone.

> Leaves pubescent or tomentose beneath. Trees.
> Fruit 2-celled, orange or red when ripe ...   . *Z. jujuba.*
> Fruit 3-celled, green when ripe, woody ...   . *X. xylopyrus.*
> Straggling shrubs or climbers.

Leaves oblique, brown silky beneath. Fruit size of a
pea, black ...   . *Z. œnoplia.*
Leaves with white velvetty tomentum beneath. Fruit
size of a cherry ...   . *Z. nummularia.*
Leaves pubescent beneath when young, glabrous when
old.
Fruit with a white mealy pulp and a crustaceous stone. *Z. rugosa.*
Leaves glabrous. A profusely armed shrub. Fruit
woody, 3-celled ...   . *Z. horrida.*

Z. jujuba, Lamk. Dict. III. 318; Fl. Br. I. 1. 632; Dalz. & Gibs.
Bomb. Fl. 49; Bedd. Fl. Sylv. t. 149; Brandis For. Fl. 86. *Ilanji,* K.;
bor, ber. M. Throughout the presidency and Sind, usually planted;
thrives in very dry situations. Fl. Apl., June. Fr. Dec.-Mch.

Z. œnoplia, Mill. Gard. Dict. No. 3; Fl. Br. I. 1. 634; Dalz. &
Gibs. Bomb. Fl. 49; Brandis For. Fl. 86; Bedd. Fl. Sylv. 69. Through-
out the Konkan and North Kánara in moist forests, also found in the
Deccan (Dalz.). Fruit edible. Fl. Sept., Oct. Fruit ripe Feb. in North
Kánara.

Z. nummularia, W. & A. Prod. 162; Fl. Br. I. 1. 633; Brandis For.
Fl. 88; Bedd. Fl. Sylv. 69. Dalz. & Gibs. Bomb. Fl. 49. *Gangr, jangra,*
Sind. Throughout the presidency and Sind in the driest situations, absent
from the moist region near the coast. Fl. Mch., June. Fr. Nov., Jan.

Z. xylopyrus, Willd. Sp. Pl. 1104; Fl. Br. I. 1. 634; Dalz. & Gibs.
Bomb. Fl. 49; Brandis For. Fl. 90; Bedd. Fl. Sylv. 68. *Mullu kare,* K.;
ghát, kanta gotti, M.; *guti,* Vern. A common tree in the moist
forests of North Kánara and Konkan, a straggling shrub throughout the
drier parts of the presidency. Fl. May. Fr. ripe end of rainy season.

Z. horrida, Roth. Nov. Sp. 159; Fl. Br. I. 1. 636; DC. Prod. II. 20.
Konkan, Stocks. I have not seen this species.

Z. rugosa. Lamk. Dict. III. 319; Fl. Br. I. 1. 663; Dalz. & Gibs.
Bomb. Fl. 49; Brandis For. Fl. 89; Bedd. Fl. Sylv. 68. *Turan,* M.
Very common in open places on the North Kánara ghâts and elsewhere
throughout the presidency. Fl. Nov., Mch. Fr. May.

RHAMNUS, L.

Shrubs or trees. Leaves alternate. Flowers hermaphrodite or
polygamous. Calyx-tube urceolate, limb 4-5-lobed, lobes keeled
within. Petals 4-5, or 0, inserted on the calyx-tube above the
ovary, hood-shaped or flat. Stamens 4-5. Disc thin, lining the
calyx-tube. Ovary free, 3-4-celled. Fruit a berry-like drupe, girt
at the base by the small calyx-tube, 2-4-seeded.

R. Wightii, W. & A. Prod. 164; Fl. Br. I. 1. 639; Bedd. Fl. Sylv. 70;
Dalz. & Gibs. Bomb. Fl. 50. " *Rugt rorar,*" Vern. Highest hills of
the Konkan ghâts.

SCUTIA, Comm.

A scandent, prickly shrub. Leaves opposite or alternate, penninerved. Flowers hermaphrodite, in axillary fascicles or subumbellate. Calyx 5-fid; tube hemispherical or turbinate. Petals 5, clawed,

flat or hooded. Disc filling the calyx-tube. Stamens 5. Ovary sunk in the disc, 2-4-celled. Fruit girt at the base by the calyx, pyrenes 2-4.

S. indica, Brogn. in Ann. Sc. Nat. X. 363; Dalz. & Gibs. Bomb. Fl. 50; Fl. Br. I. 1. 640; Bedd. Fl. Sylv. 70. *Chimat*, Vern. On the higher ghâts of the Konkan; common on the Supa ghâts of North Kánara, in evergreen forests, at about 2,000 ft. elevation. Fl. Oct. Fr. June.

SAGERATIA, Brogn.

Unarmed or spinous shrubs. Leaves subopposite with deciduous stipules. Flowers small, pentamerous, bisexual, in sessile clusters. Calyx 5-fid. Petal. Petals 5, clawed, hooded. Stamens 5. Disc cup-shaped, lining the calyx-tube, margin free, 5-lobed. Ovary 3-celled; style short, 3-grooved. Fruit globose, 3-seeded, indehiscent.

S. oppositifolia, Brogn. in Ann. Nat. Sc. Ser. 1. X. 360; Fl. Br. I. 1. 641; Brandis For. Fl. 95. Konkan, Brandis. Flowers at various seasons.

COLUBRINA, Rich.

Erect unarmed shrubs. Leaves alternate. Flowers in short axillary cymes. Calyx 5-fid, tube hemispherical. Petals 5, hooded, inserted on the margin of the disc. Disc fleshy, filling the calyx-tube. Ovary immersed in the disc, 3-celled; stigmas reflexed. Fruit obsoletely 3-lobed, surrounded at the base by calyx-tube, 3-celled, cells 1-seeded, tardily dehiscent.

C. asiatica Brogn. in Ann. Sc. Nat. Ser. 1. X. 369; Fl. Br. I. 1. 642; Dalz. & Gibs. 50; Bedd. Fl. Sylv. 69. *Guti*, M. On the Konkan ghâts; common near the sea at Ratnágiri, Dalz.

GOUANIA, Linn.

Unarmed climbing shrubs. Leaves alternate. Flowers in axillary or terminal spikes, polygamous. Calyx 5-lobed, the tube adherent to the ovary. Petals 5, hooded. Stamens 5. Disc filling the calyx-tube, 5-angled or stellate. Ovary immersed in the disc, 3-celled. Fruit 3-winged. The flowering rachis is often cirrhose.

Flowers sessile, densely pubescent   ...   . *G. microcarpa.*
Flowers shortly pedicelled, glabrous   ...   . *G. leptostachya.*

G. microcarpa, DC. Prod. II. 40; Fl. Br. I. 1. 643. From the Konkan southwards, very common on the North Kánara ghâts, in evergreen forests. Fl. Nov. Fr. Jan.

G. leptostachya, DC. Prod. II. 40; Fl. Br. I. 1. 643; Dalz. & Gibs. Bomb. Fl. 50. At Bánda in the Warri country, Dalz.

ORDER 30. AMPELIDEÆ.

Erect shrubs or small trees with jointed branches or woody climbers. Leaves alternate or opposite, simple or compound. Flowers regular, hermaphrodite, rarely unisexual, cymose. Calyx 4-5 toothed or entire. Petals 4-5, free or cohering. Stamens 4-5, opposite petals, on outside of disc. Ovary 2-6-celled. Ovules 1-2 in each cell. Fruit a berry. Seeds 1-6. Albumen ruminate.

Scandent shrubs, usually with tendrils. Ovary 2-celled, cells 2-ovuled ... ... VITIS.

Erect shrubs, no tendrils. Ovary 3-6-celled. Cells 1-ovuled ... ... LEEA.

VITIS.

Shrubs climbing by means of tendrils. Leaves simple or 3-9-foliate. Flowers variously cymose, hermaphrodite, unisexual or polygamous. Calyx 4-5-lobed. Petals 4-5 free or cohering at apex. Stamens 4-5. Ovary 2, rarely 3-4-celled. Style 0 or short. Ovules 2 in each cell. Fruit a berry 1-4-seeded.

Leaves simple.
Flowers tetramerous, in umbels.
Stems and branches acutely winged ... ... *V. quadrangularis.*
Stems cylindric or obscurely angled.
Leaves glabrous.
Branchlets terete, white mealy ... ... *V. repens.*
Branchlets 6-angled... ... ... *V. discolor.*
Branches fleshy, hollow ... ... *V. pallida.*
Branches glaucous, cylindric, striate ... ... *V. glauca.*
Leaves pubescent beneath.
Subtomentose beneath ... ... *V. gigantea.*
Adpressed woolly tomentum beneath ... ... *V. repanda.*
Rusty pubescent beneath ... ... *V. adnata.*
Flowers pentamerous in umbels.
Cymes woolly, umbellate ... ... *V. tomentosa.*
Cymes glabrous, paniculate ... ... *V. latifolia.*
Cymes tomentose, paniculate ... ... *V. vinifera.*
Cymes spicate ... ... *V. indica.*
Leaves trifoliate.
Flowers tetramerous.
Plant glabrous ... ... *V. Rheedii.*
Plant covered with bristly hairs ... ... *V. setosa.*
Plant pubescent. Fruit black, 4-seeded. Seeds
tubercled ... ... *V. carnosa.*
Stems woody. Fruit globose or ovoid. Petals 5 ... *V. canarensis.*
Stems slender. Petals 5 ... ... *V. araneosus.*
Leaves 5-foliate, digitate.
Quite glabrous. Leaflets nearly sessile. Fruit black *V. elongata.*
Pubescent. Leaflets long-stalked. Fruit red ... *V. auriculata.*
Leaves pedately 5-foliate.
Flowers subcorymbose.
Stigma subpeltate ... ... *V. tenuifolia.*
Stigma 4-lobed. ... ... *V. lanceolaria.*
Stigma simple ... ... *V. pedata.*

V. quadrangularis, Wall. Cat. 5992; Fl. Br. I. 1. 645; Brandis For. Fl. 100; *Cissus quadrangularis,* Linn.; Dalz. & Gibs. Bomb. Fl. 39. *C. edulis,* Dalz. & Gibs. Bomb. Fl. 40. *Hursanker,* Vern. Throughout the driest districts of the presidency from Gujarát to Dhárwár. Fl. R.S. Fr. C.S.

V. repens, W. & A. Prod. 125; Fl. Br. I. 1. 646; *Cissus repens*, Lam. Dalz . & Gibs. Bom. Fl. 39. Very common on the North Kánara and Konkan gháts. Fl. March. Fr. May.

V. discolor, Dalz. in Hook. Kew Jour. Bot. II. 39 ; Fl. Br. I. 1. 647 ; *Cissus discolor,* Dalz. & Gibs. Bom. Fl. 40. *Telitsayel,* M. Throughout the moist forests of the Konkan and North Kánara. Fl. Aug. Fr. Oct., Nov.

V. pallida, W. & A. Prod. 125 ; Fl. Br. I. 1. 647 ; *V. Linnæi,* Wall? *Cissus vitiginea,* Dalz. & Gibs. Bom. Fl. 40 (not of Roxb.). In the dry districts of the Deccan, common on the Kupnt hills of the Dhárwár district, usually an erect shrub with hollow glaucous stems. Fruit 1-seeded. Fl. June, July. Fr. Aug.

V. glauca, W. & A. Prod. 126 ; Fl. Br. I. 1. 648. In the evergreen forests of the Konkan and North Kánara, not common. Fl. Fr. Apl., May.

V. gigantea, Bedd. in Trans. Linn. Soc. XXV. 212 ; Fl. Br. I. 1. 648. Common throughout the moist forests of North Kánara from the sea-level upwards, also probably in the Konkan. Fl. Fr. Aug.

V. repanda, W. & A. Prod. 125 ; Fl. Br. I. 1. 648; *Cissus repanda,* Vahl. Dalz. & Gibs. Bomb. Fl. 39. Common throughout the presidency along the border of the heavy rainfall zone. Fl. Mch., Apl. Fr. Apl. May.

V. adnata, Roxb. Fl. Ind. 1. 405; Dalz. & Gibs. Bom. Fl. 39; Brandis For. Fl. 100. In the moist forests near Goond, North Kánara. Fl. Dec. Fr. Feb.

V. tomentosa, Heyne in Roth. Nov. Sp. 157 ; Fl. Br. I. 1. 650. Throughout the dry forests on the North Kánara border; common in the forests of the Dhárwár districts along the banks of streams and rivers. Fl. Mch. Fr. May.

V. latifolia, Roxb. Fl. Ind. 1. 661 ; Fl. Br. I. 1. 652; Brandis For. Fl. 99. *Nádena,* M. In the moist forests near the coast from the Konkan southwards ; common in the forests near Kárwár and on the North Kánara gháts. Stigma concave at top. Fruit edible, black. Fl. June, July, Aug. Fr. Oct.

V. vinifera, Linn. Sp. Fl. 202 ; Fl. Br. I. 1. 652 ; Brandis For. Fl. 98 ; Dalz. & Gibs. Bomb. Fl. Suppl. 15. The grape vine. *Draksha,* Vern. Cultivated in the drier districts of the presidency.

V. indica, Linn. Sp. Pl. 202 ; Fl. Br. I. 1. 653 ; Dalz. & Gibs. Bomb. Fl. 41 ; Brandis For. Fl. 100. In the evergreen forests of the North Kánara and Konkan gháts. Fl. and Fr. from Aug. till Jan.

V. Rheedii, W. & A. Prod. 127 ; Fl. Br. I. 1. 653 ; *Cissus trilobata,* Lamk. ; Dalz. & Gibs. Bom. Fl. 39. Konkan, Dalz.

V. setosa, Wall. Cat. 6009 ; Fl. Br. I. 1. 654 ; *Cissus setosa,* Dalz. & Gibs. Bomb. Fl. 41. About Juneer, Dalz. On the Kupnt hills of the Dhárwár district in dry rocky situations. Fl. Apl. Fr. June.

V. carnosa, Wall. Cat. 6018 ; Fl. Br. I. 1. 654; Brandis For. Fl. 101 ; *Cissus carnosa,* Roxb. ; Dalz. & Gibs. Bomb. Fl. 40. On the Konkan and North Kánara gháts, common in the Dhárwár district bordering on North Kánara. Fl. and Fr. May, Oct.

V. canarensis, Dalz. in Hook. Kew Jour. Bot. III. 123; Fl. Br. I. 1. 655. Supa gháts of North Kánara, in evergreen forests, common. The description of this species in the Fl. Br. I. is very short. The North Kánara plant has a globose (size of a small cherry), yellow (turning red when ripe), one-seeded, dry fruit. Seed with a crustaceous episperm and a white horny, ruminate albumen. The leaves are always trifoliate. Flowers diœcious, in short spreading cymes; it is quite distinct from *V. lanceolaria,* Roxb. Fl. C.S. Fr. April.

V. araneosus, Dalz. & Gibs. Bomb. Fl. 41 ; Fl. Br. I. 1. 657. Highest gháts of the Konkan. Fl. Br. I.

V. elongata, Wall. Cat. 6016 ; Fl. Br. I. 1. 658. In the forests on the coast near Kárwár, North Kánara ; common. A large glabrous climber. Leaves digitately 3-5-foliate; petiole 3-4 in. sulcate. Flowers small, 4-merous, in short divaricating corymbose cymes. Tendrils simple. This is *V. elongata,* Wall., or a new species. Fl. Aug. Fr. Oct.

V. auriculata, Roxb. Fl. Ind. 1. 411; Fl. Br. I. 1. 658 ; *Cissus auriculata,* D. C. ; Dalz. & Gibs. Bomb. Fl. 40. Common in the dry forests along the North Kánara frontier in the Dhárwár and Belgaum districts and probably in other places throughout the presidency. Fl. May, June. Fr. Nov.

V. tenuifolia, W. & A. Prod. 129. Fl. Br. I. 1. 660. In the moist forests of the North Kánara gháts from Ainshi southwards. A very distinct species. Fl. July, Aug. Fr. Sept.

V. lanceolaria, Roxb. Fl. Ind. I. 412; Fl. Br. I. 1. 660 ; Brandis For. Fl. 101 ; *Cissus muricata,* Dalz. & Gibs. Bomb. Fl. 40. *Kajolitsayel,* M. Common on the North Kánara gháts, in evergreen forests. Fl. Dec. Fr. Mch.

V. pedata. Vahl. in Herb. Madr. ex Wall. Cat. 6027; Fl. Br. I. 1. 661 ; *Cissus pedata,* Dalz. & Gibs. Bomb. Fl. 40. Throughout the Konkan and North Kánara ; not common. Fl. H. & R.S. Fr. Nov.

LEEA, Linn.

Small trees or erect shrubs. Leaves simple or compound, petiole dilated at base into a sheath. Inflorescence leaf opposed, corymbose. No tendrils. Calyx 5-toothed. Petals 5, united at base with staminal tube. Stamens 5, united at base into a 5-lobed tube. Ovary on disc, 3-6-celled. Ovules 1 in each cell. Berry 3-6-celled. Seeds erect.

Leaves simple, very large	... *L. macrophylla.*
Leaves pinnate.	
Stems, petioles, &c., with crisped wings ...	... *L. crispa.*
Stems without wings. Cymes small	... *L. aspera.*
Fruiting cymes, large and fleshy	... *L. coriacea.*
Leaves bi-tri-pinnate.	
Glabrous	... *L. sambucina.*
Pubescent.	
Cymes large, corymbose. Bracts large, persistent ...	*B. robusta.*
Cymes small, compact. Bracts very small	... *L. hirta.*

L. macrophylla, Roxb. Fl. Ind. 1. 653 ; Fl. Br. I 1. 664 ; Brandis For. Fl. 102 ; Dalz. & Gibs. Bomb. Fl. 41. *Dinda,* M. In the moist forests of North Kanara and the Konkan ; common in the forests near Yellápur, North Kanara. Fl. and Fr. R.S.

L. crispa, Willd. Sp. Pl. 1. 1177 ; Fl. Br. I. 1. 665. Throughout the Konkan and North Kánara from the sea-level upwards in moist forests. Testa smooth ; albumen ruminate. Fl. Aug. Fr. Oct.

L. aspera, Wall. in Roxb. Fl. Ind. II. 468 ; Fl. Br. I. 1. 665 ; Brandis For. Fl. 102. Throughout North Kánara, in moist forests ; common near Kárwár, also in the Sátpudás of Khándesh. Fl. Aug. Fr. Oct.

L. coriacea, Laws. Fl. Br. I. 1. 665. Konkan, Stocks, not seen by me.

L. sambucina, Willd. Sp. Pl. 1. 1177 ; Fl. Br. I. 1. 666 ; Brandis For. Fl. 102 ; *L. Staphylea,* Roxb. Dalz. & Gibs. Bomb. Fl. 41. *Kurkunnie,* Vern.; *dino,* Konkani. Throughout the presidency. In the moist forests of North Kánara. Where there is a heavy rainfall this species develops short aereal roots abundantly from the branches. Fl. Fr. R.S.

L. robusta, Roxb. Fl. Ind. 1. 656 ; Fl. Br. I. 1. 667. On the gháts of North Kánara, also at the sea-level near Kadra. A large shrub with bipinnate leaves ; the terminal leaflet of the pinnule is much larger than the lateral, opposite leaflets. The corymbs are very large. Fruit 6-lobed. I think this species is *L. robusta,* Roxb.; it may however be distinct. Fl. Aug. Fr. ripe Nov.

L. hirta, Roxb. Fl. Ind. 1. 655 ; Fl. Br. I. 1. 668. In the evergreen forests near Kárwár. The fruit is succulent, and when dry is very deeply lobed, resembling that of some *Grewias.* Fl. Fr. R. S.

ORDER 31. SAPINDACEÆ.

Trees or shrubs. Leaves alternate or opposite, simple or pinnate. Flowers small, generally polygamous. Sepals 4-5 or united. Petals 4-5 free, often squamate at base. Disc complete or one-sided. Stamens free, 5-10, generally 8. Ovary 2, 3, 4-celled. Ovules 1 in each cell on axile placentas. Fruit capsular or indehiscent.

Stamens inserted inside the disc. Seeds exalbuminous *(Sapindeæ)*.
Flowers irregular. Disc unilateral. Leaves pinnate.
Fruit indehiscent, not lobed   ...   . HEMIGYROSA.
Fruit lobed   ...   . ERIOGLOSSUM,
Leaves trifoliate   ..   .. ALLOPHYLUS.
Flowers regular. Disc annular.
Petals 0. Fruit 1-seeded, ovoid ..   .. SCHLEICHERA.
Petals usually present. Fruit deeply 1-3-lobed.
Sepals imbricated   ...   . SAPINDUS.
Calyx-divisions valvate.
Fruit tubercled, indehiscent   ...   . NEPHELIUM.
Fruit an orange-coloured inflated capsule   ... HARPULLIA.
Stamens inserted outside the disc. Seeds
   albuminous or exalbuminous.
Fruit a winged capsule. Leaves simple ...   ... DODONÆA.
Fruit globose, indehiscent. Leaves pinnate   ... TURPINIA.

HEMIGYROSA, Blume,

Trees. Leaves coriaceous abruptly pinnate. Flowers irregular,
polygamo-monœcious. Sepals 5. Petals 4-5, with a scale at base of
claw. Disc unilateral, cushioned-shaped. Stamens 8; in male
flowers 6-10. Ovary excentric, trigonous, 3-celled. Ovule solitary.
Fruit woody, tomentose, yellow.

H. canescens, Thwaites Enum. 56, 408; Fl. Br. I. 1. 671; Bedd. Fl.
Sylv. t. 151; *Cupania canescens*, Pers.; Dalz. & Gibs. Bom. Fl. 35.
Kurpa, lakhandi, M.; *kurpah*, K. Common on the Konkan and North
Kánara gháts in moist forests. Fl. Feb., May. Fruit ripe in June and
July.

ERIOGLOSSUM, Blume,

Trees or shrubs. Leaves alternate, pinnate, exstipulate; leaflets
opposite, entire. Flowers in terminal, erect panicles, polygamo-
diœcious. Sepals 5, unequal, imbricated. Petals 4, unequal,
clawed; scale hairy, hooded, with an appendage at the tip. Disc
1-sided, lobed. Stamens 8, unequal; filaments hairy. Ovary
stipitate obcordate, 3-lobed, 3-celled; cells 1-ovuled. Fruit lobed to
the base, indehiscent.

E. edule, Bl. Bijdr. 299; Fl. Br. I. 1. 672; *E. rubiginosum*, Brandis
For. Fl. 108; *Sapindus rubiginosa*, Roxb. Fl. Ind. II. 282; Dalz. & Gibs.
Bom. Fl. Suppl. 14; Bedd. Fl. Sylv. 73. Planted near Bombay; very
doubtfully indigenous.

ALLOPHYLUS, Linn.

Small trees or climbing shrubs. Leaves 1-3-foliate. Flowers
small, white, in simple spikes. Sepals 4. Petals 4, with a scale
above the claw. Stamens 8. Disc 1-sided. Fruit small, indehiscent, red, shining.

A. Cobbe, Bl. Rumph. III. 131; Fl. Br. I. 1. 673; *Schmidelia Cobbe*,
Bedd. Fl. Sylv. 72. Throughout the Konkan and North Kánara in
moist forests and along the banks of rivers and streams, usually a scandent
shrub. Fl. June. Fr. Nov.

SCHLEICHERA, Willd.

Trees. Leaves alternate, pinnate. Calyx 4-8 cleft. Petals 0.
Stamens 6-8, on centre of disc. Ovary 3-4-celled. Fruit dry.
Seeds arillate.

S. trijuga, Willd. Sp. Pl. IV. 1096 ; Dalz. & Gibs. Bom. Fl. 35 ; Bedd.
Fl. Sylv. t. 119 ; Brandis For. Fl. 105. *Kusumb, koon, kohan*, M.;
sajala, K. Throughout the presidency in dry and moist forests;
common. Fl. Mch. Fr. May.

SAPINDUS, Linn.

Trees. Leaves alternate, usually paripinnate. Flowers regular.
Sepals and petals 4-5. Disc complete, annular. Stamens 8-10.
Fruit 1-3-lobed, rusty tomentose.

S. trifoliatus, Linn. Sp. Pl. ed. 1. 367 ; Fl. Br. I. 1. 682 ; *S. laurifolia*,
Vahl.; Dalz. & Gibs. Bom. Fl.34 ; Bedd. Fl. Sylv. 73 ; Brandis For. Fl.106.
S. emarginata, Roxb. ; Bedd. Fl. Sylv. t. 154 ; Dalz. & Gibs. Bom. Fl. 35;
Brandis l. c. 107. *Aratala*, K. ; *rita*, M. Throughout the presidency ;
usually planted about villages. Both the varieties are found in North
Kánara. Var. *laurifolia* is indigenous on the ghàts, in evergreen forests.
Fl. Oct., Dec. Fr. Feb., Apl.

NEPHELIUM, Linn.

Trees. Leaves pinnate, glaucous beneath. Flowers polygamous.
Calyx 4-8-lobed. Petals small. Stamens 5-10, within disc. Fruit
1-3-coccous, indehiscent. Seeds globose, arillate.

N. longana, Camb. in Mém. Mus. Par. XVIII. 30; Fl. Br. I. 1. 689 ;
Dalz. & Gibs. Bom. Fl. 35 ; *Euphoria longana*, Lamk. Bedd. Fl. Sylv.
t. 156. *Wumb*, M.; *kanakindali*, K. In the evergreen forests of the Konkan
and North Kánara ghàts. Common in the forests near the Ainshi ghàts.
Fl. Mch. Fr. R.S.

N. Litchi, Cam. Fl. Br. I. 1. 687. *Litchi*, Vern., is cultivated near
Bombay.

HARPULLIA, Roxb.

Trees. Leaves pinnate. Sepals 4-5. Petals 4-5, obovate, clawed
Stamens 5-8 elongate. Disc obscure. Capsule inflated, 2-lobed.
Cells 1-2-seeded.

H. cupanoides, Roxb. Fl. Ind. II. 442 ; Fl. Br. I. 1. 692 ; *H. imbricata*,
Bedd. Fl. Sylv. t. 158. In the evergreen forests of the Konkan and North
Kánara, very common in the forests near Goond. Fl. Dec., Jany. Fr.
Mch., April.

DODONÆA, Linn.

Shrubs. Leaves alternate. Sepals 5-2. Petals 0. Stamens 10-5,
usually 8, on outer side of disc. Capsule 2-6-sided ; valves winged,
Cells 1-2-seeded.

D. viscosa, Linn. Mant. Pl. Alt. 228; Fl. Br. I. 1. 697; Brandis For. Fl. 113; Bedd. Fl. Sylv. 75; *D. Burmanniana,* Dalz. & Gibs. Bom. Fl. 36. *Lutchmi, paorki,* M.; *bundurgi,* K. Throughout the dry forests of the presidency and in open situations in the driest districts, absent from the heavy rainfall zone. Fl. Aug. Fr. C.S.

TURPINIA, Vent.

Trees. Leaves opposite, oddpinate. Calyx 5-partite. Petals 5, imbricated. Stamens 5, outside the disc. Fruit globose, indehiscent, 3-celled; seeds angular; albumen fleshy.

T. pomifera, DC. Prod. II. 3; Fl. Br. I. 1. 698; *T. nepalensis,* Wall. Bedd. Fl. Sylv. t. 159. In the evergreen forests of the Konkan and North Kánara; a very large tree. Fl. Jany. Fr. April.

ORDER 32. SABIACEÆ.

Shrubs or trees, rarely climbers. Leaves alternate, simple or pinnate. Flowers small, hermaphrodite or polygamous; inflorescence usually a panicle. Calyx 4-5-partite, imbricate. Petals 4-5, equal or unequal, opposite to or alternating with the sepals. Disc small, annular. Stamens 4-5, opposite the petals, inserted at the base of or on the disc, usually 2 only perfect and 3 without anthers. Ovary 2-3-celled, compressed or 2-3-lobed; styles 2-3, free or connate or 0, stigmas punctiform; ovules 1-2 in each cell. Ripe carpels 1-2 drupaceous or dry, endocarp crustaceous or bony, albumen 0 or scanty.

MELIOSMA, Blume.

Trees or shrubs, usually pubescent or tomentose. Flowers minute in thyrsoid panicles. Sepals 4-5. Petals as many very unequal, the 2 smaller, interior, behind the fertile stamens, sometimes 2-cleft. Stamens 2 fertile and 3 deformed, broad opposite the larger petals, 2-fid with 2 empty cells. Disc annular toothed. Ovary sessile, 2 rarely 3-celled, with 2 horizontal or pendulous ovules in each cell. Drupe obliquely globular, stone bony or crustaceous, 1 rarely 2-celled.

M. Wightii, Planch in Herb. Hook.; Fl. Br. I. 2. 4; Brandis For. Fl. 116; *M. pungens,* Wall.; Bedd. Fl Sylv. 77. From the Konkan southwards, Fl. Br. I.

ORDER 33. ANACARDIACEÆ.

Trees or shrubs. Leaves simple or compound, usually alternate. Flowers unisexual or hermaphrodite. Calyx 3-5-partite. Petals as many as calyx divisions. Disc usually annular. Stamens as many as petals, inserted under the base of the disc. Ovary unilocular (in *Spondias* 5-celled), superior or half inferior. Styles 1-4. Fruit a drupe, 1-5-celled and seeded. Seeds exalbuminous.

Fruit 1-celled, 1-seeded.
Ovules pendulous from a basal funicle.
Leaves compound. Stamens 4-10.... ... ... Rhus.
Leaves simple. Disc lobed.
Stamens 1-5. Peduncle not enlarged in fruit ... Mangifera.
Stamens 8-10. Peduncle enlarged in fruit ... Anacardium.
Stamens 10. Carpels 5-6, 1 fertile... ... ... Buchanania.

Ovules pendulous from top of cell or from the
ovarian walls above the middle.
Leaves pinnate
Style 1. Petals valvate ... ... Solenocarpus,
Styles 3-4. Petals imbricate ... ... Odina.
Leaves simple.
Stamens 5.
Drupe on enlarged disc and calyx... ... Semecarpus.
Disc and calyx not enlarged in fruit ... Holigarna.
Stamens 4. Drupe superior ... ... Nothopegia,
Fruit 2-5-celled, 2-5-seeded ... ... Spondias.

RHUS, Linn.

Trees. Leaves 1-3-foliate or pinnate. Flowers small, paniculate,
polygamous. Calyx 4-6-partite. Petals 4-6. Stamens 4-10, free.
Ovary 1-celled. Styles 3. Drupe small, dry with a pendulous seed.

R. mysorensis, Heyne ; W. & A. Prod. 172 ; Fl. Br. I. 2. 9 ; Brandis
For. Fl. 119 ; Bedd. Fl. Sylv. 78. In the dry forests of the Deccan and
in Sind ; common in the Dhárwár forests ; a small tree or large shrub.
Fl. Feb. Fr. Apl.

MANGIFERA, Linn.

Trees. Leaves alternate. Flowers in terminal panicles. Clayx 4-5 partite. Petals 4-5. Stamens 1-5, 1 more perfect than the others. Ovary
sessile, oblique. Drupe large ; stone compressed, fibrous.

M. indica, Linn. Fl. Br. I. 2. 13 ; Dalz. & Gibs. Bomb. Fl. 51 ;
Brandis For. Fl. 125 ; Bedd. Fl. Sylv. t. 162. Mangoe. *Mavina mara,* K ;
amba, M. In the evergreen forests of the Konkan and North Kánara
gháts and in the ravines of the Sátpudás of Khándesh, also cultivated
throughout the presidency and Sind. Fl. Feb., Apl. Fr. ripe June.

ANACARDIUM, Rottl.

Trees or shrubs. Leaves petioled, simple. Panicle terminal.
Calyx 5-partite. Petals 5. Disc erect. Stamens 8-10 ; filaments
connate and adnate to disc. Ovary obovoid ; ovule 1. Nut kidney-
shaped, on enlarged disc and peduncle.

A. occidentale, Linn. Fl. Br. I. 2. 20 ; Dalz. & Gibs. Bomb. Fl. Suppl.
18 ; Bedd. Fl. Sylv. t. 163. *Godámbe,* K.; *kaju,* M.; *geru mavu* in
Dhárwár ; "*Hijuli-badam*" or cashewnut. Naturalized from America
throughout the presidency ; common near the sea-coast in open situations.

BUCHANANIA, Roxb.

Trees. Leaves simple, villous or glaucous. Panicles terminal.
Flowers hermaphrodite. Calyx 3-5-lobed. Petals 4-5. Disc orbicular, 5-lobed. Carpels 5-6, in cavity of disc, only 1 fertile. Drupe
with a bony stone, 2-valved.

Leaves and panicles villous .. *B. latifolia*
Leaves and panicles glabrous .*B. angustifolia.*

B. **latifolia,** Roxb. Fl. Ind. II. 385 ; Fl. Br. I. 2. 23 ; Bedd. Fl. Sylv. t. 165 ; Brandis For. Fl. 127 ; Dalz. & Gibs. Bomb. Fl. 52. *Char, chirauli.* M.; *nurkal,* K.; *payal,* Vern. Dalz. Throughout the presidency in dry, deciduous forests. Fl. Jan., Mch. Fr. Apl., May.

B. **angustifolia,** Roxb. Fl. Ind. II. 386 ; Fl. Br. I. 2. 23 ; Grah. Cat. Bomb. Pl. 41, Konkan from the Adjunta jungles, southwards. Graham.

SOLENOCARPUS, Wt. & Arn.

Tree. Leaves oddpinnate, leaflets opposite, crenulate. Panicles terminal. Calyx 5-toothed. Petals 5. Disc annular. Stamens 10, at base of disc. Ovary sessile, 1-celled. Drupe small, compressed, pericarp full of oil.

S. **indica,** Wt. & Arn. Prod. 1. 171 ; Fl. Br. I. 2. 27 ; Bedd. Fl. Sylv. t. 233. In the evergreen forests of the Yellápur táluka of North Kánara, rare. Fl. Nov. Fr. Jany.

ODINA, Roxb.

Trees. Leaves oddpinnate deciduous, leaflets opposite. Flowers small, mon or diœcious, fascicled, in terminal panicles. Calyx 4-5-lobed. Petals 4-5. Disc annular, lobed. Ovary 4-5 partite. Drupe small, red ; stone hard.

O. **Wodier,** Roxb. Fl. Ind. II. 293 ; Fl. Br. I. 2. 29; Bedd. Fl. Sylv. t. 123 ; Dalz. & Gibs. Bomb. Fl. 51 ; Brandis For. Fl. 123. *Gujel,* K.; *moce, shembat, shimti,* M.; *moina,* Vern. Gujarát. Common throughout the presidency in deciduous forests. Fl. Feb., Apl. Fr. July, Aug.

SEMECARPUS, Linn. F.

Trees. Leaves simple. Flowers polygamous or diœcious, in terminal panicles. Calyx 5-6-fid. Petals 5-6. Disc annular. Stamens 5-6. Ovary 1-celled. Drupe fleshy, seated on thickened disc and calyx base.

S. **Anacardium,** Linn. f. Fl. Br. I. 2. 30; Dalz. & Gibs. Bomb. Fl. 52 ; Bedd. Fl. Sylv. t. 166 ; Brandis For. Fl. 124. Marking nut tree ; *bibha,* M.; *ger,* K. Throughout the presidency in dry forests ; locally abundant. Flowers at various times. Fruit from November to February. Var. *cuneifolia* ; DC. *Bibu,* M. Gháts near Khandála. Fl. Dec. Graham.

HOLIGARNA, Ham.

Trees. Leaves simple ; petiole with 2-4 deciduous appendages. Panicles axillary or terminal. Flowers polygamous. Calyx 5-toothed. Petals 5, villous in front. Disc lining calyx-tube. Stamens 5, on edge of disc. Ovary inferior, 1-celled. Styles 3-5, terminal. Ovule pendulous from top of cell. Drupe ovoid with resinous pulp, stone coriaceous.

Leaves quite glabrous beneath   ...   . *H. Arnottiana.*
Leaves pubescent beneath   ...   . *H. Grahamii.*

H. Arnottiana, Hook. f. Fl. Br. I. 2. 36 ; Dalz. & Gibs. Bomb. Fl. 51 ; Bedd. Fl. Sylv. t. 167. *Holigar, hoolyeri,* K. ; *sudrabilo,* M. In the evergreen forests of the Konkan and North Kánara, common. Fl. Jan., Feb. Fr. June, July.

H. Grahamii Hook. f. Fl. Br. I. 2. 37 ; *Semecarpus Grahamii,* Dalz. & Gibs. Bomb. Fl. 52. In the evergreen forests of the Konkan and North Kánara. Mira hills and near Thul, Dalz.; common in the North Kánara ghát forests from Ainshi southwards. Fl. Jany. Fr. ripe June.

NOTHOPEGIA, Blume.

Tree. Leaves alternate or opposite, entire. Racemes axillary Calyx small, 4-5 lobed. Petals 4-5. Disc annular, 4-5-lobed. Stamens 4-5, filaments hairy. Ovary 1-celled. Ovule pendulous from top of cell. Drupe fleshy striate, seed pendulous.

N. Colebrookiana, Blume Mus. Bot. 1. 203 ; Fl. Br. I. 2. 40 ; Bedd, Fl. Sylv. t. 164 ; *Glycycarpus racemosus,* Dalz. & Gibs. Bomb. Fl. 51. *Amberi,* Vern. Common in the evergreen forests of the Konkan and North Kánara ghát. Fl. March. Fr. May.

SPONDIAS, Linn.

Trees. Leaves odd-pinnate, leaflets caudate, acuminate. Flowers small, in terminal panicles. Calyx 4-5-fid. Petals 4-5, spreading. Disc cupular, crenate. Stamens 8-10, inserted beneath disc. Ovary free, 4-5-celled. Ovules solitary, pendulous in the cells. Drupe fleshy with a hard thick stone, 1-5-celled. Seeds pendulous.

> Leaflets not caudate. Fruit rough, fibrous ... *S. mangifera.*
> Leaflets caudate, acuminate. Fruit smooth ... *S. acuminata.*

S. mangifera, Willd. ; DC. Prod. II. 75, Bedd. Fl. Sylv. t. 169 ; Dalz. & Gibs. Bomb. Fl. Suppl. 19 ; Brandis For. Fl. 128. Hog plum. *Ambada,* M. ; *amate,* K. Throughout the presidency, usually in dry forests, common in North Kánara from the sea-coast inland, often planted. Fl. Feb., Apl. Fr. ripe next Nov., Dec.

S. acuminata, Roxb Fl. Ind. II. 453 ; Fl. Br. I. 2. 42 ; Grah. Cat. Bomb. Pl. 42. *Ambut, ambada,* M. In the Konkan hills near the Kennery Caves, Graham. North Kánara.

Order 34. MORINGEÆ.

Trees. Leaves alternate, impari-bi or tri-pinnate; pinnæ and leaflets opopsite. Flowers bisexual in axillary panicles. Calyx cup-shaped, 5-cleft; segments unequal, petaloid. Petals 5, unequal. Stamens inserted on the edge of the disc, declinate, 5 perfect, opposite the petals, alternating with 5-7 filaments without anthers ; anthers 1-celled. Ovary stipitate, lanceolate, 1-celled, with 3 parietal placentas; style simple, slender ; ovules numerous. Capsule pod-shaped, rostrate, 3-6-angled, torulose, 1-celled, 3-valved, corky and pitted within. Seeds many, in pits of the valves, testa corky, winged or 0. Albumen 0 ; embryo with a many leaved plumule.

MORINGA, Lamk.

Only genus, with characters those of the order.

Leaflets small, nerves obscure. Petals white ... ... *M. pterygosperma.*
Leaflets large, nerves 4-6 pairs, distinct. Petals
streaked with pink... ... ... *M. concanensis.*

M. pterygosperma, Gærtn. DC. Prod. II. 478; Fl. Br. I. 2. 45; Brandis For. Fl. 129; Bedd. Fl. Sylv. t. 80; Dalz. & Gibs. Bomb. Fl. 314. *Shevgi*, M.; *nuggi mara*, K.; *segava*, *segata*, Vern. Horse-radish tree. Cultivated throughout the presidency. Fl. Jany., Apl. Fr. Apl. onwards.

M. concanensis, Nimmo in Grah. Cat. Bomb. Pl. 43; Fl. Br. I. 2. 45; Brandis For. Fl. 130; Dalz. & Gibs. Bomb. Fl. 311. *Mhūa*, Sindhi. Throughout the Konkan and Sind, in dry forests. Fl. Nov., Dec. Fr. Dec. onwards.

ORDER 35. CONNARACEÆ.

Shrubs or trees, climbing or erect. Leaves alternate, 1-3-foliate or odd-pinnate. Flowers bisexual, in racemes or panicles. Calyx 5-divided. Petals 5. Stamens 5-10, sometimes declinate, those opposite the petals often shorter and imperfect. Carpels 5, rarely more or less, hairy, 1-celled; ovules 2, collateral. Fruit of 1, rarely 2-3, sessile or stalked usually 1-seeded follicles. Seed arillate or not, aril various.

Calyx accrescent. Capsule sessile ... ... ROUREA.
Calyx not accrescent. Capsule pedicellate ... ... CONNARUS.

ROUREA, Aubl.

Trees or shrubs. Leaves odd-pinnate. Panicles axillary. Sepals 5, orbicular, imbricate, accrescent in fruit. Petals 5, stamens 10; filaments connate at the base. Ovaries 5-4, usually imperfect. Fruit a capsule, apiculate.

R. santaloides, W. & A. Prod. 144; Fl. Br. I. 2. 47; Dalz. & Gibs. Bomb. Fl. 53. In the evergeeen forests of the Konkan and North Kánara; abundant on the hills near Kárwár. Fl. Oct. Fr. Apl.-July There is a variety of the above, found on the Ainshi ghát of North Kánara, the sepals of which are ciliate and the capsules small, $\frac{1}{3}$-$\frac{1}{2}$ in. on slender pedicels.

CONNARUS, L.

Trees or shrubs. Leaves odd-pinnate; leaflets usually 5. Flowers small in axillary and terminal branched panicles. Sepals 5. Petals 5. Stamens 10, 5 shorter and sometimes without anthers. Ovaries 5, densely pubescent, 4 usually imperfect or obsolete, the fifth with a slender style, stigma capitellate. Capsule oblique, inflated, broader upwards; valves glabrous or pubescent within; seeds arillate, albumen 0, testa shining.

Capsule not veined or shining, contracted into the stalk... *C. monocarpus.*
Capsule strongly striate, shining, narrowly keeled ... *C. Wightii.*
Capsule suddenly contracted into the stalk, base cordate. *C. Ritchiei.*

C. monocarpus, Linn.; Fl. Br. I. 2. 50 ; *C. pinnatus,* Lamk. Dalz. & Gibs. Bomb. Fl. 53. *Sundar,* M. Common in the ghát forests of the Konkan and North Kánara.

C. Wightii, Hook. f. Fl. Br. I. 2. 51. A. lofty climber common in the moist evergreen forests of North Kánara along the gháts. Agrees with the specimens of *C. pentandrus,* Roxb. in the Herb. Calcutt. The capsules are, however, quite glabrous within. Fl. Feb. Fr. Apl., May.

C. Ritchiei, Hook. f. Fl. Br. I. 2. 51. Forests of the Konkan and on the Rám ghát near Belgaum. Fl. Br. I. Also on the Supa gháts of North Kánara. Fl. Apl. Fr. May, June.

ORDER 36. LEGUMINOSÆ.

Herbs, shrubs or trees. Leaves stipulate, simple or compound. Inflorescence racemose or panicled. Flowers regular or irregular, hermaphrodite or polygamous. Sepals 5. Petals 5. Stamens usually 10 ; filaments free or combined. Ovary free. Ovules 1 or more on ventral suture. Fruit a pod, dehiscent or indehiscent. Seeds usually exalbuminous.

PAPILIONACEÆ—Cerolla papilionaceous. Stamens definite.

Genisteæ—Stames monadelphous. Pod dehiscent. Leaves simple or digitately trifoliate.

Anthers dimorphous	...	...	...	CROTALARIA.

Galegeæ—Stamens diadelphous. Pod not jointed. Leaves imparipinnate.

Hairs fixed by the centre	...	...	...	INDIGOFERA.
Hairs basifixed.				
Pod few seeded				
Filaments filiform...		...	...	MILLETTIA.
Filaments dilated ...		...	...	MUNDULEA.
Pod many seeded ...		...	...	SESBANIA.

Hedysareæ—Stamens diadelphous or monadelphous. Pod jointed if more than 1 seeded. Leaves odd-pinnate.

Leaves exstipellate			
Stamens diadelphous, 9-1	...	...	ALHAGI.
Stamens monadelphous	...	...	STYLOSANTHES.
Leaves stipellate.			
Racemes in fascicles from old wood	...	...	OUGEINIA.
Racemes simple or panicled from the year's shoots	...	DESMODIUM.

Phaseoleæ—Stamens monadelphous or diadelphous. Leaves pinnately trifoliate.

Leaves not gland-dotted. Stamens monadelphous.			
Climbing shrubs with irritating bristles	...	...	MUCUNA.
Prickly trees	...	...	...	ERYTHRINA.
Stamens monadelphous.			
Glabrous climbers. Calyx 2-lipped, glabrous	...	CANAVALIA.
Pubescent climbers. Calyx not lipped, silky	...	PUERARIA.
Stamens diadelphous. Pod like a samara reversed.			
Flowers small, panicled	...	...	...	SPATHOLOBUS,
Flowers large, racemose	...	...	...	BUTEA.

Leaves gland-dotted beneath.
Ovules 3 or more.
Seeds arillate ... ... ATYLOSIA.
Seeds not arillate ... ... CAJANUS.
Ovules 1-2.
Calyx teeth accrescent ... ... CYLISTA.
Calyx teeth not accrescent.
Leaves pinnate ... ... RYNCHOSIA.
Leaves digitate ... ... FLEMINGIA.

Dalbergieæ.—Stamens monadelphous.   Leaves odd-pinnate.

Leaflets alternate.
Flowers, small, white ... ... DALBERGIA.
Flowers, large, yellow ... ... PTEROCARPUS.
Leaflets opposite.
Pod compressed, not winged ... ... PONGAMIA.
Pod thin winged ... ... DERRIS.

CÆSALPINIEÆ.—Petals imbricate, slightly unequal.
Stamens definite.   Leaves bipinnate.

Sepals imbricate.
Sutures of pod not winged ... CÆSALPINIA.
Upper suture of pod winged ... MEZONEURON.
Sepals valvate.
Unarmed trees. Pod thin flat ... ... POINCIANA.
Spinous tree. Pod turgid ... ... PARKINSONIA.

Cassieæ.—Leaves pinnate. Calyx-tube short. Disc sub-basal.

Petals 5.
Pod variable usually many seeded.   Seeds exalbumin-
ous... ... ... CASSIA.
Pod thick turgid, 1-seeded ... ... CYNOMETRA.
Petals 0.
Pod samaroid, 1-seeded ... ... HARDWICKIA.

Amherstieæ.—Leaves equally pinnate. Disc at top of prolonged calyx-tube.

Petals 0 ... ... SARACA.
Petals 3-5 ... ... TAMARINDUS.
Leaves simple bilobed ... ... BAUHINIA.

MIMOSEÆ.—Petals regular, valvate often united. Stamens definite or indefinite.

Mimoseæ.—Stamens usually 10.

Anthers gland-crested.
Flowers in globose heads ... ... XYLIA.
Flowers in slender spikes. A climber ... ... ENTADA.
Tree erect, not prickly ... ... ADENANTHERA.
Prickly trees. Pod turgid ... ... PROSOPIS.
Spinous trees. Pod thin... ... ... DICHROSTACHYS.
Anthers not gland-crested.
Pod continuous ... ... LEUCÆNA.
Pod jointed... ... ... MIMOSA.

Acacieæ.—Stamens indefinite.

Stamens free ... ... ACACIA.
Stamens monadelphous.
Pod thin, ligulate ... ... ALBIZZIA.
Pod circinate ... ... PITHECOLOBIUM.

CROTALARIA, Linn.

Herbs or shrubs. Leaves simple or 3-5-foliate. Flowers racemose, often showy. Corolla included or exserted; keel incurved, distinctly beaked. Pod turgid, many seeded.

C. **Burhia**, Hamilt. in Wall. Cat. 5386; Fl. Br. I. 2. 66; Brandis For. Fl. 144. *Sis, sissai,* Vern. Common in the plains of Sind in dry sandy places. Fl. Nov.-Mch. C. **retusa**, L. *Gayri,* M. C. **soricea**, Retz. C. **Leschenaultii**, DC. *Dyli, dingala,* M. C. **Heyneana**, Grah. C. **leptostachya**, Benth. C. **juncea**, L. (*Sann, taag,* Vern.) C. **fulva**, Roxb. C. **striata**, DC. and C. **laburnifolia**, L., are all undershrubs found in the forests of the western ghats of Bombay. *C. juncea,* L., is cultivated throughout North Kánara and also in the Deccan and Konkan.

INDIGOFERA, Linn.

Herbs or shrubs. Flowers in axillary racemes. Corolla caducous; keel straight, spurred on each side near the base. Anthers apiculate. Pod usually linear cylindrical.

Leaflets 3-5	... *I. paucifolia.*
Leaflets many. Flowers small.	
Pod torulose, 3-4-seeded	... *I. argentea.*
Pod not torulose, 8-12-seeded	... *I. tinctoria.*
Pod torulose, tetraquetrous, 8-12-seeded ...	... *I. constricta.*
Pod cylindrical stout, 8-12-seeded	... *I. Wightii.*
Flowers large	... *I. pulchella.*

I. **paucifolia**, Delile; DC. Prod. II. 224; Fl. Br. I. 2. 97; Dalz. & Gibs. Bomb. Fl. 59. Throughout the dry plains of the presidency and in Sind. Fl. & Fr. Sept., Oct. Leaves silvery-shining.

I. **argentea**, Linn.; DC. Prod. II. 224; Fl. Br. I. 2. 98. Var. *cærulea;* Dalz & Gibs. Bomb. Fl. 59. Brandis For. Fl. 136. Throughout the dry plains of the presidency and in Sind. Fl. June. Fr. Aug., Sept.

I. **tinctoria**, Linn.; DC. Prod. II. 224; Fl. Br. I. 2. 99; Dalz. & Gibs. Bomb. Fl. 59; Brandis For. Fl. 135. The indigo plant. *Nil,* Vern. In open places near villages throughout the Konkan and North Kánara, doubtfully indigenous. Cultivated in Sind. Fl. Fr. Oct. Jany.

I. **constricta**, Trim. Syst. Cat. Fl. Pl. Ceylon. *I. flaccida* var. *constricta,* Thw. Enum. 411; Fl. Br. I. 2. 99. In the moist forests of North Kánara; common on the Supa gháts. Fl. Oct., Nov. Fr. Dec., Jany.

I. **Wightii**, Grah. in Wall. Cat. 5458; Fl. Br. I. 2. 99; Dalz. & Gibs. Bomb. Fl. 59. Plains of the Deccan, Belgaum.

I. **pulchella**, Roxb. Hort. Beng. 57; Fl. Ind. III. 382; Fl. Br. I. 2. 101; Dalz. & Gibs. Bomb. Fl. 60; Bedd. Fl. Sylv. 85. *Chimnati,* Vern. Throughout the Konkan and North Kánara from the sea-level upwards, usually in moist forests along the gháts. Fl. Dec. Fr. ripe Feb.

MILLETTIA, Wt. & Arn.

Climbers. Leaves imparipinnate. Flowers showy. Corolla much exserted; standard broad, keel not beaked. Pod turgid, sometimes torulose.

M. racemosa, Benth. Pl. Jung. 249; Fl. Br. I. 2. 105. *Wisteria pallida,* Dalz. & Gibs. Bomb. Fl. 61; *W. racemosa,* Dalz. & Gibs. Bomb. Fl. 61. Throughout the deciduous forests of the Konkan and North Kánara. Fl. Apl., May. Fr. ripe next cold season.

MUNDULEA, DC.

Shrubs. Leaves odd-pinnate. Corolla much exserted; standard with a long claw, keel incurved, obtusely pointed. Pod large, linear with thickened sutures.

M. suberosa, Benth. Pl. Jung. 248; Fl. Br. I. 2. 110; Bedd. Fl. Sylv. 85; *Tephrosia suberosa,* Dalz. & Gibs. Bomb. Fl. 60. *Supti,* Vern. On the rocky hills east of Belgaum, Dalz. Common in the forests near Bádámi. Fl. Aug., Sept. Fr. Oct., Nov.

SESBANIA, Pers.

Shrubs, herbs or trees. Leaves long, abruptly pinnate, leaflets mucronate. Flowers showy, in axillary racemes. Corolla much exserted; petals with long claws. Pod long, narrow with septa between the numerous seeds.

Flowers small ... ... *S. œgyptiaca.*
Flowers very large ... ... *S. grandiflora.*

S. œgyptiaca, Pers.; DC. Prod. II. 264; Fl. Br. I. 2. 114; Brandis For. Fl. 137; Bedd. Fl. Sylv. 86. *Shewarie,* M. Along nálás and water-courses in North Kánara, but not indigenous; commonly planted in gardens. Fl. R. & C.S.

S. grandiflora, Pers. Syn. II. 316; Fl. Br. I. 2. 115; Bedd. Fl. Sylv. 86; Brandis For. Fl. 137. *Agati grandiflora,* Dalz. & Gibs. Bom. Fl. Suppl. 22. *Augusta,* Vern. Commonly planted throughout the presidency; often as a support for piper betel. An ornamental tree. Fl. Fr. at various times throughout the year.

ALHAGI, Desv.

Spinous shrub. Leaves simple. Flowers axillary. Corolla exserted; standard broad, keel obtuse. Pod linear, moniliform, falcate or straight.

A. maurorum, Desv.; DC. Prod. III. 352; Fl. Br. I. 2. 145; Dalz. & Gibs. Bomb. Fl. 67; Brandis For. Fl. 144. The camel-thorn. *Kas,* Sind; *jowassi,* Vern. Throughout the dry plains of the presidency and Sind. Fl. Mch., Apl. Fr. Aug.

STYLOSANTHES, Sw.

Shrub. Leaves rigid, pinnately trifoliate. Flowers in dense heads ; corolla not exserted, keel subrostrate. Pod flattened, 1-2-jointed, joints rugose.

S. mucronata, Willd. ; DC. Prod. II. 318 ; Fl. Br. I. 2. 148. Common on the dry hills near Dhárwár. Shores of the Western peninsula, Fl. Br. I. Fl. Fr. R.S.

OUGEINIA, Benth.

Tree. Leaves pinnately trifoliate. Flowers in fascicled axillary racemes. Corolla much exserted ; standard broad, keel obtuse. Pod linear, flat, smooth, 2-5-jointed.

O. dalbergioides, Benth. Pl. Jungh, 216 ; Fl. Br. I. 2. 161 ; Bedd. Fl. Sylv. t. 36 ; Brandis For. Fl. 146 ; *Dalbergia oojeinensis*, Roxb. ; Dalz. & Gibs. Bomb. Fl. 78. *Kuri mutal*, K.; *tewas*, M.; *kula phulas*, Hind.; *telus*, Khándesh Dangs ; *tunuj, sandan, timsa*, Vern. Throughout the presidency in decidnous forests ; common in the forests of the Yellápur sub-division of North Kánara where large trees are numerous ; yields a valuable elastic wood ; not attacked by white ants. Fl. Mch.-May. Fr. June.

DESMODIUM, Desv.

Shrubs or herbs. Leaves simple or trifoliate, stipellate. Flowers small, racemose. Corolla exserted ; standard broad. Pod of several 1-seeded indehiscent joints.

Leaves trifoliate. Flowers in axillary umbels.		
Bracts minute, deciduous.		
Branches terete. Joints of pod large	*D. umbellatum.*	
Branches triquetrous. Joints of pod small ...	*D. Cephalotes.*	
Leaves trifoliate. Umbels in continuous rows.		
Bracts large persistent	*D. pulchellum.*	
Leaves 1-foliate, petiole winged	*D. triquetrum.*	
Leaves 1-3-foliate. Flowers in simple or panicled racemes.		
Joints of pod indehiscent.		
Leaves trifoliate	*D. laxiflorum.*	
Leaves 1-foliate.		
Leaflet glabrescent above	*D. gangeticum.*	
Leaflet scabrous above	*D. latifolium.*	
Joint of pod dehiscent	*D. polycarpum.*	
Pod dehiscent, not jointed	*D. gyrans.*	

D. umbellatum, DC. Prod. II. 325 ; Fl. Br. I. 2. 161 ; Bedd. Fl. Sylv. 87. Dalz. & Gibs. Bomb. Fl. 66. South-east of Surat ; near Belgaum, Dalz.

D. Cephalotes, Wall. Cat. 5721 ; Fl. Br. I. 2. 161 ; Bedd. Fl. Sylv. 87. *D. congestum*, Dalz. & Gibs. 66. Common in the moist forests of North Kánara and the Konkan. Fl. July, Sept. Fr. C.S.

D. pulchellum, Benth. MSS.; Fl. Br. I. 2. 162. Throughout the presidency in moist places along nálás and water-courses, common in North Kánara as undergrowth in the high timber deciduous forests of the Yellápur gháts. Fl. Fr. R.S.

D. triquetrum, DC. Prod. II. 326; Fl. Br. I. 2. 163; Dalz. & Gibs. Bomb. Fl. 66. Common in the moist forests of the Konkan and North Kánara from the coast inwards, a very distinct species. Fl. Sept.-Jany. Fr. C.S.

D. laxiflorum, DC. Prod. II. 335; Fl. Br. I. 2. 164. *Jungly ganga*, Vern. Common in the forests along the Supa gháts of North Kánara. Fl. Fr. R.S. C.S.

D. gangeticum, DC. Prod. II. 327; Fl. Br. I. 2. 168; Dalz. & Gibs. Bomb. Fl. 66. Throughout the presidency both in the plains and in the moist forests of the Konkan and North Kánara. Fl. May, June. Fr. July, Aug.

D. latifolium, DC. Prod. II. 327; Fl. Br. I. 2. 168; Dalz. & Gibs. Bomb. Fl. 66. In the moist forests of the Konkan and North Kánara. Fl. Fr. R.S.

D. polycarpum, DC. Prod. II. 334; Fl. Br. I. 2. 171; Dalz. & Gibs. Bomb. Fl. 66. Common throughout the presidency, scarcely worth including in the present list. Fl. R.S. Fr. C.S.

D. gyrans, DC. Prod. II. 326; Fl. Br. I. 2. 174. The semaphore plant. Throughout the moist forests of the Konkan and North Kánara, very common. Fl. R.S. Fr. Oct., Nov.

The species of *Desmodium*, although undershrubs, are included in this list, as they are common in the undergrowth of many of the forests of the Konkan and North Kánara.

ABRUS, Linn.

Climbers. Leaves pinnate; leaflets deciduous. Flowers in racemes on axillary peduncles or short branches. Corolla exserted, keel ovate. Stamens 9. Pod thin or turgid, septate.

Pod turgid, 3-5-seeded   ...   . *A. precatorius.*
Pod thin, flat, 6-8-seeded ...   . *A. pulchellus.*

A. precatorius, Linn.; DC. Prod. II. 381; Fl. Br. I. 2. 175; Dalz. & Gibs. Bomb. Fl. 76; Brandis For. Fl. 139. *Gunchi*, Vern.; *gunja*, Sans. Throughout the presidency; common in the moist forests of the Konkan and North Kánara. Fl. R.S. Fr. Jany.

A. pulchellus, Wall. Cat. 5819; Fl. Br. I. 2. 175. In the moist forests of North Kánara; very common near Kárwár and along the coast southwards. Fl. R.S. Fr. Nov., Jany.

MUCUNA, Adans.

Twining plants. Leaves trifoliate, stipellate. Flowers large, in racemes, often pendulous. Corolla exserted; keel rostrate, exceeding the wings. Anthers dimorphous. Pod clothed with irritating bristles.

Pod obliquely plaited, 1-seeded   ...   ... *M. monosperma.*
Pod without plaits, several seeded ...   ... *M. pruriens.*

M. monosperma, DC. Prod. II. 406; Fl. Br. I. 2. 185; Dalz. & Gibs. Bomb. Fl. 70. In the moist forests along the Konkan and North Kánara gháts; locally common. Fl. C.S. Fr. H. and R.S.

M. prurions, DC. Prod. II. 405; Fl. Br. I. 2. 187; *M. prurita,* Dalz. & Gibs. Bomb. Fl. 70. Cowitch. *Hasaguni gidda,* K.; *kivanch* in Gujarát; *kuhila,* Vern. Throughout the presidency from the coast inland, common in hedges, sometimes cultivated. Fl. Oct., Nov. Fr. Dec, Feb.

M. atropurpurea, DC., with broad 2-seeded, plaited, bristly pods, and *M. gigantea,* DC., with broad, winged. 2-6-seeded pods, not plaited, are stated by Nimmo to be found in the Konkan, Grah. Cat. Bo. Pl. 53.

ERYTHRINA, Linn.

Prickly trees. Leaves trifoliate. Flowers large, coral red, in racemes. Calyx spathaceous, or campanulate and 2-lipped. Standard much exserted, exceeding keel.

Pod linear, turgid, torulose.
Calyx spathaceous, not 2-lipped.
  Calyx 5 cleft at tip. Pod 6-8-seeded ... ... *E. indica.*
  Calyx entire at tip, 2-3-seeded ... ... *E. stricta.*
Calyx campanulate, 2-lipped ... ... *E. suberosa.*

E. indica, Lam.; DC. Prod. II. 412; Fl. Br. I. 2. 188; Bedd. Fl. Sylv. 87; Dalz. & Gibs. Bomb. Fl. 70; Brandis For. Fl. 139. Indian coral tree. *Pangara,* M.; *mullu mutala,* K. In the deciduous forests of the Konkan and North Kánara; commonly planted throughout the presidency as a support for pepper vines. Fl. Apl., May. Fr. June.

E. stricta, Roxb. Fl. Ind. III. 251; Fl. Br. I. 2. 189; Dalz. & Gibs. Bomb. Fl. 70; Bedd. Fl. Sylv. t. 175. In the deciduous forests of North Kánara and the Konkan. Fl. May. Fr. June.

E. suberosa, Roxb. Fl. Ind. III. 253; Fl. Br. I. 2. 189; Bedd. Fl. Sylv. 87; Dalz. & Gibs. Bomb. Fl. 70. *Pangra,* Vern. Throughout the dry forests of the presidency, does not extend into the heavy rainfall zone. Fl. May. Fr. June.

SPATHOLOBUS, Hassk.

Woody climbers. Leaves trifoliate stipellate. Flowers in terminal panicles. Corolla exserted, keel straight. Pod 1-seeded, winged.

Pod broad, rusty tomentose ... ... *S. Roxburghii.*
Pod glabrescent, narrower ... ... *S. purpureus.*

S. Roxburghii, Benth. Pl. Jungh. 238; Fl. Br. I. 2. 193; Brandis For. Fl. 143; *Butea parviflora,* Dalz. & Gibs. Bomb. Fl. 71. *Phalsun,* M. Throughout the forests of the Konkan and North Kánara. This lofty climber does great damage to the teak and other timber trees in North Kánara. Fl. C.S. Fr. April.

·**S. purpureus,** Benth. MSS. Fl. Br. I. 2. 194. Evergreen forests of North Kánara on the Supa gháts, rare. Fl. C.S. Fr. Apl., May. Fruit like the samara of the sycamore reversed, narrow, glabrescent.

BUTEA, Roxb.

Trees or shrubs. Leaves large, trifoliate, stipellate. Flowers large, in racemes or panicles. Corolla exserted, keel semi-circular. Pod stalked, compressed, thin, membranous, 1-seeded.

Erect tree ...   . *B. frondosa.*
Climbing shrub   ...   . *B. superba.*

B. frondosa, Roxb. Fl. Ind. III. 244 ; Dalz. & Gibs. Bomb. Fl. 71 ; Fl. Br. I. 2. 194 ; Brandis For. Fl. 142. Bedd. Fl. Sylv. 176. Bastard teak. *Tésú-ká-jhar,* H.; *muttala,* K.; *phulas,* M.; *phullas kakria,* Guj. Throughout the presidency, common in deciduous forests. Fl. March. Fr. June, July.

B. superba, Roxb. Fl. Ind. III. 247 ; Fl. Br. I. 2. 195. Dalz. & Gibs. Bomb. Fl. 71 ; Brandis For. Fl. 143. *Palasvél, beltivas,* M. In the moist forests of the North Konkan. Fl. Mch.

CANAVALIA, DC.

Twining shrubs. Leaves 3-foliate, stipellate. Flowers large, in racemes. Calyx 2-lipped. Corolla exserted, standard large, keel short, incurved. Pod large, ribbed along upper suture.

Racemes many flowered. Pod many seeded ...   ... *C. ensiformis.*
Racemes few flowered. Pod few seeded   ...   ... *C. obtusifolia*

C. ensiformis, DC. Prod. 2. 404 ; Fl. Br. I. 2. 195. *C. Stocksii,* Dalz. & Gibs. Bomb. Fl. 69 ? *C. virosa,* Dalz. & Gibs. Bomb. Fl. 69. *Gowara, arsambal,* Vern. In hedges throughout the presidency, common. Fl. Feb. Fr. June.

C. obstusifolia, DC. Prod. 2. 404 ; Fl. Br. I. 2. 196. Along the coast of North Kánara and Konkan. Fl. Mch. Fr. June, July.

PUERARIA, DC.

Herbs or twining shrubs. Leaves trifoliate, stipellate. Flowers fascicled, in long racemes. Corolla exserted ; standard spurred at base, equalling in length the keel and wings. Pod linear, flattish. Root immense, tuberous, in *P. tuberosa.*

P. tuberosa, DC. Prod. II. 240 ; Fl. Br. I. 2. 197 ; Dalz. & Gibs. Bomb. Fl. 67. Brandis For. Fl. 141. Throughout the Konkan and North Kánara, in open situations. Fl. Mch., Apl.; leafless when in flower. Fr. May.

ATYLOSIA, W. & A.

Erect or twining shrubs. Leaves trifoliate, gland-dotted beneath. Flowers axillary or racemed. Corolla more or less exserted, keel not beaked. Pod linear or oblong, turgid ; seeds with a divided strophiole.

Petals marcescent.   Pod not lineate.   Erect shrubs.
Flowers in peduncled pairs   ...   . *A. geminiflora.*
Flowers in sessile pairs.
Pod thinly pilose...   ...   . *A. lineata.*
Pod densely pilose   ...   . *A. sericea.*
Twiners. Pod lineate between seeds.
Pod grey canescent   ...   . *A. mollis.*
Pod densely hairy   ...   . *A. kulnensis.*
Petals calucous. Pod bristly, recurved   ... *A. rostrata.*
Pod pilose straight   ...   . *A. barbata.*

A. gominiflora, Dalz. in Jour. Linn. Soc. XIII. 185 ; Fl. Br. I. 2. 212. In the moist forests at the foot of the ghâts in the Yellápur táluka of North Kánara, probably throughout the Konkan, but nowhere abundant. Fl. R.S.

A. lineata, W. & A. Prod. 258 ; Fl. Br. I. 2. 213 ; *Cajanus lineatus,* Grah. *Atylosia Lawii,* Wgt. ; Dalz. & Gibs. Bomb. Fl. 74. *Ran-toor,* Vern. Very common in the moist forests along the Konkan and North Kánara ghâts. Fl. Nov., Dec. Fr. Jany.

A. sericoa, Benth. MSS. Fl. Br. I. 2. 213. Moist forests of the Konkan. Fl. Br. I.

A. mollis, Benth. Pl. Jungh, 243 ; Fl. Br. I. 213 ; *Cajanus glandulosus,* Dalz. & Gibs. Bomb. Fl. 73. South Konkan, Fl. Br. I.

A. kulnonsis, Dalz. in Jour. Linn. Soc. XIII. 186 ; Fl. Br. I. 2. 214. *Cajanus kulnensis,* Dalz. & Gibs. Bomb. Fl. 72. In the moist evergreen forests of the Konkan and North Kánara, common. Fl. Fr. Dec., Mch.

A. rostrata, Baker. Fl. Br. I. 2. 216. Konkan, Stocks.

A. barbata, Baker. Fl. Br. I. 2. 216 ; *A. goensis,* Dalz. & Gibs. Bomb. Fl. 73. Konkan ghâts, Belgaum Collectorate, Dalz.

CAJANUS, DC.

Erect shrub. Leaves trifoliate. Flowers racemed. Corolla exserted, keel truncate. Pod linear, torulose.

C. indicus, Spreng. Syst. III. 248 ; Fl. Br. I. 2. 217 ; Dalz. & Gibs. Bomb. Fl. Suppl. 24. Pigeon pea. *Toor* ; *dhal,* Vern. Cultivated throughout the presidency.

CYLISTA, Ait.

Climber. Leaves trifoliate. Flowers racemed. Calyx scariose. Corolla not exserted, keel much incurved. Pod small, oblique, enclosed in calyx.

C. scariosa, Ait. ; DC. Prod. II. 410 ; Fl. Br. I. 2. 219 ; Dalz. & Gibs. Bomb. Fl. 74. *Ranguera,* Vern. Throughout the presidency in open places in deciduous forests, often in hedges. Fl. Fr. cold season.

RYNCHOSIA, Lour.

Erect or twining shrubs. Leaves pinnately trifoliate, gland-dotted. Flowers axillary or racemed. Corolla included, or exserted ; keel incurved. Pod usually continuous between the seeds deeply torulose, flat, turgid ; seeds usually arillate.

R. cyanosperma, Benth. in Oliv. Fl. Trop. Africa II. 218 ; Fl. Br. I. 2. 222. *Cyanospermum tomentosum.* W. & A. Dalz. & Gibs. Bomb. Fl. 75. Evergreen forests of the South Konkan and North Kánara nowhere abundant Fl. Fr. C.S.

FLEMINGIA, Roxb.

Shrubs. Leaves simple or trifoliate, gland-dotted beneath. Inflorescence various. Corolla usually not exserted, keel obtuse or slightly rostrate. Pod small, turgid, usually 2-seeded.

Leaves simple. Flowers in small cymes, each in the axil
  of a large folded persistent bract ... ... F. strobilifera.
Leaves digitately trifoliate.
Flowers in axillary racemes ... ... F. congesta.
Flowers in globose heads. Bracts large persistent ... F. involucrata.

F. strobilifera, R. Br. in Ait. Hort. Kew ed. 2, IV. 350; Fl. Br. I. 2. 227; Dalz. & Gibs. Bomb. Fl. 75. *Kankuti*, Vern. Throughout the moist forests of the Konkan and North Kánara, common. Bracts often with sterile flowers. Fl. Fr. R. and C.S.

F. congesta, Roxb. Hort. Beng. 56; Fl. Ind. III. 340; Fl. Br. I. 2. 228; Dalz. & Gibs. Bomb. Fl. 75; Var. *nana. F. procumbens*, Roxb.; Dalz. & Gibs. Bomb. Fl. 75. *Dow dowla*, Vern. Throughout the presidency the var. *semialata* is very common in the moist forests of the Konkan and North Kánara. Flowers throughout the year.

F. involucrata. Benth. Pl. Jungh. 246; Fl. Br. I. 2. 229; in the deciduous forests of the Konkan and North Kánara: a gregarious shrub growing in the open glades. Fl. R.S. Fr. Nov. Feb.

DALBERGIA, Linn. f.

Trees or climbers. Leaves imparipinnate, leaflets alternate. Flowers paniculate. Corolla exserted, keel obtuse. Stamens 9-1, 10-9 or 5-5. Pod thin, flat, indehiscent, few seeded.

Trees.
Stamens 9. Leaflets large.
Leaflets acuminate ... ... D. Sissoo.
Leaflets obtuse ... ... D. latifolia.
Scandent shrubs.
Leaflets few. Pod strongly veined, 1-2-seeded ... ... D. rubiginosa.
Leaflets many, small.
Flowers in terminal panicles. Pod strap-shaped, brown-
  ish, veined ... ... D. confertiflora.
Panicles axillary.
Pod thin, oblong, very long stalked ... ... D. Stocksii.
Pod thin, oblong, short stalked ... ... D. sympathetica.
Pod strap-shaped, not veined, long stalked. Stamens 10. D. tamarindifolia.
Stamens 5-5, equally diadelphous.
Trees.
Leaflets with prominent lateral nerves ... ... D. lanceolaria.
Leaflets with reticulate venation ... ... D. paniculata.
Climbers.
Pod straight, thin, 1-2-seeded ... ... D. volubilis.
Pod crescent-shaped, 1-seeded ... ... D. monosperma.
Pod reniform, flat. A spinous, erect shrub ... ... D. spinosa.

D. Sissoo, Roxb. Fl. Ind. III. 223. Fl. Br. I. 2. 231. Bedd. Fl. Sylv. t. 25; Brandis For. Fl. 149; Dalz. & Gibs. Bomb. Fl. Suppl. 24. The Sisso tree. *Sissu*, Vern. Planted throughout the presidency; believed to be indigenous in Gujarát. Fl. Mch.-June. Fr. Nov.-Feb.

D. latifolia, Roxb. Cor. Pl. II. 7, t. 113; Fl. Ind. III. 221; Fl. Br. I. 2. 231; Bedd. Fl. Sylv. t. 24; Dalz. & Gibs. Bomb. Fl. 77. The black-

wood or rosewood tree of Southern India. *Biti*, K.; *shisham*, *sisu*, M.
Throughout the deciduous forests of the presidency, associated with teak
in the high timber forests of North Kánara. Fl. Aug. Fr. Jany. in
North Kánara.

D. **rubiginosa**, Roxb. Fl. Ind. III. 231. Fl. Br. I. 2. 232. In the
evergreen forests of North Kánara from the Ainshi ghát southwards.
Leaflets unequal, largest 2½ in. by 1½ in. Pod strongly veined. 1-2-
seeded. Fl. and Fr. at different times throughout the year.

D. **confertiflora**, Benth. Pl. Jungh. 1. 255; Fl. Br. I. 2. 232.
Konkan.

D. **Stocksii**, Benth. in Jour. Linn. Soc. IV. Suppl. 42. Fl. Br. 1.
2. 231. Konkan.

D. **sympathetica**, Nimmo in Grah. Cat. Bomb. Pl. 55 ; Fl. Br. I. 2.
231; Dalz. & Gibs. Bomb. Fl. 78. *Pendguliyel*, *yekyel*, M. Throughout the Konkan and North Kánara from the coast inland in deciduous
forests. Stems armed with large curved spines. Fl. and Fr. Jany.-
Feb.

D. **tamarindifolia**, Roxb. Fl. Ind. III. 233 ; Fl. Br. I. 2. 231. In
evergreen forests on the southern gháts of North Kánara, common. Fl.
and Fr. Apl., May.

D. **lanceolaria**, Linn.; DC. Prod. II. 417 ; Fl. Br. I. 2. 235 ; Dalz. &
Gibs. Bomb. Fl. 78 ; Bedd. Fl. Sylv. 88; Brandis For. Fl. 151. *Dandous*,
dandoshi, M.; *harráni*, *gengri*, Vern. In the deciduous forests of North
Kánara from Mundgod to Siddápur, common. Konkan and Khándesh,
Dalz. Panicles large, diffuse. Fl. Mch.-May. Fr. ripe next cold season
and remains long on tree.

D. **paniculata**, Roxb. Fl. Ind. III. 227; Fl. Br. I. 2. 236; Dalz. &
Gibs. Bomb. Fl. 78 ; Brandis For. Fl. 151 ; Bedd. Fl. Sylv. 88. *Pussi*,
padri, Vern.; *phansa*, M. In the deciduous forests of the Konkan and
North Kánara, common in the forests of the Kalghatgi táluka of Dhárwár. Panicles terminal, congested. Wood of an abnormal structure,
annual rings separated by concentric layers of soft tissue. Fl. April,
May.

D. **volubilis**, Roxb. Fl. Ind. III. 231 ; Fl. Br. I. 2. 235; Dalz. & Gibs.
Bomb. Fl. 78; Brandis For. Fl. 152. *Alei*, Vern. Throughout the
forests of the presidency; common. Fl. Feb.-Mar. Fr. May.

D. **monosperma**, Dalz. in Kew Jour. Bot. II. 36; Fl. Br. I. 2. 237 ;
Dalz. & Gibs. Bomb. Fl. 78. Hills of the Konkan, Málvan, Dalz. Fl.
June.

D. **spinosa**, Roxb. Fl. Ind. III. 233 ; Fl. Br. I. 2. 232. Coast of the
Konkan.

PTEROCARPUS, Linn.

Trees. Leaflets alternate coriaceous. Flowers large yellow in
racemes or lax panicles. Petals exserted, with long claws, those of
keel free. Stamens 5-5. Pod orbicular winged, 1-seeded.

P. marsupium, Roxb. Fl. Ind. III. 234; Fl. Br. I. 2. 239; Bedd. Fl. Sylv. t. 21; Dalz. & Gibs. Bomb. Fl. 76; Brandis For. Fl. 152. *Bijasal, hond, honne,* K.; *asan, bibla,* M. Throughout the presidency in deciduous forests; common in North Kánara. Yields a valuable timber which contains a gum resin, kino of commerce. Fl. May-June. Fr. Dec.-March.

PONGAMIA, Vent.

Tree. Leaves imparipinnate, leaflets opposite. Flowers racemed. Corolla exserted, standard broad, keel obtuse. Pod woody, indehiscent, 1-seeded. Seed reniform, thick.

P. glabra, Vent. Jard. Malm. t. 28; Fl. Br. I. 2. 240; Dalz. & Gibs. Bomb. Fl. 77; Bedd. Fl. Sylv. t. 177; Brandis For. Fl. 153. Throughout the presidency in moist situations along rivers and nálás; often planted as a road-side tree; very common near the sea-coast. Fl. May, June. Fr. next April.

DERRIS, Lour.

Climbing shrubs or trees. Leaves imparipinnate, exstipellate. Flowers in simple or paniculate racemes. Corolla much exserted, standard broad, keel petals slightly cohering. Pod indehiscent, few seeded, winged along 1 or both sutures.

Flowers in axillary racemes.
Pod thin, strap-shaped, several seeded.
Climber   ...   . *D. scandens.*
Tree ...   . *D. robusta.*
Pod broad, flat, 1-2-seeded, winged along upper suture.
Leaflets few ...   . *D. uliginosa.*
Leaflets many...   ...   . *D. oblonga.*
Flowers paniculate.   Leaflets large, 5-9.
Pod silky   ...   . *D. brevipes.*
Pod glabrous.
Leaflets 5, obtuse.   Panicles downy   ...   . *D. Heyneana.*
Leaflets 5-9, acute.   Panicles brown, pubescent   ... *D. thyrsiflora.*
Leaflets 15-21, small   ...   . *D. canarensis.*

D. scandens, Benth. in Jour. Linn. Soc. IV. Suppl. 103; Fl. Br. I. 2. 210; Brandis For. Fl. 154; *Brachypterum scandens,* Dalz. & Gibs. Bomb. Fl. 76. Throughout the Konkan and North Kánara from the sea-coast inland, nowhere common. Fl. R. S. June-Aug. Fr. Sept.

D. robusta, Benth. in Jour. Linn. Soc. IV. Suppl. 104; Fl. Br. I. 2. 241; *Brachypterum robustum,* Dalz. & Gibs. Bomb. Fl. 77. Konkan. Dalz.

D. uliginosa, Benth. Pl. Jungh. I, 252; Fl. Br. I. 2. 241; Dalz. & Gibs. Bomb. Fl. 77. Common in North Kánara and the Konkan, along the banks of tidal rivers and near the coast. Fl. C.S.

D. oblonga, Benth. in Jour. Linn. Soc. IV. Suppl. 112. Fl. Br. I. 2. 242. Konkan. Stocks.

D. brevipes. Baker; Fl. Br. I. 2. 214. On the higher ghàts of the Konkan. On the top of Dursing, the highest peak of the North Kanara ghâts, 3,400 feet high. Fl. Feb.

D. Heyneana, Benth. Pl. Jungh. I. 252; Fl. Br. I. 2. 244; Dalz. & Gibs. Bomb. Fl. 77. Moist forests of the Konkan and North Kánara; in the Ainshi ghát forests at 500 feet elevation above the sea-level. Fl. and Fr. C. S.

D. canarensis, Baker. Fl. Br. I. 2. 246; *Brachypterum canarense,* Dalz. & Gibs. Bomb. Fl. 76. Along the ghát of North Kánara from the Ainshi ghát to the Mysore frontier in moist forests, locally common. Fl. H. S. Apl.-May. Fr. R. S; remains long on the stem.

D. thyrsiflora, Benth. in Jour. Linn. Soc. IV. Suppl. 114; Fl. Br. I. 2. 243. Common on the southern ghát of North Kánara. Fl. Dec. Fr. Feb.

SOPHORA, Linn.

Trees or shrubs. Leaves odd-pinnate. Flowers showy, racemed or panicled. Corolla much exserted, petals equal with long claws. Pod moniliform with turgid joints.

S. Wightii, Baker. Fl. Br. I. 2. 250; *S. heptaphylla,* Dalz. & Gibs. Bomb. Fl. 79; Bedd. Fl. Sylv. 89. Konkan ghát, hills east of Belgaum. Dalz.

CÆSALPINIA, Linn.

Trees or woody climbers, armed with prickles. Leaves bipinnate. Flowers showy, yellow, in axillary racemes. Calyx deeply cleft, the lowest lobe cuculate. Petals spreading, orbicular, clawed. Pod oblong or ligulate, armed or smooth.

Pod dry, armed with prickles ... ... *C. Bonducella.*
Pod dry, unarmed.
　Pod broad, glabrous, 1-seeded ... ... *C. Nuga.*
　Pod ligulate, glabrous, 4-8-seeded ... ... *C. sepiaria.*
　Pod inflated, bristly, 2-seeded ... ... *C. mimosoides.*

C. Bonducella, Fleming in Asiat. Res. XI, 159; Fl. Br. I. 2. 254; *Guilandina Bonducella,* Linn. *G. Bonduc,* Dalz. & Gibs. Bomb. Fl. 79, in part. Fever nut plant. *Sagurgota,* Vern.; *karbat* in Sind. Throughout the presidency in hedges and waste places, common in the forests near Kárwár. Fl. R. S. Fr. C. S.

C. Nuga, Ait. Hort. Kew III. 32 ; Fl. Br. I. 2. 255; *C. paniculata,* Roxb.; Dalz. & Gibs. Bomb. Fl. 79. Common along the banks of tidal rivers and creeks near the coast of the Konkan and North Kánara. Fl. C. & R.S. Fr. remains long on plant.

C. sepiaria, Roxb. Hort. Beng. 32 ; Fl. Ind. II. 360 ; Fl. Br. I. 2. 256; Dalz. & Gibs. Bomb. Fl. 80 ; Brandis For. Fl. 156. Mysore thorn. *Chillur, chillari,* M. In hedges throughout the dry districts of the presidency, very common in the Southern Marátha Country. Fl. Feb. May. Fr. Dec., Jan.

C. mimosoides, Lam. Ill. t. 335, fig. 2 ; Fl. Br. I. 2. 256 ; Dalz. & Gibs. Bomb. Fl. 80. Throughout the moist forests of North Kánara and the Konkan, not found in the dry zone. Fl. C.S. Fr. May. *C. coriaria,*

Willd., the dividivi tree, and *C. pulcherrima*, Sw., an ornamental shrub, are cultivated throughout the presidency in plantations ; along road sides and in gardens.

MEZONEURON, Desf.

Climbers. Leaves bipinnate. Flowers in large panicles, yellow. Calyx oblique, lowest lobe cucullate. Petals spreading, subequal, obovate. Stamens free, exserted. Pod large, flat, thin, winged down the upper suture, indehiscent.

M. cucullatum, W. & A. Prod. 283 ; Fl. Br. I. 2. 258 ; Dalz. & Gibs. Bomb. Fl. 80. *Ragi.* Vern. In the moist evergreen forests of the Konkan and North Kánara, abundant in the extreme south of North Kánara. Fl. Nov. Feb. Fr. March.

POINCIANA, Linn.

Trees. Leaves bipinnate. Flowers in corymbose racemes, large. Petals orbicular, clawed, margin of blade fimbriate-crisped. Stamens 10, free, exserted. Pod large, flat, ligulate.

Flowers bright scarlet   ...   . *P. regia.*
Flowers yellow   ...   . *P. elata.*

P. regia, Bojer. Bot. Mag. t. 2884. Fl. Br. I. 2. 260 ; Dalz. & Gibs. Bomb. Fl. Suppl. 27. Goolmohr tree. Cultivated in gardens throughout the presidency, very ornamental, a native of Madagascar. Fl. H. S. Fr. remains long on tree.

P. elata, Linn. ; DC. Prod. II. 484 ; Fl. Br. I. 2. 260 ; Bedd. Fl. Sylv. t. 178 ; Brandis For. Fl. 157 ; Dalz. & Gibs. Bomb. Fl. Suppl. 28. Indigenous in the forests of the western coast as far north as Gujarát. Brandis. Cultivated in gardens throughout the presidency. Fl. R.S. Fr. C. S.

PARKINSONIA, Linn.

Spinous trees. Leaves bipinnate, pinnæ with a flattened rachis, leaflets very small, often absent. Flowers yellow. Petals exserted, broad, with long claws. Stamens 10, villous, included. Pod turgid, dry, moniliform.

P aculeata, Linn. DC. Prod. II. 486 ; Fl. Br. I. 2. 260 ; Bedd. Fl. Sylv. 91 ; Brandis For. Fl. 158 ; Dalz. & Gibs. Bomb. Fl. Suppl. 28. *Vilayate kikar.* Cultivated as a hedge plant or as a road-side tree throughout the presidency, a native of the West Indies. Fl. throughout the year.

WAGATEA, Dalz.

Prickly climber. Leaves abruptly bipinnate. Flowers in simple or panicled spikes, orange-coloured. Corolla little exserted. Pod ligulate, oblong, indehiscent, few seeded, sutures thickened.

W. spicata, Dalz. in Kew Jour. III. 90 ; Fl. Br. I. 2. 260 ; Dalz. & Gibs. Bomb. Fl. 80. *Wagati,* M. Common throughout the presidency both in dry and in moist forests. Fl. C.S. Fr. H.S.

CASSIA, Linn.

Trees, shrubs or herbs. Leaves abruptly pinnate. Flowers yellow, in axillary racemes and terminal panicles. Calyx-tube very short; sepals imbricated. Petals 5, nearly equal, imbricate. Stamens 10, all perfect or 3-5 more or less abortive. Pod indehiscent or 2-valved.

Stamens all perfect. Trees.
Pod long, cylindrical ... ... *C. fistula.*
Pod flat, strap-shaped ... ... *C. glauca.*
Perfect stamens 7.
Stipules large, lunate. Pod not winged. Shrub. ... .. *C. auriculata.*
Stipules deltoid. Pod winged. Shrub. ... ... *C. alata.*
Stipules minute, caducuons.
Pod ligulate, not winged, sutures thickened. Tree. ... *C. siamea.*
Sutures of pod not thickened. Glabrescent. Shrub. ... *C. montana.*
Stipules persistent. All parts brown pubescent. Tree.... *C. timoriensis.*

C. fistula, Linn.; DC. Prod. II. 490; Fl. Br. I. 2. 261; Bedd. Fl. Sylv. 91; Dalz. & Gibs. Bomb. Fl. 80; Brandis For. Fl. 164. The Indian laburnum. *Boya, bahawa,* M.; *kakkai,* K.; *chimkani,* Sind; *girmala,* Deccan, Gujarát. Throughout the presidency in deciduous forests. Fl. Apl., May. Fr. C. S.

C. auriculata, Linn.; DC. Prod. II. 496; Fl. Br. I. 2. 263; Bedd. Fl. Sylv. 92; Dalz. & Gibs. Bomb. Fl. 81; Brandis For. Fl. 165. *Turwar, arsul,* Vern. Common in the dry zone throughout the presidency. Fl. throughout the year.

C. siamea, Lam.; DC. Prod. II. 499; Fl. Br. I. 2. 264. *C. florida,* Vahl. Bedd. Fl. Sylv. t. 179. *C. sumatrana,* W. & A. Dalz. & Gibs. Bom. Fl. Suppl. 29. *Kassod,* M. Introduced, common throughout the presidency, often planted along road-sides.

C. alata, Linn.; DC. Prod. II. 492; Fl. Br. I. 2. 264; Dalz. & Gibs. Bom. Fl. Suppl. 29; Bedd. Fl. Sylv. 92. Introduced from the West Indies into the Deccan and Konkan.

C. montana, Heyne; DC. Prod. II. 499; Fl. Br. I. 2. 264; Dalz. & Gibs. Bomb. Fl. 265. Moist forests of the Belgaum and North Kánara gháts. Fl. July, Aug. Fr. C.S.

C. timoriensis, DC. Prod. II. 499; Fl. Br. I. 2. 265; Bedd. Fl. Sylv. 92. Indigenous in the forests of North Kánara from the gháts in Goa southwards, common in the moist forests of the Ankola and Kumta talukas. Fl. Oct. Fr. ripe C. S. (a small tree or large shrub).

C. glauca, Lam.; DC. Prod. II. 495; Fl. Br. I. 2. 265; Bedd. Fl. Sylv. 91. Dalz. & Gibs. Bom. Fl. Suppl. 30. In the forests of the Deccan and North Kánara. Common in the forests of the Dhárwár district near the North Kánara border at Devikope. Fl. Fr. Oct. Nov.

C. Goensis, Dalz.; Bedd. Fl. Sylv. 92. A large shrub; foot of the Bombay gháts and several cultivated species (*C. lanceolata,* W. & A.,

C. grandis, C. bicapsularis, W. & A.) I am unacquainted with; see Dalz. & Gibs. Bom. Fl. Suppl. p. 29.

CYNOMETRA. Linn.

Trees or shrubs. Leaves odd-pinnate, leaflets few. Flowers minute, axillary. Petals 5, equal, not exserted. Pod turgid, indehiscent, fleshy, rugose, 1-seeded.

C. ramiflora, Linn.; DC. Prod. II. 509; Dalz. & Gibs. Bom. Fl. 83 : Bedd. Fl. Sylv. t. 315; Fl. Br. I. 2. 267. In the forests of the Konkan and North Kánara. The characteristic fruit of this species is carried down the Kálánadi during the rainy season and is thrown up on the sea-shore near Karwár. Fl. H. S. Fr. ripe Aug., Sept.

HARDWICKIA, Roxb.

Trees. Leaves abruptly pinnate. Flowers small in racemose panicles Sepals 5, petaloid. Petals 0. Stamens 10, alternately shorter. Pod dry, 1-seeded near top.

H. binata, Roxb. Fl. Ind. II. 423; Fl. Br. I. 270; Dalz. & Gibs. Bomb. Fl. 83; Bedd. Fl. Sylv. t. 26; Brandis For. Fl. 162. *Anjan,* M. In the dry forests of the presidency; a gregarious tree, but very local, common on the trap of the Sátpudás in Khandesh; a small forest of this specie is found in the Ránebennur taluka of Dhárwár. Fl. C.S. Fr. Apl., May.

SARACA, Linn.

Trees. Leaves abruptly pinnate. Flowers in dense corymbose axillary panicles. Calyx-tube long, limb 4-cleft. Petals 0. Stamens 3-8, exserted. Pod oblong, woody 2-valved.

S. indica, Linn. Mant. 98; Fl. Br. I. 2. 271; Bedd. Fl. Sylv. t. 57; Brandis For. Fl. 166; *Jonesia Asoca,* Roxb. Fl. Ind. II. 218; Dalz. & Gibs. Bomb. Fl. 82. *Ashok, jassundie,* Vern. In the evergreen forests of the Konkan and North Kánara, common, sometimes planted. Fl. Mch., Apl. Fr. Aug., Sept.

TAMARINDUS, Linn.

Tree. Leaves abruptly pinnate, stipules caducous. Flowers racemed. Calyx-tube turbinate, divisions 4, membranous. Petals 3. Stamens 3 perfect, united in a sheath; remaining stamens reduced to staminodes at top of the sheath. Pod 3-10-seeded, mesocarp fleshy.

T. indica, Linn. DC. Prod. II. 488; Fl. Br. I. 2. 273; Dalz. & Gibs. Bomb. Fl. 82. Bedd. Fl. Sylv. t. 184; Brandis For. Fl. 163. The tamarind. *Hunuse,* K.; *chinch,* M.; *amli,* Vern. Self-sown near villages and in waste lands; also cultivated throughout the presidency, indigenous in tropical Africa. Fl. May, June. Fr. ripe Feb., Mch.

BAUHINIA, Linn.

Trees, shrubs or climbers. Leaves simple, cleft at tip into two more or less connate leaflets. Flowers in paniculate racemes. Calyx-tube short, limb spathaceous or 2-5 cleft. Petals 5, subequal clawed. Stamens 10, 3-5 or all perfect. Pod linear, indehiscent or 2-valved.

Trees. Flowers large ; calyx spathaceous ;
stamens 10.
Leaflets obtuse. Calyx-limb broad, ovate. ... ...*B. tomentosa.*
Leaflets acute. Calyx-limb long pointed. ... ...*B. acuminata.*
Trees. Flowers small, white ; calyx spathaceous or 5-cleft ; stamens 10 ; pod glabrous.
Pod turgid, not veined. Calyx spathaceous ... ...*B. racemosa.*
Pod reticulate-veined. Calyx 5-cleft ... ...*B. malabarica.*
Pod red, tomentose, turgid. Calyx 5-cleft. ... ...*B. Lawii.*
Twining shrubs with circinate tendrils.
Leaflets distinct. Stamens 10. ... ...*B. diphylla.*
Leaflets connate. Stamens 3 ... ...*B. Vahlii.*
Trees. Stamens 3-5.
Stamens 3. Calyx 2-cleft to base ... ...*B. purpurea.*
Stamens 5. Calyx spathaceous ... ...*B. variegata.*

B. tomentosa, Linn.; DC. Prod. II. 514 ; Bedd. Fl. Sylv. 92. Dalz. & Gibs. Bom. Fl. Suppl. 31. Cultivated in gardens. Flowers yellow with a purple centre. Fl. C.S.

B. acuminata, Linn. ; DC. Prod. II. 513 ; Fl. Br. I. 2. 276 ; Dalz. & Gibs. Bomb. Fl. Suppl. 30. A cultivated species. Fl. H. & R.S.

B. racemosa, Lam. Dict. I. 390 ; Dalz. & Gibs. Bomb. Fl. 82 ; Bedd. Fl. Sylv. t. 182 ; Brandis For. Fl. 159. *Apta*, M.; *bonne*, K. ; *wanu rajah*, Vern. Throughout the presidency in deciduous forests. Fl. Mch. June. Fr. Nov., Mch.

B. malabarica, Roxb. Fl. Ind. II. 321 ; Fl. Br. I. 2. 277 ; Dalz. & Gibs. Bomb. Fl. 82 ; Brandis For. Fl. 159 ; Bedd. Fl. Sylv. 92. *Amli*, Vern. Throughout the moist forests of the Konkan and North Kánara, common. Fl. Oct., Nov. Fr. Apl., May.

B. Lawii, Benth. MSS. Fl. Br. I. 277 ; *B. foveolata*, Dalz. in Jour. Linn. Soc. 13, p. 188. *Buswanpad*, K. ; *kanchin*, M. Throughout the moist forests of the Konkan and North Kánara, common. This large, diœcious tree has the white flowers sweetly scented. Pods 1 foot long, twisted, turgid, red tomentose. Fl. Oct. Fr. ripe Jany. Male and female flowers on separate stems ; this I have also found to be usually the case in *B. malabarica* in North Kánara.

B. diphylla, Hamilt. in Symes It. Avens. t. 24 ; Fl. Br. I. 2. 278. Konkan, Stocks.

B. Vahlii, W. & A. Prod. 297 ; Fl. Br. I. 2. 279 ; Dalz. & Gibs. Bomb. Fl. 83 ; Brandis For. Fl. 161. *Chambuli, chambil*, M. In the moist forests of the Konkan ghâts. A gigantic climber. Fl. Apl. Fr. ripe next Apl.

B. purpurea, Linn.; Roxb. Fl. Ind. II. 320; Fl. Br. I. 2. 284; Bedd. Fl. Sylv. 92; Brandis For. Fl. 160; Dalz. & Gibs. Bomb. Fl. 30. *Dewakunchun, ragtahanchun, atmatti,* Vern. Cultivated throughout the presidency, also wild in the dry forests of the Deccan and Konkan. Fl. Sept., Nov. Fr. Jan., April.

B. variegata, Linn.; DC. Prod. II. 514; Bedd. Fl. Sylv. 92; Brandis For. Fl. 160; Fl. Br. I. 2. 284; Dalz. & Gibs. Bomb. Fl. Suppl. 30. *Kanaraj, koridara, kanchan,* Vern. Cultivated throughout the presidency, often as a road-side tree. Fl. Feb., April. Fr. Nov.

XYLIA, Benth.

Tree. Leaves bipinnate, pinnæ 1 pair. Flowers in globose, tomentose heads. Petals linear, valvate. Stamens 10, free, anthers with deciduous, stipitate glands. Pod woody, falcate, dehiscent.

X. dolabriformis, Benth. in Hook. Jour. Bot. IV. 417; Fl. Br. I. 2. 286; Brandis For. Fl. 171; Bedd. Fl. Sylv. t. 186; Dalz. & Gibs. Bomb. Fl. 85. The iron wood of Pegu and Arracan. *Pynkado,* Burm. *Jamba, yerul, suria,* M.; *jambe,* K. Throughout the deciduous forests of the presidency. Fl. Mch., April. Fr. C.S. Yields a hard, durable, reddish brown timber, suitable for railway sleepers.

ENTADA, Linn.

Climber with tendrils. Leaves bipinnate. Flowers minute, yellow in long spikes. Calyx minute, 5-toothed. Stamens 10, free exserted; anthers glandular. Pod very large, constricted between the seeds; seeds flat ovate brown shining.

E. scandens, Benth. in Hook. Jour. Bot. IV. 332; Fl. Br. I. 2. 287; Brandis For. Fl. 167; *E. puswtha,* DC.; Dalz. & Gibs. Bomb. Fl. 83. *Gardul,* Vern; *gárambi, garbe,* M. Throughout the forests of the North Kánara and Konkan gháts, often along river banks. Fl. Mch., May. Fr. Dec., Jan.

ADENANTHERA, Linn.

Trees. Leaves abruptly bipinnate. Flowers in slender axillary or paniculate racemes. Petals free or connate at the base. Stamens 10, free, anther cells with a deciduous gland at top. Pod linear, 2-valved, seeds with a red or bi-coloured testa.

A. pavonina, Linn.; DC. Prod. II. 446; Fl. Br. I. 2. 287; Bedd. Fl. Sylv. t. 46; Brandis For. Fl. 168. *Munjuti,* K.; *val, thorla-guny,* M. In the moist forests of the Konkan and North Kánara, nowhere abundant. Fl. Mch., May. Fr. Aug.-Oct.

PROSOPIS, Linn.

Thorny trees. Leaves bipinnate, pinnæ 2 pair, leaflets obliquely oblong, 3-nerved. Flowers small, yellow, in the axils of ovate membranous bracts. Stamens 10, free, anthers tipped with a gland. Pod coriaceous, indehiscent, linear, pendulous. Mesocarp mealy.

P. spicigera, Linn.; DC. Prod. II. 446; Fl. Br. I. 2. 288; Bedd. Fl. Sylv. t. 56; Brandis For. Fl. 169; Dalz. & Gibs. Bomb. Fl. 84. *Sumri, hamra* Guj.; *shema, saunder, savandal,* Deccan; *kandi,* Sind. In the dry forests of the Deccan and Sind. Fl. Feb.-Apl. Fr. May, Aug.

DICHROSTACHYS, DC.

Shrub or small tree. Leaves bipinnate, pinnæ 8-10 pairs, with stipitate glands at the base of each pair. Flowers in axillary pedunculate spikes, upper flowers of spike bisexual, lower neuter, with long exserted filiform staminodes. Stamens 10, free, exserted; anthers tipped with a globose stipitate gland. Pod linear, twisted, indehiscent or irregularly opening.

D. cinerea, W. & A. Prod. 271; Fl. Br. I. 2. 288; Bedd. Fl. Sylv. t. 185; Brandis For. Fl. 171; Dalz. & Gibs. Bomb. Fl. 84. *Vurtuli,* Vern.; *sigamkáti,* M. Common on the dry stony hills of the Deccan, cultivated in North Kánara and the Konkan. Fl. H.S.

LEUCÆNA, Benth.

Trees. Leaves bipinnate. Flowers in dense, globose heads. Stamens 10, free, exserted. Anthers not gland-crested. Pod flat, strap-shaped, dehiscent, many seeded.

L. glauca, Benth. in Hook. Jour. Bot. IV. 416; Fl. Br. I. 2. 290; Brandis For. Fl. 172. Naturalised throughout the presidency from the coast inland. Fl. June-Aug. Fr. remains long on tree. A tropical American species.

MIMOSA, Linn.

Shrubs or herbs, prickly or not. Leaves bipinnate, often sensitive. Flowers in globose heads or cylindrical spikes. Calyx shortly toothed. Petals connate towards the base. Stamens twice the number of petals, exserted, free. Pod oblong or linear, valves membranous or coriaceous, made up of 1-seeded joints that separate, when mature, from the sutures.

> Pod small, sutures prickly   ...   . *M. pudica*
> Pod large, sutures without prickles   ...   . *M. rubicaulis.*
> Pod large, sutures armed with large hooked prickles . *M. hamata.*

M. pudica, Linn.; DC. Prod. II. 426; Fl. Br. I. 2. 291; Dalz. & Gibs. Bomb. Fl. Suppl. 25. *Lajalu,* M. Sensitive plant. Native of Brazil. Run wild in the Konkan and North Kánara; usually cultivated in gardens. Fl. R. and C.S.

M. rubicaulis, Lam.; DC. Prod. II. 429; Fl. Br. I. 2. 291; Dalz. & Gibs. Bomb. Fl. 85; Brandis For. Fl. 172. *Hajeru,* Sind. In the dry districts of the presidency and Sind, in open jungles and plains. Fl. Aug., Sept. Fr. Nov., Jan.

M. hamata, Willd.; DC. Prod. II. 427; Fl. Br. I. 2. 291; Dalz. & Gibs. Bomb. Fl. 85. *Arkur*, Vern. Throughout the dry Deccan districts in open jungles, common on the Kuput range of hills of the Dhárwár district. Fl. R.S. Fr. C.S.

ACACIA, Willd.

Erect or climbing, spinose, prickly shrubs or trees. Leaves bipinnate, leaflets minute. Flowers in globose heads or cylindrical spikes, usually pentamerous. Calyx shortly toothed. Petals exserted. Stamens free, indefinite, exserted. Pod ligulate or oblong, dehiscent or indehiscent.

Erect shrubs or trees. Stipular spines long, straight.
 Flowers in globose heads.
 Pod thick, straight, short, cylindrical ... ... *A. Farnesiana.*
 Pod thick, grey-downy, moniliform, straight ... *A. arabica.*
 Pod thin, linear, ¼ in. broad, straight; sutures
  slightly repand. Flowers unpleasantly scented ... *A eburnea.*
 Pod thin, broad, linear, sutures straight ... ... *A. Jacquemontii.*
 Pod ligulate-falcate.
 Pod thinly downy, dry dehiscent. ... ... *A. tomentosa.*
 Pod tomentose, subindehiscent ... ... *A. leucophloea.*
Erect shrubs or trees. Pairs of stipular spines short,
  hooked. Flowers in spikes.
 Bark white. Rachis pubescent. Pinnæ 10-20 pairs,
  leaflets 30-50 pairs. Calyx long, tomentose. Pod
  strap-shaped, subindehiscent ... ... *A. Suma.*
 Bark brown; rachis downy. Pinnæ 20-40 pairs,
  leaflets 30-58 pairs. Calyx short, downy. Pod
  thin, narrow, often repand .. .. *A. Catechu.*
 Rachis dark, shining, glabrous. Pinnæ 15-20 pairs,
  leaflets 30-40 pairs. Calyx short, glabrous. Pod
  smaller than in either of the above 2 species ... *A. Sundra.*
 Pod strap-shaped, veined, narrowly winged along
  upper suture, distinctly stalked ... ... *A. ferruginea.*
 Spines often ternate. Pod firm, indehiscent... ... *A. Senegal.*
 Spines long, straight. Pod thin, broadly falcate ... *A. latronum.*
Climbing shrubs with scattered prickles. Flowers
  in globose heads.
 Pod red, strap-shaped, thick, fleshy ... ... *A. concinna.*
 Pod thin, dry.
 Leaflets subfalcate, pale beneath, 16 to 60 ... ... *A. Intsia.*
 Leaflets linear, not pale beneath, 80 to 100 ... ... *A. pennata.*

A. Farnesiana, Willd. DC. Prod. II. 461; Fl. Br. I. 2. 292; Brandis For. Fl. 180; Bedd. Fl. Sylv. 52. *Vachellia Farnesiana*, W. & A. Dalz. & Gibs. Bomb. Fl. Suppl. 26. *Vikiyati babul, gu-kikar, iri-babul*, Vern.; *jalli*, K., *kankar*, M. Run wild near villages and in moist situations throughout the presidency and Sind; indigenous in Central America. Fl.

A. arabica, Willd.; W. & A. Prod. 277; Fl. Br. I. 2. 293; Bedd. Fl. Sylv. 95; Dalz. & Gibs. Bomb. Fl. 86; Brandis For. Fl. 180. *Jali*, K.; *babul*, M.; *kikar, ramakanta*, Vern.; *babbar, kálikikar*, Sind. Throughout the dry districts of the presidency, forms extensive forests along the Indus in Sind and also in the black soil country of the presidency proper; it is an excellent road-side tree wherever there is a scanty rainfall. Fr. R.S. Fr. C.S.

A. eburnea, Willd. ; DC. Prod. II. 461; Fl. Br. I. 2. 293 ; Dalz.
& Gibs. Bomb. Fl. 85; Bedd. Fl. Sylv. 95 ; Brandis For. Fl. 183.
Marmat, M. Throughout the dry Deccan districts, common in the
Southern Marátha Country, in stony places. Fl. Nov., Jan. Fr. May,
June.

A. Jacquemontii., Benth. in Hook. Jour. Bot. 1. 490 ; Fl. Br. I. 2.
293; Brandis For. Fl. 183. *Ratobauli,* Guj. Plains of Sind and Gujarát.
Fl. Feb., May.

A. tomentosa. Willd. ; DC. Prod. II. 462; Fl. Br. I. 2. 294. Dalz.
& Gibs. Bomb. Fl. 86; Bedd. Fl. Sylv. t. 48. Rare ; in the Deccan and
Khándesh jungles, Dalz.   •

A. leucophlœa, Willd. ; DC. Prod. II. 462 ; Fl. Br. I. 2. 294; Dalz.
& Gibs. Bomb. Fl. 86; Bedd. Fl. Sylv. t. 49. *Hewar,* Vern.; *haribával,* Guj,
Throughout the dry Deccan districts, common in the Southern Marátha
Country, in dry open forests. Fl. Aug., Sept. Fr. C.S.

A. Suma, Kurz. in Brandis For. Fl. 187 ; Fl. Br. I. 2. 294. *Kamtiya*
M. In North Kánara, near the Dhárwár frontier and in the Southern
Marátha Country ; usually in moist places, along nálás, not common.
Fl. H. S. Fr. Sept., Oct.

A. Catechu, Willd. Sp. Pl. IV. 1070 ; Fl. Br. I. 2. 295 ; Bedd.
Fl. Sylv. t. 49 ; Dalz. & Gibs. Bomb. Fl. 86. *Khair, kaderi,* M.
Common along the coasts of the Konkan and North Kánara, often on
laterite ; also scattered throughout the presidency in dry or moist forests.
Fl. July-Sept. Fr. C. S. The principal catechu yielding tree.

A. Sundra, DC. Prod. II. 458; Fl. Br. I. 2. 295; Dalz. & Gibs.
Bomb. Fl. 86; Bedd. Fl. Sylv. t. 50 ; *A. Catechu,* Brandis For. Fl. 186.
Khair, lál khair, M. Common throughout the presidency. Also yields
catechu. Fl. Aug., Sept. Fr. Dec.

A. ferruginea, DC. Prod. II. 458; Bedd. Fl. Sylv. t. 51; Brandis
For. Fl. 185. *Pándhra khair,* M.; *kaiger,* Panch Maháls, Konkan,
Gujarat. Fl. R.S. Fr. Jany., Feb.

A. Senegal, Willd. DC. Prod. II. 459 ; Fl. Br. I. 2. 295 ; *A. rupestris,*
Stocks, Brandis For. Fl. 184. *Khor,* Sind. Dry rocky hills of Sind.
Fl. Fr. Dec.

A. latronum, Willd. ; DC. Prod. II. 460 ; Fl. Br. I. 2. 296 ; Bedd.
Fl. Sylv. 95; Dalz. & Gibs. Bomb. Fl. 87; Brandis For. Fl. 180.
Donu mullina jali, K. ; *dev babul, bhes,* M.; *tumbuti* in Dhárwár. Com-
mon in the dry plains of the Deccan. Fl. Jany., Mch.

A. concinna, DC. Prod. II. 464 ; Fl. Br. I. 2. 296 ; Dalz. & Gibs.
Bomb. Fl. 87 ; Brandis For. Fl. 188. *Sigekai,* K. Common in the
Konkan and North Kánara moist forests. Fl. Mch., July. Fr. ripe C. S.

A. Intsia Willd. in DC. Prod. II. 464 ; Fl. Br. I. 2. 297 : Dalz.
& Gibs. Bomb. Fl. 88. *A. cæsia,* W. & A. Prod. 1. 278 ; Dalz. & Gibs.

Bomb. Fl. 87; Brandis For. Fl. 189. *Chilar*, M. Common throughout the presidency from the coast inland in deciduous forests. Fl. Apl., Aug. Fr.

A. pennata, Willd. ; Sp. Pl. IV. 1090. Fl. Br. I. 2. 297; Brandis For. Fl. 189; *A. pinnata*, Dalz. & Gibs. Bomb. Fl. 87. *Shembi, shemberti*, M. Throughout the presidency, common in the Konkan and North Kánara forests. Fl. June, Ang.

ALBIZZIA, Durazz.

Unarmed trees or shrubs. Leaves bipinnate. Flowers in large globose heads. Calyx toothed. Petals 4-5 united below, valvate. Stamens indefinite exserted, filaments united at the base. Pod flat, broad, indehiscent or 2-valved. Seeds compressed. Pinnæ few, leaflets oblong, comparatively large and few.

Heads fasciculate, not panicled. Pinnæ 2-4 pairs, leaflets
  3-9 pairs   ...   . *A. Lebbek.*
Heads copiously panicled.
Tomentose. Pinnæ 3-8 pairs, leaflets 10-25 pairs   ... *A. odoratissima.*
Glabrous, young leaves pubescent. Pinnæ 3-4 pairs,
  leaflets 6-8 pairs   ...   . *A. procera.*
Pinnæ numerous, leaflets linear small and in numerous
  pairs. Stipules broad cordate. Midrib of leaflet
  near upper edge   ...   . *A. stipulata.*
Stipules minute. Midrib of leaflet central   ...   ... *A. amara.*

A. Lebbek, Benth. in Hook. Lond. Jour. 1844, 87. Fl. Br. I. 2. 298; Dalz. & Gibs. Bomb. Fl. 88; Bedd. Fl. Sylv. t. 53. Brandis For. Fl. 176. *Sirsul*, K.; *chichola*, M., *siris, harréri*, Vern. Throughout the presidency in dry and moist forests. Commonly planted along road sides. Fl. Apl., May. Fr. Sept.

A. odoratissima. Benth. in Hook. Jour. Bot. 1844, 88. Fl. Br. I. 2. 299; Dalz. & Gibs. Bomb. Fl. 88; Bedd. Fl. Sylv. t. 54; Brandis For. Fl. 175. *Godhunchi*, K. ; *siris*, M. Common in the moist forests of North Kánara and the Konkan, also in the dry Deccan. Fl. Apl.-June. Fr. C.S.

A. procera, Benth. in Hook. Jour. 1844, 89; Fl. Br. I. 2. 229 ; Bedd. Fl. Sylv. 98. *Acacia procera*, Willd. ; Dalz. & Gibs. Bomb. Fl. 87. Brandis For. Fl. 175. *Bellati*, K.; *kinhai*, M. ; *karalla, kilai, tihiri, gurar*, Vern. Throughout the presidency ; common in the North Kánara and Konkan moist forests. In the dry districts, along ravines and nálás. Fl. May. Fr. Jan., Feb.

A. stipulata, Boiv.; Benth. in Hook. Jour. Bot. 1844, 92 ; Fl. Br. I. 2. 300; Brandis For. Fl. 178 ; Bedd. Fl. Sylv. t. 55 ; Dalz. & Gibs. Bomb. Fl. 88. *Bagana, kalbage*, K. ; *laeli, ulul, kasir*, M. Common in the evergreen forests of the Konkan and North Kánara gháts. Fl. Apl., June. Fr. C. S.

A. amara, Boiv. Benth. in Hook. Jour. Bot. 1844, 90. Fl. Br. I. 2. 301. Brandis For. Fl. 178; Bedd. Fl. Sylv. t. 61. Dalz. & Gibs.

Bomb. Fl. 83. *Tugli*, K.; *lullei*, Vern. Throughout the presidency, in dry forests. Fl. Apl., June. Fr. C.S.

PITHECOLOBIUM, Mart.

Trees. Leaves bipinnate. Flowers in globose heads or cylindrical spikes, pentamerous, bisexual. Calyx with short teeth. Corolla segments valvate. Stamens indefinite, much exserted. Pod flat, circinate or falcate, 2-valved. Seeds in a scanty pulp.

Spinose tree. Leaflets 1 pair. Pod turgid, twisted,
   linear   ...   .   *P. dulce.*
Unarmed tree. Pinnæ, 1-2 pair. Pod flat, spirally contorted. *P. bigeminum.*
Pod thick curved. Pinnæ more than 1 pair   ...   ... *P. Saman.*

P. dulce, Benth. in Hook. Lond. Jour. 1844, 199 ; Fl. Br. I. 2. 302 ; Bedd. Fl. Sylv. t. 183. A cultivated species. Fl. R.S. Fl. C.S.

P. bigeminum, Benth. in Hook. Lond. Jour. 1844, 206 ; Dalz. & Gibs. Bomb. Fl. 89 ; Fl. Br. I. 2. 303 ; Bedd. Fl. Sylv. 96. Iron wood. Common in the evergreen forests of the Konkan and North Kánara. Fl.. C.S. Fr. ripe June, July.

P. Saman, L. Introduced into North Kánara. Fl. C.S. Fr. R.S.

ORDER 37. ROSACEÆ.,

Herbs, shrubs or trees. Leaves alternate, simple or compound. Flowers bisexual, regular. Calyx superior or inferior, gamosepalous. Petals 5, rarely 0, inserted under the margin of the disc, imbricate. Stamens perigynous, indefinite, inserted with the petals, or on the disc. Carpels 1 or more, free or connate or adnate to the calyx-tube. Styles free or connate. Ovules 1 or more in each carpel. Fruit variable, of achenes berries or drupes.

Leaves simple. Carpel 1 ...   . PYGEUM.
Leaves compound. Carpels many...   ...   ... RUBUS.

PYGEUM, Gærtn.

Trees or shrubs. Leaves alternate, persistent ; basal glands 2 or 0. Flowers small, racemose. Calyx-tube obconic, urceolate, deciduous ; Limb 10-15-toothed. Petals minute, 5-6 or 0 villous or tomentose, rarely glabrous. Stamens 10-50, at mouth of calyx-tube. Fruit a transversely oblong drupe. Cotyledons very thick, hemispheric.

P. Wightianum, Bl. Melanges Bot. 1855, Av. 2 (ex Walp. Ann. IV. 642) ; Fl. Br. I. 2. 319 ; *P. ceylanicum*, Bedd. Fl. Sylv. t. 59. Excl. synon. Common on the southern ghâts of North Kánara, in evergreen forests. Fl. Nov., Dec. Fr. H. S.

P. Gardneri, Hook. f. Fl. Br. I. 2. 321 ; *P. zeylanicum*, Dalz. & Gibs. Bomb. Fl. 891. Konkán ghâts, Mahábaleshvar, Munohur and Párwár, Dalz.

RUBUS, Linn.

Prickly, trailing shrubs or creeping herbs. Leaves alternate, simple or compound ; stipules adnate to petiole. Flowers usually in terminal or axillary panicles. Calyx cleft into 5 persistent lobes.

Petals 5. Stamens numerous. Carpels many, on a convex receptacle ; ovules 2, collateral, pendulous. Drupes many 1-seeded, crowded upon a dry or spongy receptacle.

Leaves simple, palmately lobed   ...   . *R. moluccanus.*
Leaves trifoliate. A suberect bush   ...   ... *R. ellipticus.*
Leaves 5-9-foliate   ...   . *R. lasiocarpus.*

R. moluccanus, Linn.; DC. Prod. II. 566; Fl. Br. I. 2. 330. *R. rugosus,* Dalz. & Gibs. Bomb. Fl. 89. Along the higher ghats from Mahábaleshvar southwards ; also on the Nilkund and southern ghats of North Kánara at 1,800 to 2,000 ft. elevation.   Fl. & Fr. C.R.S.

R. ellipticus, Smith in Rees' Cyclop. XXX., Rubus 16 ; Fl. Br. I. 2. 336 ; *R. flavus,* Brandis For. Fl. 197 ; *R. Wallichianus,* Dalz. & Gibs. Bomb. Fl. 89. From Mahábaleshvar southwards along the highest ghats ; 4-5,000 ft. Fl. Fr. R. S.

R. lasiocarpus, Smith Fl. Br. I. 2. 339 ; Dalz. & Gibs. Bomb. Fl. 89. *Gariphul,* M. Highest ghats to the southwards, Dalz. Common on the Bababudon hills of Mysore above 5,000 ft. Fl. Fr. R. S.

ORDER 38. **RHIZOPHORACEÆ.**

Trees or shrubs. Leaves entire, opposite, coriaceous. Stipules caducous. Flowers regular, bisexual. Calyx adherent to ovary, persistent, 4-14-toothed. Petals as many as calyx-lobes. Stamens twice the number of petals, in pairs, opposite and embraced by them or indefinite. Ovary 2-5-celled, ovules geminate, pendulous. Fruit coriaceous, 1-celled, 1-seeded.

Ovary inferior. Seed germinating on tree with
  a long, exserted radicle.
Petals 4. Stamens 8. Ovary 2-celled   ...   ... RHIZOPHORA.
Petals 5-6. Stamens 10-12. Ovary 3-celled   ... CERIOPS.
Petals 5-6. Stamens indefinite. Ovary 1-celled... KANDELIA.
Petals 8-14, bifid. Stamens 16-28. Ovary 2-4-
  celled...   ...   . BRUGUIERA.
Ovary ½ inferior. Seed not germinating on tree ... CARALLIA.

RHIZOPHORA, Linn.

Trees. Leaves leathery, ovate, mucronate. Flowers in dichotomous cymes. Calyx 4-lobed, subtended by the united bractlets. Corolla of 4 entire petals, inserted at the base of a fleshy disc. Stamens 8-12, filaments short. Anthers multilocellate. Ovary 2-celled, produced in a fleshy cone ; cells 2-ovuled. Fruit 1-celled, 1-seeded. Radicle elongated, clavate. Albumen 0.

Peduncles longer than the petioles. Flowers
  pedicellate ...   . *R. mucronata.*
Peduncles shorter than the petioles. Flowers
  sessile...   ...   . *R. conjugata.*

R. mucronata, Lam. Dict. IV. 169; Fl. Br. I. 2. 436 ; Bedd. Fl. Sylv. Anal. Gen. t. XIII. fig. 4 ; Dalz. & Gibs. Bomb. Fl. 95 ; Brandis For. Fl. 217. Mangrove. *Kamo,* Sind ; *kandal,* M. On the muddy tidal flats of the coast, also along the coast of Sind. Fl. R. S. Fr. Aug., Sept.

R. conjugata,. Linn.; DC. Prod. III. 33; Fl. Br. I. 2. 436; Brandis For. Fl. 218. Along the coast in tidal marshes. Fl. R.S. Fr. Oct., Nov.

CERIOPS, Arn.

Trees, or shrubs. Leaves simple, coriaceous. Flowers subcapitate, on axillary peduncles. Calyx 5-6-divided, surrounded by the connate bracts. Petals 5-6, inserted at the base of the 10-12-lobed disc; lobes of the petals with clavate bristles. Stamens 10-12, in pairs opposite the petals. Ovary ½-inferior, 3-celled; cells 2-ovuled. Fruit 1-celled, 1-seeded. Radicle as in *Rhizophora*.

C. Candolleana, Arn. in Ann. Nat. Hist. 1. 363; Fl. Br. I. 2. 436; Bedd. Fl Sylv. 99; Brandis For. Fl. 218. *Kirrari, chauri*, Sind. Tidal swamps along the coast. Fl. June-July. Fr. Aug., Sept.

KANDELIA, Wight & Arnott.

A small tree. Leaves simple, coriaceous. Flowers in axillary cymes, large, white. Calyx 5-6-divided, subtended by the united bractlets. Petals 5-6, bifid; lobes multifid, segments capillary. Ovary ½-inferior, 1-celled; ovules 6, in pairs on a central column. Fruit 1-celled, 1-seeded. Radicle fusiform.

K. Rheedii, W. & A. Prod. 1. 310; Fl. Br. I. 2. 437; Brandis For. Fl. 218; Bedd. Fl. Sylv. 100. Tidal swamps along the coast. Fl. June, July. Fr. Sept., Oct.

BRUGUIERA, Lam.

Trees. Leaves shining, coriaceous. Flowers rather large, solitary or few on recurved peduncles. Calyx 8-14-divided, without bracts or bracteoles. Petals 8-14, inserted on the margin of the calyx, 2-cleft, embracing the stamens. Stamens 16-28, in pairs opposite the petals, springing elastically from them when mature. Ovary 2-4-celled, included in the calyx-tube; cells 2-ovuled. Fruit turbinate, 1-celled, 1-seeded.

Peduncles 1-flowered. Radicle angled ... ... *B. gymnorhiza.*
Peduncles 3-flowered. Radicle subacute, subclavate ... ... *B. caryophylloides.*
Peduncle many flowered. Radicle subcylindric, truncated ... ... *B. parviflora.*

B. gymnorhiza, Lamk. Ill. t. 397; Fl. Br. I. 2. 437; Brandis For. Fl. 219; *B. Rheedii*, Dalz. & Gibs. Bomb. Fl. 95; Bedd. Fl. Sylv. 110. Littoral swamps and along creeks and tidal rivers. Fl. Fr. R.S.

B. caryophylloides, Blume. Mus. Bot. 1. 141; Fl. Br. I. 2. 438, Bedd. F. Sylv. 101. Brandis For. Fl. 219. Tidal swamps along coast. Fl. and Fr. R.S.

B. parviflora, W. & A. Prod. 1. 311; Fl. Br. I. 2. 438; Bedd. Fl. Sylv. 101; *Kanilia parviflora*, Dalz. & Gibs. Bomb. Fl. 95. Tidal swamps along the coast. Fl. Fr. R.S.

CARALLIA Roxb.

Trees or shrubs. Leaves simple, ovate. Flowers small, in short branching cymes, often crowded. Calyx with mintue bracts at the base, shortly 5-8-lobed. Petals 5-8, inserted round the thin disc. Disc 10-16-lobed. Stamens usually twice as many as petals. Ovary inferior. Fruit globose, 1-celled, 1-seeded. Seeds albuminous

C. integerrima, DC. Prod. III. 33 ; Fl. Br. I. 2. 439 ; Bedd. Fl. Sylv. t. 193 ; Brandis For. Fl. 219 ; Dalz. & Gibs. Bomb. Fl. 95. *Shenyali, panasi*, M. ; *anda-murgal*, K. ; *punschi*, Vern. Throughout the evergreen forests of the Konkan and North Kánara. Flowers in bud appear during August, but do not open until Jany. or Feb. Fr. H. S. The inflorescence is covered with resinous scales.

ORDER 39. COMBRETACEÆ.

Trees or shrubs. Leaves petiolate, entire. Flowers bracteate, usually bisexual. Calyx-tube adnate to ovary, limb 4-5-divided. Petals 4-5 or 0. Stamens 4-5 or twice as many, on the calyx. Ovary inferior, 1-celled. Ovules 1-7, pendulous from apex of cell. Fruit 1-celled, 1-seeded, often winged. Cotyledons convolute or plano-convex.

Flowers racemose or capitate. Anthers opening by a
　longitudinal slit.
Petals 0. Calyx-limb deciduous or accrescent.
Flowers spiked. Calyx-limb deciduous. ... ... TERMINALIA.
Flowers spiked. Calyx-limb accrescent. ... ... CALYCOPTERIS.
Flowers capitate. Calyx-limb deciduous. ... ... ANOGEISSUS.
Petals 4-5.
Tree. Leaves alternate. Calyx-limb persistent. LUMNITZERA.
Climbing shrubs.
Calyx-tube short ... ... COMBRETUM.
Calyx-tube long, produced beyond ovary ... ... QUISQUALIS.
Flowers cymose. Anthers opening by lateral valves. GYROCARPUS.

TERMINALIA, Linn.

Trees or shrubs. Leaves entire, petiolate, without stipules. Flowers usually bisexual, bracteolate. Calyx-tube adnate to the ovary ; limb 4-5 cleft, segments valvate. Petals 4-5 or 0. Stamens, 4-5 or 8-10, on the calyx. Ovary inferior, 1-celled ; ovules 2-3, pendulous from the apex of the cell. Fruit ovoid, angular or winged, usually indehiscent.

Fruit ovoid, not winged, obscurely ridged.
Fruit ellipsoid, 2-ridged... ... ... *T. Catappa.*
Fruit globular, tomentose ... ... *T. belerica.*
Fruit ellipsoid or obovoid, glabrous, more or less *T. Chebula.*
　5-ribbed when dry.
Fruit with 3-7 longitudinal wings.
Wings of fruit subequal.
Bark white. Fruit with 5-7, narrow, subequal
　wings ... ... *T. Arjuna.*
Fruit with 5, subequal, broad wings. Bark black ... *T. tomentosa.*
Fruit 3-winged ; wings 1 large and 2 small ... ... *T. paniculata.*

T. Catappa, Linn.; Willd. Sp. Pl. IV. 967; Fl. Br. I. 2. 444; Bedd. Fl. Sylv. t. 18; Dalz. & Gibs. Bomb. Fl. Suppl. 33. *Bengali badam*, H. Planted in the Konkan and North Kánara, a native of the Molluccas. Fl. C.S. Fr. R.S.

T. belerica, Roxb. Fl. Ind. II. 431. Fl. Br. I. 2. 445; Bedd. Fl. Sylv. t. 19; Dalz. & Gibs. Bomb. Fl. 91; Brandis For. Fl. 222. *Tare*, K.; *goting, bherda, hela, yela, balra, balda*, M. Common in the deciduous mixed forests throughout the presidency, not in Sind. Fl. Feb.-May. Fr. next cold season.

T. Chebula, Retz. Obs. V. 31; Fl. Br. I. 2. 446; Brandis For. Fl. 223; Bedd. Fl. Sylv. t. 27; Dalz. & Gibs. Bomb. Fl. 91. *Asale*, K.; *hirda, halra*, M. Throughout the presidency in deciduous forests. abundant on the laterite of North Kánara. Fl. Apl., May. Fr. Dec. to Mch. Yields the myrabolan of commerce

T. Arjuna, Bedd. Fl. Sylv. t. 28; Fl. Br. I. 2. 447; Dalz. & Gibs. Bomb. Fl. 91; Brandis For. Fl. 224. *Holematti*, K.; *savimadat*, M. *kahu, arjun* (*arjuna sadra* in Gujarát), Vern. Along banks of rivers throughout the presidency. Fl. Apl.. May. Fr. C. S.

T. tomentosa, Bedd. Fl. Sylv. t. 17; Fl. Br. I. 2. 447; Dalz. & Gibs. Bomb. Fl. 91; Brandis For. Fl. 225. *Kari matti*, K.; *ain*, M.; *sadri, hadri*, Gúj. Common throughout the presidency in mixed deciduous forests. Fl. Apl. Fr. next year's Feb.-Apl. Yields a valuable and well-known timber.

T. paniculata, Roth. Nov. Sp. 383; Fl. Br. I. 2. 448; Dalz. & Gibs. Bomb. Fl. 92; Bedd. Fl. Sylv. t. 20; Brandis For. Fl. 226. *Kinjal*; *kinjal*, M.; *honal, hongal, hanab*, K.; *kirijal*, Vern. Throughout the presidency in mixed deciduous forests; one of the most common trees in North Kánara. Fl. Aug., Sept. Fr. ripe Apl., May. Yields a good timber.

CALYCOPTERIS, Lamk.

A diffuse or scandent shrub. Leaves opposite, ovate, entire. Racemes dense, forming large panicles. Flowers small, greenish, bracteate. Calyx-tube 5-striate; limb 5-fid, enlarged in fruit-Petals 0. Stamens 10. Ovary 1-celled, inferior; ovules 3, pendulous from the top of the cell. Fruit 5-ribbed, villous, 1-seeded, crowned by the enlarged calyx.

C. floribunda, Lam. Dict. Supp. II. 41; Fl. Br. I. 2. 449; Brandis For. Fl. 220; *Getonia floribunda*, Roxb.; Dalz. & Gibs. Bomb. Fl. 91. *Ukshi*, M.; *wuksey, baguli*, Vern. Throughout the presidency, in deciduous forests Fl. Mch.-Apl, Fr. Apl., May.

ANOGEISSUS, Wall.

Trees or shrubs. Leaves alternate, entire. Calyx-tube compressed, 2 winged at the base, long, attenuated above the ovary; limb small, deciduous. Petals 0. Stamens 10 in 2 series. Ovary inferior, 1-celled ; ovules 2, pendulous from the top of the cell. Fruit small, 2-winged, in dense, globose heads.

A. **latifolia**, Wall. ; Bedd. Fl. Sylv. t, 15 ; Fl. Br. I. 2. 450 ; Brandis For. Fl. 227 ; *Conocarpus latifolia*, DC. ; Dalz. & Gibs. Bomb Fl. 91. *Dindal*, K. ; *dhaura*, M. ; *dabria*, Vern. Throughout the presidency, usually in dry forests. Fl. May-June. Fr. Nov.-Feb. Yields a useful timber.

LUMNITZERA, Willd.

Trees or shrubs. Leaves fleshy, simple, alternate. Calyx-tube elongate, bracteolate ; limb bell-shaped, 5-lobed, persistent. Petals 5, spreading. Stamens 5-10; anthers cordate. Ovary 1-celled ; ovules 2-5, pendulous. Fruit woody, obtusely angular, 1-seeded.

L. **racemosa**, Willd. ; DC. Prod. III. 22 ; Fl. Br. I. 2. 452 ; Dalz. & Gibs. Bomb. Fl. 90 ; Bedd. Fl. Sylv. 103. Brandis For. Fl. 221. South Konkan, along salt water creeks and back waters. Fl. Mch., Apl.

COMBRETUM, Linn.

Shrubs, usually climbing. Leaves opposite, seldom alternate, simple. Flowers polygamo-diœcious, bracteolate. Calyx-tube adnate to the ovary, limb 4-5-divided, segments valvate. Petals 4-5 or 0. Stamens as many or twice as many as calyx-segments Ovary inferior, 1-celled ; ovules 2-5, pendulous. Fruit 4-5-winged or angled.

     Tube of calyx equalling the ovary. Fruit with 4 papery wings ... ... *C. ovalifolium.*
     Tube of calyx 2-3 times as long as the ovary. Fruit with 4 membranous wings ... ... *C. extensum.*

C. **ovalifolium**, Roxb. Fl. Ind. 2. 256 ; Fl. Br. I. 2. 458 ; Dalz. & Gibs. Bomb. Fl. 90. *Zelloosey, madbel*, Vern. Throughout the presidency in deciduous forests. Fl. Feb. Fr. May.

C. **extensum**, Roxb. Fl. Ind 2. 229 ; Fl. Br. I. 2. 458 ; *C. Wightianum*, Wall. ; Dalz. & Gibs. Bomb. Fl. 458. *Piloka*, Vern. Throughout the moist forests of the Konkan and North Kánara. Fl. Feb. Fr. May.

QUISQUALIS, Linn.

Rambling shrubs. Leaves opposite. Flowers axillary, in short spikes. Calyx-tube long produced. Petals 5. Stamens 10. Ovary 1-celled. Fruit 5-angled, dry.

Q. **indica**, Linn. Sp. Pl. 556 ; Fl. Br. I. 2. 459 ; Brandis For. Fl. 220 : Dalz. & Gibs. Bomb. Fl. Suppl. 33. Rangoon creeper. Cultivated in gardens throughout the presidency. Fl. May, Sept.

GYROCARPUS, Jacq.

A large tree. Leaves alternate, long petioled. Flowers small, unisexual, males numerous. Male fl. Calyx 4-7-divided. Petals 0. Stamens 4-7, with alternate clavate staminodes ; anthers opening by lateral

valves. Ovary 0. Female fl. Calyx-limb 2-divided, accrescent. Petals
and stamens 0. Ovary 1-celled, ovule 1, pendulous. Nut bony, crowned
with the 2, winged calyx-lobes ; cotyledons convolute.

{ **G. Jacquini**, Roxb. Fl. Ind. 1. 415 ; Fl. Br. I. 2. 461 ; Kz. For. Fl.
Br. Burm. 1. 470 ; Grah. Cat. Bomb. Pl. 250. *Zaitun*, Hind. Deccan,
banks of the Krishna near Nalutwar, Grah. Belgaum district, in dry
forests. Fl. R.S. Fr. C.S.

ORDER 40. MYRTACEÆ.

Trees or shrubs. Leaves opposite, simple, usually gland-dotted
and with an intramarginal vein. Flowers regular, usually hermaphrodite. Calyx 4-5 or more divided. Petals as many as calyx-lobes
on disc surrounding cavity of calyx. Stamens usually many. Ovary
inferior or ½ inferior crowned by fleshy disc, 1-celled, 1-ovuled or 2,
many celled with many ovules on axile placenta. Fruit crowned
with the calyx-limb, dehiscent or indehiscent. Seeds angular,
albumen 0.

Leaves dotted with pellucid glands. Fruit a berry or
  drupe.
Berry many seeded ... ... PSIDIUM.
Berry few seeded ... ... EUGENIA.
Leaves not gland-dotted. Fruit angular fibrous, or
  globose fleshy.
Fruit angular fibrous ... ... BARRINGTONIA.
Fruit globose fleshy, many seeded ... ... CAREYA.

PSIDIUM, Linn.

Trees or shrubs. Leaves opposite, entire. Flowers on axillary
peduncles, large. Calyx-tube ovate ; limb valvately 4-5-lobed.
Petals 4-5. Stamens many. Ovary 2 or more celled ; ovules many.
Fruit a many seeded berry.

P. Guyava, Linn. Benth. Fl. Hongk. 120. Fl. Br. I. 2. 468 ; Brandis
For. Fl. 232. *P. pyriferum*, W. & A. Dalz. & Gibs. Bomb. Fl. Suppl.
34. The guyava tree. *Jam, peru*, Vern. Indigenous in Mexico, cultivated
throughout the presidency, naturalised in many places. Fl. Apl., May.
Fr. R. S.

EUGENIA, Linn.

Evergreen trees or shrubs. Leaves opposite, coriaceous or membranous, pinnate nerved. Flowers in lateral or terminal trichotomous cymes or panicles. Calyx-tube globose or more or less
elongate ; lobes 4, rarely 5. Petals usully 4, free and spreading or
united in a calyptra. Stamens many, in several series, free or
slightly collected in four bundles. Ovary 2, rarely 3-celled ; ovules
several in each cell. Fruit a berry, 1 or several seeded, crowned
with the persistant calyx-limb.

Trees. Flowers large, showy, 4-merous. Calyx with a thickened
  disc.
Petals spreading.

Flowers red, in dense lateral, short peduncled racemes ... *E. malaccensis.*
Flowers te.minal and axillary.　Calyx broadly turbinate . *E. Jambos.*
Ca'yx-tube hemispherical ... ... *E. hemisphærica.*
Ca'yx-tube slender.　Stamens crimson ... ... *E. læta.*
Flowers small. in compact cymes.　Ca'yx without a disc.
　　Petals usually ca'yptrate.
　Petals free.
　Calyx elongate.　Flowers axillary racemose ... ... *E. Wightiana.*
　Calyx shorter.　Flowers in corymbs ... ... *E. zeylanica.*
　Petals calyptrate.
　Calyx-tube shortly turbinate.
　Branches 4-angled.　Leaves lanceolate ... ... *E. lissophylla.*
　Branches terete.　Leaves obovate ... ... *E. caryophyllœa.*
　Calyx-tube slender and tapering, limb 4-5-lobed.
　Leaves petioled. narrow (near *E. lissophylla*) ... ... *E. rubicunda.*
　Leaves indistinctly nerved, dots not pellucid ... ... *E. Stocksii.*
　Leaves distinctly nerved, dots pellucid.
　Cymes short, compact beneath leaves ... ... *E. Jambolana.*
　Cymes lateral lax from the scars of fallen leaves, much
　　branched ... ... *E. Heyneana.*
　Flowers solitary or fascicled.　Petals distinct.　Shrubs.
　Branchlets, &c., pubescent.　Staminal disc broad ... *E. macrosepala.*
　Glabrous.　Staminal disc not enlarged ... ... *E. Mooniana.*

E. malaccensis, Linn. ; Lam. Dict III. 196 ; Fl. Br. I. 2. 471 ; Bedd. Fl.
Sylv. 110.　*Jambosa malaccensis,* DC. ; Dalz. & Gibs. Bomb. Fl. Suppl.
35.　Malay apple.　*Malacca jambul,* Vern.　Cultivated in the gardens of
the Haigas in North Kánara.　Also near and in Bombay. Fl. Apl , May.

E. Jambos, Linn. Fl. Br. I. 2. 474 ; Brandis For. Fl. 233 ; Bedd. Fl.
Sylv. 110.　*Jambosa vulgaris,* W. & A. Dalz. & Gibs. Bomb. Fl. 35.
Rose apple. *Gulúb-jamun, sakarajambha,* Vern.　Planted throughout the
presidency in many places, run wild near villages in North Kánara. Fl.
Feb.　Fr. July, Aug.

E. hemisphærica, Wight Ill. II. 14 ; Fl. Br. I. 2. 477 ; Bedd. Fl.
Sylv. t. 203 ; *Jambosa lanceolaria,* Wgt. ; Dalz. & Gibs. Bomb. Fl. 94.
Colonel Beddome is of opinion that *E. hemisphærica,* Wgt., and *E.
lanceolaria,* Roxb., are identical.　On the Konkan and North Kánara
gháts, in evergreen forests, common on the Nilkund and Ainshi gháts.
Fl. Mch., April.　Fr. R. S.

E. læta, Ham. in Trans. Wern. Soc. V. 338 ; Fl. Br. I. 2. 479;
Jambosa pauciflora, Wgt. Dalz. & Gibs. Bomb. Fl. 94 ; *E. Wightii,*
Bedd. Fl. Sylv. 109.　In evergreen forests, on the higher gháts of
North Kánara ; common on the Gairsoppah ghát. Fl. C.S.　Fr. R.S.

E. Wightiana, Wight. Ill. II. 15 ; Ic. t. 529 ; Fl. Br. I. 2. 485. *E.
lanceolata,* Bedd. Fl. Sylv. 110.　Konkan and North Kánara, near rivers
and along water-courses, rare. Fl. Feb., Mch.

E. zeylanica, Wgt. Ic. I. 73 ; Fl. Br. I. 2. 485 ; *E. spicata,* Bedd. Fl.
Sylv. t. 202 ; *Syzygium zeylanicum,* DC. Dalz. & Gibs. Bomb. Fl. 94.
Pitculi, M.　Throughout the Konkan and North Kánara gháts along
rivers and nálás, common in North Kánara.　Fl. Apl.　Fr. R.S.

E. lissophylla, Thw. Enum. 117 (*Syzygium*) ; Fl. Br. 1. 2. 488; Bedd. Fl. Sylv. 108. Konkan, Stocks.

E. caryophyllæa, Wgt. Ic. t. 540 ; Fl. Br. I. 2. 490 ; Bedd. Fl. Sylv. 108 ; *Syzygium caryophyllæum*, Gærtn. Dalz. & Gibs. Bomb. Fl. 93. *Radarang*, M. Throughout the Konkan and North Kánara ; in moist forests, very common near the sea. Fruit black edible. Fl. Mch., May. Fr. R. S.

E. rubicunda, Wgt. Ic. t. 538 ; Fl. Br. I. 2. 495; Bedd. Fl. Sylv. 108 ; *Syzygium rubicundum*, W. & A. Dalz. & Gibs. Bomb. Fl. 94. On the higher ghâts along streams, Dalz.

E. Stocksii, Duthie. Fl. Br. I. 2. 498. Konkan, Stocks.

E. Jambolana, Lim. Dict. III. 198 ; Fl. Br. I. 2. 499 ; Bedd. Fl. Sylv. t. 197 ; Brandis For. Fl. 233 ; *Syz. Jambolanum*, DC. Dalz & Gibs. Bomb. Fl. 93. *Jambul*, M. ; *nerlu*, K. Throughout the presidency both wild and cultivated, common in the moist forests of North Kánara and the Konkan. Fl. Mch., Apl. Fr. July-Aug.

E. Heyneana, Wall. Cat. 3599 (*Syzygium*) ; Fl. Br. I. 2. 500 ; *Syz. salicifolium*, Dalz. & Gibs. Bomb. Fl. 94 ; *E. salicifolia*, Grah. Bedd. Fl. Sylv. 112. *Pan jambul, bedas,* M. In the beds of streams and rivers ; common throughout the Konkan and North Kánara. Fl. Mch. April. Fr. ripe June.

E. macrosepala, Duthie. Fl. Br. I. 2. 501. Abundant locally on the North Kánara ghâts, in evergreen forests. Fl. Jan.-Feb. Fr. Aug. Sometimes a large shrub.

E. Mooniana, Wgt. Ic. II. 551; Fl. Br. I. 2. 505 ; var. *gracilis,* Bedd. Fl. Sylv. 110 Evergreen forests of the Konkan and North Kánara. Common in the forests near Kárwár on the coast. Fl. Feb., Mch. Fr. Aug., Sept.

BARRINGTONIA, Forst.

Trees. Leaves alternate. Flowers bracteate, in spikes or racemes. Calyx-tube ovoid or turbinate. Petals 4-5. Stamens indefinite. Ovary inferior. Disc annular. Berry fibrous, fleshy, crowned with the calyx-limb, usually 1-seeded. Embryo consisting of 2 concentric masses, one a woody rind and the other a medulla.

Calyx valvate. Fruit ovoid, large ... ... *B. racemosa.*
Calyx imbricate. Fruit quadrangular, small ... ... *B. acutangula.*

B. racemosa, Bl. DC. Prod. III. 288 ; Fl. Br. I. 2. 507 ; Dalz. & Gibs. Bomb. Fl. 94. *Nivar,* M. Along rivers and nálás near the coast of the Konkan and North Kánara. Fl. H. S. Fr. R.S.

B. acutangula, Gærtn. Fruct. II. 97. t. 101. Fl. Br. I. 2. 508 ; Brandis For. Fl. 235 ; Dalz. & Gibs. Bomb. Fl. 95 ; Bedd. Fl. Sylv. t. 204. *Tivar, ingli,* M. Banks of rivers and streams throughout North Kánara and the Konkan, common near the coast. Fl. May. Fr. Sept., Oct.

CAREYA, Roxb.

Trees or shrubs. Leaves alternate, simple. Flowers large, showy. Calyx adnate to the ovary; limb 4-divided. Petals 4-5. Stamens numerous, free. Ovary inferior, 4-5-celled; style elongate with a capitate or 4-lobed stigma. Fruit globose, fleshy, crowned with the calyx-limb, many seeded. Seeds in a fleshy pulp; cotyledons 0.

C. arborea, Roxb. Fl. Br. Ind. II. 638; Fl. I. 2. 511; Brandis For. Fl. 236; Dalz. & Gibs. Bomb. Fl. 95; Bedd. Fl. Sylv. 205. *Kaval*, K; *kumbia, kuba*, M. Common throughout the presidency in deciduous forests. Fl. March-April. Fr. ripe July.

ORDER 41. MELASTOMACEÆ.

Herbs, shrubs or trees. Leaves opposite, entire, often 3-7-nerved. Flowers, regular hermaphrodite. Calyx-tube united by vertical walls to the ovary, rarely nearly free; limb 4-6-lobed. Petals 3-6, imbricate. Stamens as many or twice as many as the petals, inserted with them; anthers opening by pores, rarely by slits; connective often appendaged near the base. Ovary 4-5-1-celled. Ovules several, rarely only 2 to each placenta. Fruit included in the calyx-tube, capsular or berried, breaking up irregularly or by slits through the top of the cells. Seeds minute, many (in *Memecylon* only 1); albumen 0.

A villous shrub. Stamens unequal, Seeds many.
Leaves 3-7-nerved, palmately from near the base   ... MELASTOMA.
Glabrous shrubs or trees. Stamens equal. Berry 1-
seeded. Leaves pinnate nerved   ...   . MEMECYLON.

MELASTOMA, L.

Strigose or villous shrubs. Leaves petioled. Flowers terminal, showy purple, 5-merous. Calyx clothed with bristles, tube bell-shaped; lobes 5. Petals usually 5. Stamens twice as many as calyx-lobes, anthers opening by a single apical pore, very unequal, 5 larger with the connective produced below in a long appendage, and 5 smaller with the basal appendage shorter or wanting. Ovary 5-7-celled more or less united to the calyx-tube, apex bristly. Fruit bursting irregularly. Seeds minute, very many, minutely punctate.

M. malabathricum, L. DC. Prod. III. 145; Fl. Br. I. 2. 523; Dalz. & Gibs. Bomb. Fl. 92. Indian rhododendron. *Paloré*, M. Throughout the Konkan and North Kánara in moist forests, often along the banks of nálás near evergreen forests. Fl. throughout the year.

MEMECYLON, L.

Leaves coriaceous, short-petioled. Flowers usually small, in axillary cymes, clusters or umbellets. Calyx-tube campanulate; limb 'lated, truncate or shortly 4-lobed   Petals 4, blue or white, rarely reddish. Stamens 8, equal, anthers opening by slits in front, con-nective ending in a horn behind. Ovary inferior, 1-celled; ovules 6-12 whorled, on a free central placenta. Berry globose, crowned with the calyx-margin, 1-seeded. Seed large, cotyledons convolute or variously folded.

Trees. Peduncles 0; pedicels under ¼ inch. Flowers clustered.

Branchlets acutely quadrangular. Leaves cordate, sub-sessile ... ... *M. Wightii.*

Branchlets round. Leaves amplexicaul ... ... *M. amplexicaule.*

Trees or shrubs. Flowers in peduncled cymes; pedicels short or long.

Peduncles long, 1-3 from.same axil; pedicels umbelled.
Shrub ... ... *M. terminale.*

Tree. Flowers purple, in a compact cyme; peduncles short; pedicels very short, slender ... . *M. edule.*

M. Wightii, Thw. Enum. 113; Fl. Br. I. 2. 554. North Kánara, in evergreen forests, not common. Fl. C. S.

M. amplexicaule, Roxb. Fl. Ind. II. 260; Fl. Br. 1. 2. 559; Wight, Ic. t. 279. Var. *malabarica. Limba,* M.; *locundi, limbtoli,* K. In the evergreen forests of North Kánara, common on the Yellápur gháts. Fl. C.S. Fr. H.S.

M. terminale, Dalz. in Hook. Kew Jour. III. 121; Fl. Br. I. 2. 558; Dalz. & Gibs. Bomb. Fl. 93. Southern gháts of the Deccan peninsula. Fl. Br. I. In the evergreen forests of North Kánara, from Ainshi southwards. Fl. Dec.-Feb. Fr. ripe Mch.-May.

M. edule, Roxb. Cor. Pl. t. 82; Fl. Br. I. 2. 563; Dalz. & Gibs. Bomb. Fl. 93; Bedd. Fl. Sylv. 113. *Anjun, anjani, kurpa,* Vern. Very common in the Koukan and North Kánara, in moist evergreen forests, and along the coast. Fl. C. S. Fr. Apl., May.

ORDER 42. **LYTHRACEÆ.**

Trees, shrubs or herbs. Leaves simple, opposite. Stipules 0. Flowers regular, hermaphrodite. Calyx 3-6 lobed. Petals as many .as calyx-lobes, rarely 0. Stamens definite or indefinite, on the calyx-tube. Ovary free, 2-6-celled, ovules many, placentas axile. Fruit a capsule, 1-2-4 or many celled. Seeds numerous, exalbuminous.

Ovary superior.

Calyx curved. Flowers secund ... ... WOODFORDIA.

Calyx straight. Flowers symmetric.

Flowers 4-fid. Capsule irregularly breaking up ... LAWSONIA.

Flowers 6-fid. Capsule 3-6-valved ... ... LAGERSTROEMIA.

Berry subglobose ... ... SONNERATIA.

Ovary inferior ... ... PUNICA.

WOODFORDIA, Salisb.

A shrub. Leaves opposite, subsessile, lanceolate, usually cordate. Flowers scarlet in axillary pedunculate cymes. Calyx long, tubular, curved, mouth oblique. Petals 6, small or 0, at the top of the calyx-tube. Stamens 12. Ovary free, sessile, 2-celled; ovules many. Capsule membranous, included in the calyx-tube. Seeds many, smooth.

W. floribunda, Salisb. Parad. Lond. t. 42 ; Brandis For. Fl. 238. *W. tomentosa.* Bedd. Fl. Sylv. 117 ; *Grislea tomentosa,* Roxb. Dalz. & Gibs. Bomb. Fl. 97. *Dayti, dhaiphal, phusati,* M. ; *dhauri, dowari,* Vern. Throughout the presidency, common in the Konkan and North Kánara near the sea-coast. Fl. Dec., Apl.

LAWSONIA, Linn.

An erect, sometimes spinous shrub. Leaves opposite, entire. Flowers in terminal panicled cymes. Calyx-tube very short; lobes 4, ovate. Petals 4, wrinkled. Stamens usually 8. Ovary free, 4-celled; ovules many, placentas axile. Seeds many, smooth, on a central placenta.

L. alba, Lamk. Ill. t. 296, fig. 2 ; DC. Prod. III. 91 ; Fl. Br. I. 2. 573. Dalz. & Gibs. Bomb. Fl. 97 ; Bedd. Fl. Sylv. 118 ; Brandis For. Fl. 238. *Henne, mendie,* Vern. Throughout the presidency in hedges, along the coast common. Fl. C. S. Fr. C. S.

LAGERSTRŒMIA, Linn.

Trees and shrubs. Leaves opposite (or the uppermost alternate), entire. Flowers in axillary or terminal panicles, bracteate. Calyx-tube funnel-shaped, cleft into 6 ovate lobes. Petals 6, clawed. Stamens numerous; filaments long exserted. Ovary sessile, 3-6-celled; ovules numerous on axile placentas. Fruit a coriaceous capsule, 3-6-celled, loculicidally dehiscent. Seeds winged.

Calyx-tube smooth.
Calyx-teeth not patent or reflexed.
  A shrub. Fruit ½ inch   ...   . *L. indica.*
  A tree. Fruit ½ to 1 inch   ...   . *L. parviflora.*
Calyx-lobes patent or reflexed. Capsule small, ¼ inch ... *L. microcarpa.*
Calyx tomentose, ribbed   ...   . *L. Flos-Reginæ.*

L. indica, Linn. DC. Prod. III. 93 ; Fl. Br. I. 2. 575. Cultivated in gardens throughout the presidency, a native of China.

L. parviflora, Roxb. Fl. Ind. II. 505 ; Bedd. Fl. Sylv. t. 31. Gamble Ind. Timbers, 200. *L. lanceolata,* Dalz. & Gibs. Bomb. Fl. 98 ; Gamble Ind. Timbers, 201 ; Brandis For. Fl. 239 ; Bedd. Fl. Sylv. t. XXXII. (var. *majuscula,* Fl. Br. I). *Chunungi, sokutia, bondarch,* Vern. Throughout the presidency in deciduous forests from the coast inland. I follow the Fl. of Br. I. in making only two species out of the 3 figured by Beddome in the Fl. Sylv. t. 30, 31, 32. *L. parviflora* is a well-defined species with leaves glaucous beneath and a large capsule with adpressed calyx-lobes. The variety *majuscula* is found in the northern parts of the presidency and is called *sokatia* in the vernacular.

L. microcarpa, Wgt. Bedd. Fl. Sylv. t. 30; *L. lanceolata*, Wall. Fl. Br. I. 2. 576; Brandis For. Fl. 210. *L. parviflora?* Dalz. & Gibs. Bomb. Fl. 98. *Bili nandi*, K.; *nana*, M. Throughout the presidency in deciduous forests, common, and attains a large size in North Kánara. I have retained Wight's and Beddome's name for our Nana tree, which has invariably a small capsule and is otherwise a very distinct species. Fl. Mch., Apl. Fr. Nov., Jany.

L. Flos-Reginæ, Retz. Obs. V. p. 25; Fl. Br. I. 2. 577; *L. reginæ*, Bedd. Fl. Sylv. t. 29; Dalz. & Gibs. Bomb. Fl. 98; Brandis Fl. 240. *Hole dásál*, K.; *taman*, M. On the North Kánara and Southern Konkan ghâts along the banks of nálás and rivers, sometimes planted as an ornamental tree. Fl H. S. Fr. R. S.

SONNERATIA, Linn. f.

Trees. Leaves opposite, thick. Flowers large, solitary or in 3-flowered cymes. Calyx thick, coriaceous, the tube adnate to the base of the ovary, lobes 4-8, valvate. Petals 4-8 or 0. Stamens many, on the calyx-tube, inflected in bud. Ovary nearly free, 10-15-celled, style long. Fruit large, supported by the persistent calyx. Seeds many, curved, angular.

Calyx-lobes 4.　Petals 0　　...　　...　　... *S. apetala.*
Calyx-lobes 6.　Petals 6　　...　　...　　... *S. acida.*

S acida, Linn. f. Suppl. 252; Fl. Br. I. 2. 579; Dalz. & Gibs. Bomb Fl. 98; Brandis For. Fl. 242; Bedd. Fl. Sylv. 118. Stigma capitate In the salt marshes and creeks of the Konkan and North Kánara. Fl. June, July Fr. Oct., Nov.

S. apetala, Ham. in Syme Emb. Ava. III. 313, t. 25. Fl, Br. I. 2. 579. Stigma umbrella-shaped, large. Koukan, Law.

Punica granatum, Linn. Roxb. Fl. Ind. II. 499; Brandis For Fl. 241. Pomegranate. Is cultivated in gardens throughout the presidency. Fl. Apl., May. Fr. July, Sept.

Order 43. SAMYDACEÆ.

Trees or shrubs. Leaves simple, alternate; stipules small, deciduous. Flowers regular, small, axillary, shortly pedicelled, fascicled or in long racemes, simple or panicled. Calyx coriaceous, persistent; tube short, free, or adnate to the ovary, limb 3-7-fid. Petals as many as calyx-lobes or 0, imbricate. Stamens definite or indefinite, usually opposite the petals, and alternating with small glands. Ovary superior or more or less inferior, 1-celled; style 1, capitate or 3-fid at the apex, or styles 2-5; ovules on usually 2-5, parietal placentas. Fruit loculicidally 2-5-valved; seeds several, usually drilled.

Petals 0, flowers in axillary fascicles   ...   ... CASEARIA.
Petals present, flowers in axillary or terminal racemes   ... HOMALIUM.

CASEARIA, Jacq.

Shrubs or small trees. Leaves alternate, distichous, often slightly serrate and transparently dotted. Calyx-tube short, limb 4-5-lobed. Petals 0. Stamens 6-15, or rarely more, alternating with as many scales or staminodes. Ovary superior, 1-celled; stigma capitate or shortly 3-lobed; ovules many, parietal. Fruit fleshy, opening into valves; seeds often with a red arillus.

Trees. Leaves glabrous.
Pedicels glabrous.
Calyx pubescent. Carpels usually 3. Leaves thin   ...*C. graveolens*:
Calyx glabrous. Carpels usually 2. Leaves thick   ...*C. esculenta.*
Shrub   Pedicels pubescent. Leaves coriaceous   ...*C. rubescens.*
Tree. Leaves tomentose...   ...   .*C. tomentosa.*

C. graveolens, Dalz. in Hook. Jour. Bot. IV. 107 ; Dalz. & Gibs. Bomb. Fl. 11 ; Fl. Br. I. 2. 592 ; Brandis For. Fl. 243. *Bohkara*, M. In the South Konkan, in open situations on the ghâts. Fl. R.S.

C. esculenta, Roxb. Fl. Ind. II. 422 ; Fl. Br. I. 2. 592; Bedd. Fl. Sylv. 119. *C. lævigata,* Dalz. & Gibs. Bomb. Fl. 11. *C. varians,* Thw. Bedd. Fl. Sylv. t. 208. *Mori,* M. Throughout the moist forests of the Konkan and North Kánara, common on the hills near Kárwár. Fl. June. Fr. Aug., Sept.

C. rubescens, Dalz. & Gibs. Bomb. Fl. 11 ; Fl. Br. I. 2. 593. On the ghâts to the south. Dalz. In the forests near the Nilkund ghát of North Kánara. Fl. C. S. Fr H.S. The leaves in the North Kánara plant are sometimes very large up to 10 in. long.

C. tomentosa, Roxb. Fl. Ind. II. 421 ; Brandis For. Fl. 243 ; Bedd. Fl. Sylv. 119; *C. Anavinga,* Dalz. & Gibs. Bomb. Fl. 11. *Bairi, chillara,* Vern. Throughout the presidency in deciduous forests, common. Fl. Jan., May. Fr. R.S.

HOMALIUM, Jacq.

Shrubs or trees. Leaves alternate, crenate. Flowers hairy, in slender, simple or panicled racemes; pedicels bracteate at the base. Calyx-tube adnate to the base of the ovary; lobes 5-7, narrow, persistent. Petals 5-7, inserted on the throat of the calyx, persistent. Disc tomentose. Stamens 1 or 1-7, fascicled, opposite the petals, with alternating glands. Ovary ½ superior, 1-celled; styles 2-5, filiform. Ovules many or several, placentas parietal. Capsule ½ superior, coriaceous, 2-5-valved at the apex. Seeds few, angular or oblong.

H. zeylanicum, Benth. in Jour. Linn. Soc. IV. 35 ; Fl. Br. I. 2.596 ;
Bedd. Fl. Sylv. t. 210. In the evergreen forests of the northern ghâts
of North Kánara, from Diggi to Ainshi. Fl. April, May. Fr. R.S.

ORDER 44. DATISCACEÆ.

Trees or herbs. Leaves simple or pinnate. Flowers small, diœcious, clustered, racemed or panicled. Male calyx-tube short, teeth
3-9; petals 0: stamens 4-25. Female calyx-tube adnate to the
ovary, lobes 3-8; petals 0 ; ovary 1-celled, open or closed at the
vertex ; styles lateral ; placentas parietal ; ovules many. Capsule
coriaceous or membranous, opening at the vertex between the styles.
Seeds many, small, albuminous.

TETRAMELES, R. Br.

Large trees. Leaves pubescent, serrate. Flowers diœcious.
Petals 0. Males : calyx-tube short; lobes 4, unequal or equal.
Stamens 4 round a depressed disc ; filaments elongate. Ovary rudimentary 4-angled or 0. Female : calyx-tube almost 4-angled ; lobes
4, short. Staminodes 0. Ovary open at apex ; ovules inserted in 3-4
rows on the parietal placentas ; styles 4 short. Capsule membranous,
open at the summit and crowned by the 4 styles.

T. nudiflora, R. Br. in Benn. Pl. Jav. Rar. 79, t. 17 ; Fl. Br. I. 2. 657;
Bedd. Fl. Sylv. t. 212 ; Brandis For. Fl. 245. *Jermali,* K. ; *uyado,* M.
Common in the evergreen forests of the ghâts of North Kánara and the
Konkan. An immense, deciduous tree, overtopping the regular evergreen
canopy. Fl. Feb.-Meh. Fr. April, May.

ORDER 45. CACTEÆ.

Herbs, shrubs or trees. Leaves reduced to tufts of spines, prickles
or small tubercles. Flowers sessile, solitary, hermaphrodite, regular.
Calyx-tube adnate to ovary, lobes 3 to many, small, imbricate. Petals
many, imbricate. Stamens many, free or adnate to the base of the
petals. Ovary 2-celled ; ovules many on parietal placentas, horizontal.
Berry 1-celled, placentas pulpy. Seeds numerous.

OPUNTIA, Mill.

Branches jointed, joints ovate flat, bearing tufts of unequal spines
and bristles. Flowers arising from the tufts or margins of the
joints, yellow or reddish. Calyx-tube not produced beyond the
ovary ; lobes numerous. Petals numerous, connate at the base
spreading. Stamens indefinite, in many series. Style cylindric,
thicker below, constricted at the base; stigma with 2-7 thick erect
branches. Fruit pyriform, fleshy, often with spines. Seeds with a
hard testa. Albumen scanty or copious.

Fl. Br. I. 2. 657 ; Brandis For. Fl. 245 ; Dalz. & Gibs. Bomb. Fl.
Suppl. 39. Prickly-pear. *Chappal, send,* Vern. Naturalized throughout
the presidency in dry waste places. Fl. throughout the year. Several
introduced species are included under the name *O. Dillenii,* which is
said to be rather rare in India.

Order 46. ARALIACEÆ.

Trees or shrubs, sometimes scandent. Leaves alternate, simple or compound long-petioled. Stipules adnate to the petiole, or 0. Flowers regular, sometimes polygamous in umbels, racemes or panicled heads ; pedicels continuous with the base of the calyx or there jointed ; bracts and bracteoles small or conspicuous. Calyx-tube adnate to ovary ; limb truncate, obsolete or with small teeth. Petals 5, rarely more, valvate, expanding or deciduous in a cap. Stamens as many as and alternate with the petals, inserted round an epigynous disc. Ovary inferior, 2-celled, or cells as many as the stamens ; styles distinct or united : ovules solitary and pendulous in each cell. Fruit small, seed pendulous, albumen uniform or ruminate.

HEPTAPLEURUM, Gærtn.

Trees or shrubs, sometimes climbing. Leaves digitate ; leaflets coriaceous, usually entire ; stipules connate within the petiole and prominent. Umbels panicled or in compound racemes ; bracts woolly ; pedicels not jointed : bracteoles few or 0. Calyx-margin toothed or truncate. Petals 5-6 or many, valvate. Stamens as many as petals. Ovary cells as many as petals, disc large or small ; styles small, separate or combined in a column. Fruit subglobose, 5-6-angled ; albumen uniform.

Leaflets with prominent nervures, on the upper surface ...*H. venulosum.*
Nervures not prominent above   ...   .*H. Wallichianum.*

H. venulosum. Seem. Rev. Heder. 44 ; Fl. Br. I. 2. 729 ; Brandis For. Fl. 249 ; Bedd. Fl. Sylv. 122. Common in the evergreen forests of the Konkan and North Kánara. Usually climbing scandent. Fl. March, June. Fr. June, Aug.

H. Wallichianum, C. B. Clarke. Fl. Br. I. 2. 730 ; Bedd. Fl. Sylv. 122 ; *Hedera Wallichiana,* Dalz. & Gibs. Bomb. Fl. 108. Konkan ghâts, pretty common, Dalz.

Order 47. CORNACEÆ.

Shrubs or trees. Leaves opposite or alternate, often unequal at the base ; stipules 0. Flowers regular, hermaphrodite or uni-sexual, in axillary or terminal cymes, panicles or heads. Calyx-tube adnate to ovary ; limb truncate or 4-5-toothed, persistent. Petals 0 or 4-5. Stamens inserted with the petals and equal to them in number. Ovary inferior, 1-4-celled ; crowned by the disc ; style single, stigma capitate or branched ; ovule solitary, pendulous. Fruit usually succulent, 1-4-celled, less often with 2 pyrenes. Seed oblong, pendulous, albumen copious, fleshy.

Petals long, narrow. Stamens 20-30   ... ALANGIUM.
Petals short. Stamens 5-4   ...   ... MASTIXIA.

ALANGIUM, Lamk.

Shrubs or small trees, armed or 0. Leaves alternate, 3-nerved
at the base, persistent. Flowers hermaphrodite, in axillary fascicles
or condensed cymes, jointed on the pedicel. Calyx-tube adnate to
ovary, limb 5-10 toothed or truncate. Petals 5-10, linear-oblong,
valvate, then reflexed. Stamens twice as many as petals or more.
Ovary inferior, 1-celled, surmounted by a disc; style long; stigma
large, capitate; ovule pendulous. Fruit a berry, crowned by the
enlarged calyx-limb. Seed oblong, albumen ruminate; cotyledons
leafy, crumpled.

A. **Lamarkii,** Thw. Enum. 133. Fl. Br. I. 2. 741; Dalz. & Gibs.
Bom. Fl 109; Brandis For. Fl. 250; Bedd. Fl. Sylv. 215. *Ankul,*
M.; *asroli,* K. Throughout the presidency in dry places, often along the
banks of nálás in North Kánara. Fl. Feb.-Apl. Fr. May-Aug.; sometimes at other seasons; erect or scandent.

MASTIXIA, Blume.

Trees. Leaves alternate or opposite, entire. Flowers hermaphrodite, 2-bracteolate, in terminal panicles; pedicels jointed under
the flower. Calyx-tube campanulate; limb 4-5-toothed. Petals
4-5, ovate, valvate, silky. Stamens 5-4. Ovary 1-celled; disc
fleshy; ovule 1 pendulous. Drupe ovoid, crowned by the calyx-
teeth; putamen grooved; endocarp protruded inwards down one
side. Seed ellipsoid; albumen fleshy.

M. **pentandra,** Blume Mus. Bot. I. 256: Fl. Br. I. 2. 746. *Bursi*
nopetalum arboreum, Dalz. & Gibs. Bomb. Fl. 28. Konkan and North
Kánara, in evergreen forests; also along nálás in North Kanara.
Dhárwár, Dalz. Fl. Feb.-Mar. Fr. May.

ORDER 48. CAPRIFOLIACEÆ.

Shrubs, small trees or herbs. Leaves opposite, simple or
pinnate. Flowers hermaphrodite, regular or irregular. Calyx-tube
adnate to the ovary, limb 3-5-divided. Corolla gamopetalous, limb
often 2-lipped, 5-lobed, lobes imbricate in bud. Stamens 5, on the
corolla. Ovary 1-6-celled, ovules solitary, pendulous, or several on
axile placentas. Fruit a drupe with 1-8, cartilaginous pyrenes, or a
many seeded berry. Albumen copious, fleshy; embryo minute.

VIBURNUM, Linn.

Shrubs or small trees. Leaves opposite. Flowers in terminal
panicles or corymbose cymes. Calyx-limb short, 5-toothed. Corolla 5-lobed. Stamens 5. Ovary 1-3-celled, with 1 pendulous
ovule in each cell. Fruit a dry or fleshy drupe, 1-seeded, 1-
celled.

V. **punctatum** Ham. in Don Prodr. 142; Fl. Br. I. 3. 5; Bedd. Fl.
Sylv. t. 217. Brandis says this is found in Kánara. (North Kánara ?)

Beddome says "common in subalpine jungles.". Var acuminata, Wall., is common on the Bababuden hills of Mysore at 4-6000 ft. alt. Fl. Sept., Oct. Fr. C.S.

ORDER 49. RUBIACEÆ.

Trees, shrubs or herbs. Leaves opposite or whorled, entire, stipulate. Inflorescence various. Calyx-tube adnate to the ovary, limb entire or toothed. Corolla gamopetalous, usually 4-5 lobed, lobes imbricate, contorted or valvate. Stamens as many as corolla lobes, alternating with them. Ovary inferior, 2 or more celled, rarely 1-celled. Fruit berried, capsular or drupaceous or of dehiscent or indehiscent cocci, 2-10-celled. Seeds with a fleshy or horny albumen.

Nauclea.—Flowers in globose beads. Corolla funnel-shaped. Stigma simple.

Ovaries confluent. Corolla lobes imbricate in bud...	ANTHOCEPHALUS.
Ovaries free or nearly so.	
Calyx 5-lobed. Corolla lobes valvate	ADINA.
Calyx-limb entire. Corolla lobes valvate	STEPHEGYNE.
Corolla lobes imbricate	NAUCLEA.

Cinchoneæ.—Corolla lobes valvate. Fruit capsular, 2-celled. Seeds winged. Bracts leafy ... ... | HYMENODICTYON.

Rondeletieæ.—Corolla lobes twisted in bud. Fruit capsular, 2-celled. Seeds angled, not winged ... ... | WENDLANDIA.

Mussændeæ.—Corolla lobes valvate in bud. Fruit indehiscent, fleshy. Calyx with a leafy lobe ... ... | MUSSÆNDA.

Gardenieæ.—Corolla lobes twisted in bud. Fruit berried or dry. Seeds large, cotyledons foliaceous.

Seeds many in each cell.

Inflorescence terminal	WEBERA.
Inflorescence axillary. Ovary 2-celled	RANDIA.
Inflorescence axillary. Ovary 1-celled	GARDENIA.
Ovules 2, in each cell...	DIPLOSPORA.

Vanguerieæ.—Corolla lobes valvate in bud. Stamens inserted in mouth of corolla. Drupe with 2 or many free or cohering pyrenes.

Ovary 2-celled	CANTHIUM.
Ovary 3-5-celled	VANGUERIA.

Ixoreæ.—Corolla lobes twisted in bud. Ovary 2-4-celled ; ovules inserted about the middle of the cell, rarely basilar. Fruit a 2-4-celled berry or drupe with 2-4 free or united pyrenes.

Flowers in trichotomous panicles.

Style short-exerted, 2-fid	IXORA.
Style long exerted, undivided	PAVETTA.
Flowers in axillary fascicles	COFFEA.

Morindeæ.—Corolla lobes valvate in bud. Stamens inserted in the mouth of the corolla. Ovules inserted below the middle of the cell and amphitropous. Fruit a 2-4 celled berry, pyrenes 2-4. Calyx-tubes cohering in a head. Fruit a fleshy syncarpium ... ... | MORINDA.

Psychotrieæ.—Corolla lobes valvate in bud. Stamens inserted near mouth of corolla. Ovules basilar, cuneate, anatropous. Drupe with 2 or more pyrenes.

Flowers cymose or cymose panicled.

Corolla-tube short, straight ...   . PSYCHOTRIA.
Corolla-tube long, curved   ...   . CHASALIA.
Flowers in axillary bracteate clusters   ...   . LASIANTHUS.
Flowers axillary and terminal, solitary or few   ...   ... SAPROSMA.
Pæderieæ.—Corolla-lobes valvate in bud.   Fruit capsular
    5-valved, or with 2 dorsally compressed pyrenes, often
    pendulous from a columella.   Albumen copious or
    scanty.
Ovary 5-celled, seeds 5.   Flowers in panicles   ...   ... HAMILTONIA.

ANTHOCEPHALUS, A. Rich.

A large tree.   Leaves opposite, coriaceous; interpetiolar stipules caducous.   Flowers in large globose heads.   Calyx-tubes confluent; limb 5-lobed.   Corolla-tube long, funnel-shaped.   Stamens 5, on throat of corolla.   Ovary 2-celled below, 4-celled above; ovules many on the 2 bifid placentas.   Fruit a fleshy mass of few seeded pyrenes.   Seeds minute, albumen fleshy.

A. Cadamba, Miq. Fl. Ind. Bat. II. 135 ; Fl. Br. I. 3. 23 ; Brandis For. Fl. 261 ; Bedd. Fl. Sylv. 127, t. 35. *Kaddam, nhyu,* Vern. North Kánara, in evergreen forests, rare.   Fl. Dec., Mch.   Fr. R.S.

ADINA, Salisb.

Trees.   Leaves opposite with large interpetiolar stipules.   Flowers bracteate in solitary or panicled globose heads.   Receptacle pilose, Calyx-tube angled, lobes 5.   Corolla-tube elongate, lobes 5. Stamens 5, on mouth of corolla, filaments short.   Capsule of 2 dehiscent cocci, many seeded.   Seeds with a winged testa.

A. cordifolia, Hook. f. in Gen. Plant. II. 30 ; Fl. Br. I. 3. 24 ; Brandis For. Fl. 263 ; *Nauclea cordifolia,* Bedd. Fl. Sylv. t. 33 ; Dalz. & Gibs. Bomb. Fl. 118. *Heddi, honangi, hedu,* M. ; *yettagal,* K.   Throughout the presidency in deciduous forests.   Yields a valuable timber.   Fl. June, July.   Fr. Dec., Mch.

STEPHEGYNE, Korth.

Shrubs or trees with petioled leaves; stipules large, caducous. Flower heads globose, solitary or paniculate often subumbelled, usually subtended by two large bracts.   Flowers surrounded with paleaceous bracteoles.   Calyx-tube short, limb truncate or 5-toothed. Corolla funnel-shaped, tube long, throat hairy.   Fruit of 2 dehiscent cocci, many seeded.   Seeds small, testa winged, albumen fleshy.

S. parvifolia, Korth. in Verh. Gesch. Nat. Bot. 161 ; Fl. Br. I. 3. 25 ; Brandis For. Fl. 262 ; *Nauclea parvifolia,* Bedd. Fl. Sylv. t. 34 ; Dalz. & Gibs. Bomb. Fl. 118. *Kalamb, kaddam,* M. ; *kadawar, kanu,* K. Throughout the presidency in deciduous forests.   Fl. May, July.   Fr. Nov., Dec.   Wood useful and ornamental.

NAUCLEA, L.

Trees or shrubs.   Leaves large, sessile or petioled ; stipules large, caducous, or subpersistent.   Flowers in globose, solitary or subpanicled heads, ebracteolate. Calyx-tube turbinate or obconic, lobes 5. Corolla-tube elongate, lobes imbricate. Stamens 5, on throat of corolla. Fruit of 2, dehiscent, many seeded cocci. Seeds imbricate, testa winged.

> Stigma capitate. Stipules flat, deciduous.
> Heads 1-3. Corolla-lobes spreading, not ridged...   .. ...*N. purpurea.*
> Heads solitary. Corolla-lobes ridged and horned near apex ...*N. elliptica.*
> Stigma elongate, acute. Stipules subpersistent..   ... ...*N. missionis.*

N. purpurea, Roxb. Fl. Ind. I. 515; Fl. Br. I. 3 26; Bedd. Fl. Sylv. 129. Konkan southwards. I. D. H. There seems to be very little difference between this and the next species, *N. elliptica*, Dalz. Good specimens sent by me of the true *N. elliptica*, Dalz., having been named *N. purpurea*, Roxb., at Kew. Both *N. purpurea* and *N. elliptica* are quite distinct from *N. missionis*, Wall., which has the stigma mitre-shaped and not globose, as is the case in the two former species.

N. elliptica, Dalz. & Gibs. Bomb. Fl. 118; Bedd. Fl. Sylv. 129. *Phuga*, M. On the Konkan and Supa ghâts of North Kánara, along streams and water-courses. Fl. Feb.-Mch. Fr. May-June.

N. missionis, Wall. Cat. 6099; Fl. Br. I. 3. 27. Konkan and North Kánara, along rivers and water-courses, common near Yellápur and in several other localities. Fl. Apl.-May. Fr. R. S. An ornamental tree when in flower.

HYMENODICTYON, Wall.

Trees or shrubs. Leaves simple, petioled. Stipules deciduous. Flowers small, paniculate with 1-2 large leafy bracts. Calyx-tube short, lobes 5-6. Corolla funnel-shaped, lobes 5, short. Stamens 5. Stigma spindle-shaped, ovules numerous. Fruit a dehiscent capsule. Seeds numerous, broadly winged.

> Flowers paniculate. Capsule deflexed...   ...   ...*H. excelsum.*
> Flowers spicate. Capsule erect...   ...   ...*H. obovatum.*

H. excelsum, Wall.; Roxb.; Fl. Ind. II. 149; Fl. Br. I. 3. 35; Dalz. & Gibs. Bomb. Fl. 117; Brandis For. Fl. 267; Bedd. Fl. Sylv. 130. *H. utile*, Bedd. Fl. Sylv. 130. *Kala kurwah*, Vern.; *dondru, dandelo, bhorsal*, M. Ghâts of the Deccan and Konkan, Panch Maháls. Fl. June. July. Fr. Oct.

H. obovatum, Wall., in Roxb. Fl. Ind. II. 153; Fl. Br. I. 3. 36; Dalz. & Gibs. Bomb. Fl. 117; Bedd. Fl. Sylv. 219; Brandis For. Fl. 268. *Kurwei, sirid*, Vern. Throughout the moist forests of the Konkan and North Kánara. Fl. Aug., Sept. Fr. Nov., Dec. There are two varieties of this species, one with large obovate pubescent leaves and capsules 1 in.; the other with small, ovate, nearly glabrous leaves and capsules, ½ inch.

WENDLANDIA, Bartl.

Shrubs or small trees. Leaves opposite or ternately whorled' F lowers in terminal panicled cymes, 2-3 bracteolate. Corolla-lobes imbricate in bud. Capsule small, dehiscent, many seeded. Seeds obscurely winged.

Leaves opposite ... ...*W. exserta.*
Leaves whorled in 3s ... ...*W. Notoniana.*

W. exserta, DC. Prod. IV. 411 ; Fl. Br. I. 3. 37 ; Brandis For. Fl. 268 ; Bedd. Fl. Sylv. 130. In the deciduous forests of North Deccan and Konkan. Fl. Mch.-Apl. Fr. May, June.

W. Notoniana, Wall. in W. & A. Prod. 403 ; Fl. Br. I. 3. 40 ; Dalz. & Gibs. 117 ; Bedd. Fl. Sylv. t. 224. *Showla,* M. In the moist forests of the Konkan and North Kánara, very common on the Supa gháts, growing on laterite, Fl. Feb. Fr. Apl.

MUSSÆNDA, L.

Shrubs. Leaves simple, sessile or petioled. Flowers yellow in terminal cymes. Calyx-lobes 5, 1 frequently developed into a white or coloured leaf. Corolla tubular with a villous throat ; lobes 5, valvate in bud. Stamens 5. Ovary 2-celled, ovules numerous on fleshy placentas. Berry fleshy, areolate at the top, many seeded. Seeds minute, testa pitted.

M. frondosa, Linn. DC. Prod. IV. 370 ; Fl. Br. I. 3. 89 ; Dalz. & Gibs. Bomb. Fl. 111. Bedd. Fl. Sylv. 131. *Bhútkes, lavasat, sherwod,* M. ; *bebana,* Vern. Throughout the moist forests of North Kánara and the Konkan where there is a heavy rainfall. Fl. Fr. Aug.-Sept.-Oct.

WEBERA, Schreb.

Trees or shrubs. Leaves entire glabrous, stipules deciduous. Flowers in terminal corymbose cymes. Corolla-tube cylindrical, lobes 5, spreading or reflexed, imbricate and usually twisted in bud. Ovary 2-celled, ovules 2 or more in each cell on fleshy peltate placentas. Berry globose 2-celled, cells 1 many seeded.

W. corymbosa, Willd.; Boxb. Fl. Ind. 1. 696 ; Fl. Br. I. 3. 102.

W. asiatica, Bedd. Fl. Sylv. 133 ; *Stylocoryne Webera,* A. Rich., Dalz. & Gibs. Bomb. Fl. 119. *Káré,* M. South Konkan and North Kánara, in moist forests on the gháts, common in the evergreen forests near the Ainshi ghát of North Kánara. Fl. at different times from December to July. Fr. R S.

RANDIA, L.

Small trees or shrubs, usually spinous. Leaves entire, stipules free or connate. Flowers large or small, in axillary cymes or clusters, white or yellowish. Calyx-tube various. Corolla from salver-shaped to nearly rotate, limb 5-lobed, lobes twisted in bud. Stamens 5, anthers subsessile. Ovary 2-celled, ovules many, attached to a fleshy peltate placenta. Berry globose, 2-celled, many seeded, usually crowned by the calyx-limb.

Spinous trees.
Flowers solitary, large, white. Berry ellipsoid, smooth, yellow, 2 in. long.

Spines short decussate, at the ends of branchlets   ...   .*R. uliginosa.*
Flowers small, white, yellow.   Fruit 1 in. long. Spines long,
    axillary ...   .*R. dumentorum.*
Climbing shrubs, unarmed.   Berry small   ...   . *R. rugulosa.*

R. uliginosa, DC. Prod. IV. 386; Fl. Br. I. 3. 110; Brandis For. Fl.
273; Dalz. & Gibs. Bomb. Fl. 119; Bedd. Fl. Sylv. 132. *Pandri,* K..;
kantha goting, Vern. Throughout the presidency, often in abandoned
rice fields, common in deciduous forests.   Fl. May-June. Fr. Dec.-Feb.

R. dumentorum, Lamk. Ill. t. 156, f. 4; Fl. Br. I. 3. 110; Dalz. &
Gibs. Bomb. Fl. 119; Brandis For. Fl. 273; Bedd. Fl. Sylv. 132;
R. longispina, DC. Dalz. & Gibs. Bomb. Fl. 119. *Ghela, peralu, mind-
hal, monigeli,* M.; *karigidda.* K. Common throughout the Presidency in
deciduous forests.   Fl. Mch.-**May.** Fr. Nov.-Mch.

R. rugulosa, Thw. Enum. 159; Fl. Br. I. 3. 113; Bedd. Fl. Sylv. 133.
Var. *speciosa,* Bedd. Fl. Sylv. 133. On the Konkan and North Kánara
ghâts, in evergreen forests, common in the forests near the Tinai ghát.
Fl. Jan., Feb. Fr. Nov., Dec.

GARDENIA, L.

Shrubs or trees, spinous or 0. Leaves simple, rarely ternately
whorled. Flowers often showy. Calyx-limb variously cleft or lobed.
Corolla-lobes 5-12, twisted in bud. Stamens as many as corolla-
lobes, not exserted, or more or less exserted. Ovary 1-celled,
ovules numerous, 2 seriate on 2-6 parietal placentas. Fruit fleshy,
indehiscent, seeds numerous, immersed in pulp.

Unarmed shrubs.
Flowers 5-6 merous. Calyx-lobes short, ovate, acute. Placentas 4-5 ...*G. gummifera.*
Calyx-lobes 5 long subulate Placentas 2.   ...   .*G. lucida.*
Small unarmed tree. Flowers 9-merous. Placentas 4-5   ...   ...*G. latifolia.*
Small spinous tree. Flowers dimorphic. Calyx with 5 short teeth
in male flowers ...   .*G. turgida.*

G. lucida, Roxb. Fl. Ind. 1. 707. Fl. Br. I. 3. 115; Dalz. & Gibs.
Bomb. Fl. 120; Bedd. Fl. Sylv. 134. Brandis For. Fl. 271. *Decamali,*
Vern. North Kánara, Konkan. Fl. Mch., June. Fr. C. S.

G. gummifera, Linn. f. DC. Prod. IV. 381; Fl. Br. I. 3. 116; Dalz. &
Gibs. Bomb. Fl. 120. Bedd. Fl. Sylv. 134-1. *Decamali,* Vern.. In the
moist forests of North Kánara and the Konkan, in open situations, common
near Siddápur. Fl. Mch., May. Fr. next Mch.-May.

G. latifolia, Ait. Hort. Kew I. 294; Fl. Br. I. 3. 116; Dalz. & Gibs.
Bomb. Fl. 120; Brandis For. Fl. 271; Bedd. Fl. Sylv. 134-1. *Pandru,
ghogari, papur,* M. Common in the dry forests of the presidency. Fl.
Apl., May. Fr. next cold season.

G. turgida, Roxb. Fl. Ind. 1. 711; Fl. Br. I. 3. 118; Brandis For. Fl.
270; Bedd. Fl. Sylv. 134-1. Var. *montana, Roxb.,* Dalz. & Gibs. Bomb.
Fl. 120. *Kurphondra,* M.; *pendra,* Vern. In dry forests throughout
the presidency, common in the Dhárwár and North Kánara jungles. Fl.
H.S. Fr. R & C.S. Foliage of young plants quite distinct from the
leaves of mature trees.

DIPLOSPORA, DC.

Shrubs or trees, evergreen. Leaves simple, stipules acuminate. Flowers in axillary fascicles or cymes, polygamo-diœcious, white; bracts free or connate in a cup. Corolla-tube short, lobes 4-5, spreading, twisted in bud. Ovary 2-3-celled; ovules 2-3 in each cell. Berry ovoid or globose. Seeds few imbricate.

Fruit sessile ellipsoid....   ...   . *D. apiocarpa.*
Fruit pedicelled globose   ...   . *D. sphærocarpa.*

D. apiocarpa, Dalz. in Hook. Kew Jour. II. 257 ; Fl. Br. I. 3. 123. *Dicospermum apiocarpum,* Dalz. & Gibs. Bomb. Fl. 120 ; Bedd. Fl. Sylv. t. 223. Evergreen forests on the higher ghâts of the Konkan and North Kánara. Fl. R.S.

D. sphærocarpa, Dalz. in Hook. Kew Jour. II. 257 ; Fl. Br. I. 3. 123 ; *Dicospermum sphærocarpa,* Dalz. & Gibs. Bomb. Fl. 120 ; Bedd. Fl. Sylv. 134. On the ghâts from the Konkan southwards. Fl. Oct. Fr. May.

CANTHIUM, Lam.

Unarmed or spinous shrubs. Leaves simple, stipules interpetiolar, pointed, with a broad base. Flowers small, axillary, fascicled or cymose. Corolla-tube with a ring of deflexed hairs within|; throat villous ; lobes valvate in bud. Ovary 2-celled, ovules 1 in each cell. Drupe didymous with 1-2 pyrenes. Seeds with fleshy albumen.

Unarmed tree ...   . *C. umbellatum.*
Spinous shrubs.
Scandent. Spines short, recurved. Leaves shin-
ing above   ...   . *C. Rheedii.*
Rigid shrub. Spines long, straight. Leaves not
shining   ...   . *C. parviflorum.*

C. umbellatum, Wgt. Ic. t. 1034 ; Fl. Br. I. 3. 132 ; Dalz. & Gibs. Bomb. Fl. 113 ; *C. didymum,* Bedd. Fl. Sylv. t. 221 ; *Plectronia didyma,* Brandis For. Fl. 276. *Tupo, arsul,* M.; *yellal,* K. Common in the evergreen forests of the Konkan and North Kanara. Fl. Nov.-Jany. Fr. Apl.-June.

C. Rheedii, DC. Prod. IV. 474 ; Fl. Br. I. 3. 134 ; Dalz. & Gibs. Bomb. Fl. 113 ; *Plectronia Rheedii,* DC. Bedd. Fl. Sylv. 134-5. *C. angustifolium.* Roxb. Fl. Br. I. 3, 135 ; *C. Leschenaultii,* W. & A. Dalz. & Gibs. Bomb. Fl. 113. *Chapyel,* M. Common in the evergreen forests of the Konkan and North Kánara. A scandent shrub. Fl. & Fr. at different times throughout the year. I am unacquainted with *C. angustifolium,* W. & A., if it is distinct from *C. Rheedii.* DC. Hooker in the Fl. Br. I. says *C. angustifolium* is probably only a glabrous variety of *C. Rheedii,* DC.

C. parviflorum, Lamk. Dict. I. 602 ; Fl. Br. I. 3. 136 ; Dalz. & Gibs. Bomb. Fl. 113 ; *Plectronia parviflora,* Roxb.; Bedd. Fl. Sylv. 134-5. *Kirna,* M. Very common in dry, open situations throughout the presidency, also along and near the coast; often semiscandent in hedges. Fl. Fr. at different times throughout the year. Fr. yellow edible.

VANGUERIA, Juss.

Small spinous tree. Leaves and flowers as in *Canthium*. Ovary usually 5-celled. Drupe large, globose, 1 inch in diameter, smooth; pyrenes, 4-5 woody, smooth.

V. spinosa, Roxb. Fl. Ind. 1. 537; Fl. Br. I. 3. 136. *Alu*, Vern. Common in the deciduous forests of North Kánara, usually a small tree. Fruit large, globose. Fl. Mch., Apl. Fr. R.S. Roxb. in the Fl. Ind. says the fruit of his species is small, size of a cherry. The fruit of the North Kánara tree is often more than 1 inch in diameter and is green when ripe. I have not observed it turning yellow.

IXORA, L.

Shrubs or small trees. Leaves with interpetiolar stipules. Flowers often showy in axillary and terminal corymbose cymes. Calyx 4-5-toothed or lobed. Corolla-tube long slender, lobes 4, spreading. Ovary 2-celled; style filiform, exserted; stigma slender, fusiform, or 2-lobed; 1 ovule in each cell. Fruit a small berry or drupe with 2 plano-convex coriaceous pyrenes. Flowers white or pink. Fruit dark-coloured. Leaves green when dry.

Shrubs.
Calyx teeth linear, much longer than ovary. Corolla-tube
   very slender. Fruit glabrous, calyx teeth persistent   ... *I. lanceolaria.*
Calyx teeth longer and broader than ovary. Fruit ¼ inch,
   hairy, striate, 1-2 seeded   ...   . *I. polyantha.*
Calyx teeth shorter than ovary. Fruit smooth, 2-lobed. ... *I. elongata.*
Trees, cymes brachiate.
Cyme branches long. Flower buds globose ...   . *I. brachiata.*
Cyme branches short. Flower buds ovoid ...   . *I. parviflora.*
Shrubs.
Flowers scarlet. Fruit red fleshy globose   ...   . *I. coccinea.*
Flowers white. Fruit black didymous, fleshy. Leaves
   turning black when dry   ...   . *I. nigricans.*

I. lanceolaria, Colebr. in Roxb. Fl. Ind. I. 387; Fl. Br. I. 3. 138; Common in the evergreen forests of North Kánara, particularly on the southern gháts. Fl. R. and C. S. Fr. Aug.

I. polyantha, Wgt. Ic. t. 1066; Fl. Br. I. 3. 140; Bedd. Fl. Sylv. 134-7. From the Konkan southwards, common on the southern gháts of North Kánara, in evergreen forests. Fruit ½ in. ovoid, striate. Sparsely hairy, crowned with the calyx-lobes; 1-2-seeded; pedicels ½ in. densely hairy, Fl. Mch. Fr. ripe Nov.

I. elongata, Heyne. in Wall. Cat. 6131; Fl. Br. I. 3. 141; *I. pedunculata,* Dalz. & Gibs. Bom. Fl. 113. Konkan and North Kánara, on the gháts. Fl. Feb. Fl. H S.

I. brachiata, Roxb. Fl. Ind. 1. 381; Fl. Br. I. 3. 142. From the Konkan southwards in evergreen forests, common on the North Kánara gháts. Fl. C. S. Fr. Apl., May.

I. parviflora, Vahl. Symb. III. 11, t. 52; Fl. Br. I. 3. 142; Bedd. Fl. Sylv. t. 222; Dalz. & Gibs. Bomb. Fl. 113; Brandis For. Fl. 275. The torch tree. *Kurat, kura,* M.; *heunnu, gorvi,* K.; *makrichijhar,* Hind. Throughout the presidency, in deciduous forests. Fl. Mch., Apl. Fr. May-June.

I. **coccinoa,** Linn. Roxb. Fl. Ind. 1. 375 ; Fl. Br. I. 3. 145 ; Dalz. & Gibs. Bomb. Fl. 112 ; Brandis For. Fl. 275 ; Bedd. Fl. Sylv. 134-7. *Bakora, pendgul,* M. Very common in the Konkan and North Kánara moist forests near the sea-coast, also on the ghâts along river banks. Fl. throughout the year. Fruit red ; seeds plano-convex ; more or less ventrally concave.

I. **nigricans,** Br. in Wall. Cat. 6154 ; Fl. Br. I. 3. 148 ; Dalz. & Gibs. Bomb. Fl. 113. *Lokhandi, katkura,* M. Common in the evergreen forests of North Kánara, also in the Konkan. Fl. apparently throughout the year. Fruit glabrous, size of a pea. Seeds 2, deeply ventrally concave.

PAVETTA, L.

Shrubs and small trees. Leaves usually membranous. Flowers in trichotomous corymbs. Corolla-tube slender, lobes 4-5. Ovary 2-celled, style long exserted ; stigma fusiform or 2-dentate ; ovule 1 in each cell. Fruit a 2-seeded berry. Albumen horny.

> Corolla white, tube ½ inch   ...   . *P. indica.*
> Corolla yellowish, tube more than 1 inch. Leaves
>   turning black when dry   ...   . *P. hispidula.*

P. **indica,** Linn.; DC. Prod. IV. 490 ; Fl. Br. I. 3. 150 ; Var. *indica proper;* leaves glabrous. Dalz. & Gibs Bomb. Fl. 112 ; Brandis For. Fl. 275. Bedd. Fl. Sylv. 134-7. Var. *tomentosa,* Roxb. Leaves and cymes tomentose or villous. Brandis For. Fl. 275 ; Bedd. Fl. Sylv. 134-7 ; *P. Brunonis,* Dalz. & Gibs. Bomb. Fl. 112. Common throughout the presidency. Fl. Mch., May. Fr. R.S.

P. **hispidula,** W. & A. Prod. 431 ; Fl. Br. I. 3. 151 ; *P. siphonantha,* Dalz. & Gibs. Bomb. Fl. 112. Bedd. Fl. Sylv. 134-8. In the evergreen forests of the Supa ghâts of North Kánara, rare. Fl. May. Fr. R.S.

COFFEA, L.

Shrubs. Leaves simple, stipules broad. Flowers in axillary fascicles or cymes. Corolla-lobes 4-7, spreading, twisted in bud. Ovary 2-celled, ovules 1, peltately attached, in each cell. Fruit with two plano-convex, ventrally grooved seeds. Albumen horny.

C. **arabica,** L. Fl. Br. I. 3. 153 ; Brandis For. Fl. 276. Bedd. Fl. Sylv. 13-48. Dalz. & Gibs. Bomb. Fl. Suppl. 44. Coffee. *Boon,* M.; *kawa,* Vern. Cultivated throughout the presidency. Fl. March. Fr. Oct., Jany.

MORINDA, L.

Shrubs or trees. Leaves usually membranous. Flowers sessile on a globose receptacle. Corolla tubular, lobes 4-7, valvate in bud. Stamens inserted in the mouth of the corolla, filaments short.

Ovary 4-celled, cells 1-ovuled.   Fruit a compound succulent berry containing a number of hard pyrenes, 2-4 from each flower.

Leaves pubescent, tomentose or glabrous, not shining   .. *M. citrifolia.*
Leaves glabrous, shining   ...   . *M. tinctoria.*

M. citrifolia, Linn. ; DC. Prod. IV. 446 ; Fl. Br. 1. 3. 155 ; Dalz. & Gibs. Bomb. Fl. 114; Brandis For. Fl. 277. Bedd. Fl. Sylv. t. 220. *Aavl, bartundi,* Vern. ; *aak,* M. ; *ainshi,* K. Var. 1, *citrifolia* proper. Cultivated.   Var. 2, *bracteata,* is common along the coast of the Konkan quite near the sea.   Fruit a white syncarpium.   Fl. and Fr. R.S. Var. 3, *elliptica.*   Konkan, Stocks.

M. tinctoria, Roxb. Fl. Br. I. 3. 156; Brandis For. Fl. 277 ; *M. citrifolia,* Bedd. Fl. Sylv. t. 220. *M. tomentosa,* Dalz. & Gibs. Bomb. Fl. 114. *M. multiflora,* Roxb., Brandis For. Fl. 227 ; *M. exserta,* Brandis For. Fl. 276. *Al, ack, alleri, alludi, ainshi,* Vern. Var. 1, *tinctoria* proper, is cultivated.   Var. 2, *tomentosa,* common in the deciduous forests of the presidency.   Fl. May.   Fruit ripe June, July.   Var. 4, *forma exserta,* Panch Maháls.   Brandis.

PSYCHOTRIA, L.

Shrubs or small trees. Leaves petioled, attenuated at the base. Stipules solitary or in pairs, often with glandular hairs. Flowers in terminal corymbose cymes, bracteate or not. Corolla-tube short, straight ; throat hairy or glabrous. Anthers included or 0. Ovary 2-celled, cells 1-ovuled. Fruit small, ovoid, globose or oblong, rarely didymous. Seeds plano-convex, ventrally flat or grooved Albumen hard, sometimes ruminate.

Seeds plano-convex, no ventral groove. Albumen
   ruminated.
Cyme branches whorled. Flowers bracteolate. Corolla-
   tube very short.
Flowers mixed with rufous hairs. Calyx-limb not en-
   larged in fruit. Stipules small ...   . *P. Thwaitesii.*
Flowers   without   rufous hairs. Cymes   congested,
   branches short. Stipules large...   ...   . *P. truncata.*
Cymes with long stout branches. Fruit crowned with
   the enlarged calyx lobes...   ...   . *P. Dalzellii.*
Cyme branches opposite, yellow in fruit   ...   ... *P. flavida.*
Seeds dorsally ridged, furrowed. Albumen equable   ... *P. canarensis.*

P. Thwaitesii Hook. f. Fl. Br. I. 3. 162.   On the southern gháts of North Kánara, in evergreen forests, common near the Nilkund and Gairsoppah gháts at about 2,000 feet elevation.   Fl. Mch., Apl. Fr. ripe Nov., Dec.   A large gregarious shrub.

P. truncata, Wall. in Roxb. Fl. Ind. II. 162 ; Fl. Br. I. 3. 163. *Grumilea vaginans,* DC.; Dalz. & Gibs. Bomb. Fl. 111.   In the ever-green forests of the Konkan and North Kánara gháts ; common in the forests near the Tinai ghát.   Fl. Jany. Fr. ripe Oct.

P. Dalzellii, Hook f. Fl. Br. I. 3. 163. Throughout the evergreen forests of North Kánara ; very common on the ghâts near Yellápur. Fl. June, July. Fr. Jany. Mch. Fruit black, succulent, crowned with the calyx-tube. Albumen ruminate.

P. flavida, sp. nov.—A small shrub. Leaves elliptic, glabrous, abruptly, shortly acuminate, coriaceous, narrowed into the petiole, 10·4 by 4·2 in., nerves about 12 pairs, much stronger than the veins, stipules ovate, acuminate, deciduous. Flowers numerous, very small in terminal, peduncled cymes ; peduncle 1-2 in. ; branches opposite, flattened, dichotomously divided, shorter than the peduncle. Bracts 2, leafy, deciduous. Calyx short, truncate. Corolla ½ in. ; tube very short ; throat villous, Fruit ¼ in., ellipsoid, compressed, crowned with the calyx, black, scarcely pulpy, 2-seeded ; seeds with a crustaceous testa, plano-convex with a narrow dorsal ridge, rugose, black ; albumen strongly ruminate. The cyme branches turn bright yellow in fruit. Abundant in the evergreen forests of the Sirsi and Siddápur talukas of North Kánara. Fl. June, July. Fr. Jany., Mch. The following about this species has just been received from Kew :—This is certainly *P. nudiflora*, W. & A. It is identical with the following specimens which have not been taken up in the Fl. Br. I. No. 735 A. Pl. Ind. Or. (Terr. Canara et confin.) R. F. Hohenacker. *Psychotria ambigua*, W. & A. Pt. 433—Miq.—North and South Konkan, Mr. Law.—No. 6, Gibson 20 Dec. 1845. From the general habit it comes very near *P. Johnsoni*, Hook. f. As it does not agree with the short description ·given in the Fl. Br. I., I have described and named it as above.

P. canarensis, sp. nov.—An erect shrub, branches smooth. Leaves glabrous, lanceolate, or ovate. acute, attenuate at the base, minutely punctate, very pale beneath ; petiole short or 0, blade 2-5 inches long by ½ inch to 1 inch broad ; stipules small, short acuminate, caducuons. Lateral nerves 8 pairs. Cymes terminal or axillary, few flowered, 2-3 inches, branches as long as peduncle, slender. Fruit ridged, sulcate, black, shining, crowned with the toothed calyx ½ in. long. by 2 lines broad at base, narrowed upwards. Seeds ridged along back, ventrally flat. Albumen equable. This shrub is found in the moist evergreen forests situated near the Falls of Gairsoppah in North Kánara. It is closely alied to both *P. bisulcata*, W. & A., and *P. filipes*, Hook. f. Flowers during the rainy season. Fruit ripe Dec., Jan. I have specimens, in fruit, of another, probably a new species from North Kánara. Leaves obovate, cuneate, acuminate, pale beneath, glandular in the axils of the nerves. Fruit dry, ½ in. 10-ribbed, 2-seeded.

CHASALIA, Comms.

Shrubs. Leaves membranous. Corolla-tube slender and curved. Fruit globose, didymous seeds orbicular, dorsally much compressed, ventrally deeply concave ; albumen equable.

C. curviflora, Thw. Enum. 150, 421 ; Fl. Br. I. 3. 176. *Psychotria longifolia*, Dalz. in Hook. Lond. Jour. Bot. II. 133. In the evergreen forests of North Kánara and the Konkan, common. Fl. May-Aug.

C. virgata sp. nov.—An erect, virgate shrub. Leaves ovate, lanceolate, acuminate scaberulous or glabrous shining, dark green above; paler beneath; blade narrowed into the petiole, membranous, 2-5 in. by ¾-2 in.; petiole 0-½ in.; nerves 7 pairs, strong beneath. Stipules connate, long acuminate. Inflorescence many flowered, in pedundled, panicled cymes; branches opposite or obscurely whorled; peduncle 1 in, or less; pedicels short, slender. Calyx 4-toothed. Corola tubular, ¼ in. long, 4-lobed, glabrous within. Seeds very concave. This shrub comes near *C. curviflora*, but it differs in the long pedicels and in the regular suppression of one of the ovary cells; it is found in the evergreen forests of North Kánara. Fl. Apl., May.

LASIANTHUS, Jack.

Shrubs or small trees. Leaves distichous. Stipules broad. Flowers in axillary bracteate clusters, cymes or heads. Calyx-tube short, limb 3-7-toothed. Corolla throat villous; lobes 3-7-valvate in bud. Stamens 4-6, filaments short. Ovary 4-9-celled with 1 erect, basal ovule in each cell. Drupe small, pyrenes triquetrous, 1-seeded.

L. sessilis. sp. nov.—Branchlets adpressed hairy. Leaves opposite, shortly petioled elliptic-acuminate, glabrous above, hairy on the nerves beneath; nerves 6 pairs, prominent beneath, strongly curved upwards, blade 2-5-in. long by 1-2-in. broad; petiole ¼-in. hairy. Stipules broad hairy, acuminate. Flowers small, white, axillary, sessile, fascicled, with triangular, hairy bracts. Calyx-tube short, teeth triangular, hairy outside. Corolla tubular, slightly hairy without, villous within. Stamens inserted in throat, nearly sessile. Ovary 4-celled, cells 1-ovuled; stigma 4-divided. Fruit a black fleshy drupe, containing 4 triangular pyrenes, dorsally tubercled. Leaves and fruit fœtid when bruised. A stout evergreen shrub nearly allied to *L. strigosus*, Wight, but differing in the calyx-teeth and nervation of leaves.

In the evergreen forests of North Kánara from Yellápur southwards to Gairsoppah. Fl. Aug.-Sept. Fr. Oct.

SAPROSMA, Blume.

Fœtid shrubs, often with bristles at the ends of the branches. Leaves opposite or whorled, membranous. Stipules connate, 1-3-cuspidate, deciduous. Flowers small, axillary or terminal. Calyx-tube obconic; limb 4-6-divided, persistent. Corolla with a villous throat; lobes 4-5, valvate. Stamens 4-5. Ovary 2-celled; style filiform with two short arms; ovules 1, erect, basal in each cell. Drupe with 1-2, crustaceous pyrenes.

S. indicum, Dalz. & Gibs. Bomb. Fl. 112; Fl. Br. I. 3. 192. In the evergreen forests of the Konkan and North Kánara. Common on the southern gháts of North Kánara. Fruit bright blue, fœtid. Seed solitary. Fl. C.S. Fr. ripe March.

HAMILTONIA, Roxb.

Shrubs. Leaves petioled, many-nerved. Flowers in large trichotomous panicles. Calyx-tube ovoid, limb 5-divided. Corolla funnel-
shaped; tube long, lobes 5, valvate. Stamens 5, inserted in the
throat of the corolla. Ovary 5-celled and furrowed. Capsule 1-
celled, 5-seeded. Seeds triquetrous, outer coat of testa reticulate.
Cotyledons foliaceous, cordate.

H. suaveolens, Roxb. Hort. Beng. 15; Fl. Ind. 1. 554; Fl. Br. I. 3.
197; Brandis For. Fl. 278; Bedd. Fl. Sylv. 134-12. *H. mysorensis*,
Dalz. & Gibs. Bomb. Fl. 115. *Gulesa*, Vern. On the higher ghâts of
the Konkan, also in North Kánara on the highest hills of the Supa
táluka. Fl. Oct., Feb.

ORDER 50. COMPOSITÆ.

Herbs or shrubs, rarely trees. Leaves alternate or opposite.
Stipules 0. Inflorescence capitate, bracteate, on the receptacle
which is either naked or with hairs, scales or bristles between the
flowers. Flowers all tubular (head discoid), or the outer or all
ligulate (head rayed), all 2-sexual, or the inner 2-sexual or male,
the outer female or neuter, sometimes diœcious. Calyx superior;
limb 0, or of hairs (pappus) or scales. Corolla of two forms: 1st,
tubular or campanulate, 4-5-lobed, lobes valvate; 2ndly, ligulate,
lobes elongate and connate. Disc epigynous. Stamens 4-5, filaments
usually free: anthers connate; connective produced; cells simple or
tailed at the base. Ovary 1-celled; style slender 2-fid; ovule solitary, basal, erect. Fruit a dry achene usually crowned with the
calyx (pappus); albumen 0.

VERNONIA, Schreb.

Herbs, shrubs, climbers or small trees. Leaves alternate. Heads
homogamous, solitary, cymose or panicled. Involucre equalling or
shorter than the flowers; bracts in many series. Corollas equal,
tubular, slender. Anther bases obtuse. Style-arms subulate.
Achenes usually ribbed or angled; pappus of many hairs, often with
an outer row of flattened shorter hairs.

V. arborea, Ham. in Trans. Linn. Soc. Fl. Br. I. 3. 239. *V. javanica*
DC.; Bedd. Fl. Sylv. 135. On the ghâts from the Konkan southwards.
Fl. Br. I. I have not seen this tree in the North Kánara or Konkan
forests; it is cultivated in gardens near Bombay. Graham. *V. divergens*,
Benth. (*bundar*, M.) and *V. indica*, C.B.C., are common undershrubs in
the Konkan and North Kánara moist forests. Fl. Fr. C.S.

ORDER 51. GOODENOVIEÆ.

Shrubs or herbs. Leaves alternate; stipules 0. Flowers axillary
or terminal, racemed or panicled, irregular or regular. Calyx-tube

adnate to the ovary ; limb 5-fid or nearly obsolete. Corolla-lobes 5,.
valvate in bud. Stamens 5 ; anthers free or connate in a ring round
the style. Ovary 1-2-celled. Style simple, with a cup-shaped
indusium, including the stigma. Ovules in each cell one or many,.
on the dissepiment. Fruit a drupe or capsule. Seeds-albuminous..

SCÆVOLA, Linn.

Shrubs. Leaves entire or toothed. Flowers axillary, in short
cymes or solitary. Corolla oblique, split at the base behind. Anthers free. Ovary 1-2-celled, with two erect ovules. Fruit a drupe.

Calyx-lobes linear-lanceolate, enlarged in fruit   ... *S. Koenigii.*
Calyx-lobes very short, obtuse or subobsolete·   .. *S. Lobelia.*

S. Koenigii, Vahl. Symb. III. 36 ; Fl. Br. I. 3.421 ; *S. Taccada,* Dalz.
& Gibs. Bomb. Fl 134. *Bhadrak,* M. On the sea-coast of the Konkan
near Ratnágiri, cultivated in gardens, Bombay.

S. Lobelia, Linn.; Fl. Br. I. 3. 421. *S. uvifera,* Stocks. Wight Ic..
1613. Sea-coast of Sind and South Deccan. Fl. Br. 1. Mouths of the
Indus near Karáchi.

Order 52. MYRSINEÆ.

Trees or shrubs. Leaves alternate, entire or serrate. Flowers·
small, regular, in axillary clusters, racemes or panicles, rarely termi-.
nal. Calyx free (in *Mæsa* adnate to the ovary) ;. limb 4-6-lobed,
lobes persistent, often enlarged in fruit. Corolla regular, tube
short or 0, lobes 3-7, contorted or imbricate. Stamens epipetalous,
as many as divisions of the corolla and opposite to them. Ovary.
1-celled ; ovules few or many, inserted on the free central placenta..
Fruit an indehiscent berry or drupe (dehiscent in *Ægiceras*), 1-
several seeded. Seeds often with more than one embryo ; albumen.
pitted or ruminate.

Calyx adnate to ovary,.2-bracteolate. Berry many seeded. MÆSA.
Calyx free. Fruit 1-seeded, indehiscent.
Petals free. Flowers racemed or panicled ...   . EMBELIA.
Corolla with a short tube; limb deeply lobed.
Flowers in dense axillary fascicles. Corolla imbricate   ... MYRSINE.
Flowers in umbels, corymbs or panicles. Corolla lobes
    contorted in bud   ...   . ARDISIA.
Fruit cylindric, curved, dehiscing longitudinally ...   ... ÆGICERAS.

MÆSA, Forsk.

Trees or shrubs. Leaves entire or serrate. Flowers small,.
2-bracteolate, 1-2-sexual, 4-5-merous. Calyx ½-inferior, in fruit
½-adnate; teeth small. Corolla campanulate; lobes imbricate.
Stamens 5. Ovary with numerous ovules on a globose placenta.
Berry globose.

Leaves and branchlets glabrous ...   . *M. indica.*
Leaves and branchlets rusty tomentose ...   . *M. dubia.*

M. indica, Roxb. F'. Ind. 1. 558 (Bæobotrys) ; Fl. Br. I. 3. 509 ; Dalz. & Gibs. Bomb. Fl. 136 ; Brandis For. Fl. 283 ; Bedd. Fl. Sylv. 137. *Atki*, Vern. In the moist forests of the Konkan and North Kánara along the gháts, abundant in the evergreen forests of the Sirsi and Siddápur tálukas of North Kánara. Fl. at different times throughout the year.

M. dubia. Wall. in Roxb. Fl. Ind. ed. Carey & Wall. II. 235 ; *M. indica*, var., Bedd. Fl. Sylv. 137, t. XVIII. 4, fig. 1 only. Throughout the evergreen forests of the Konkan and North Kánara. Fl. C. S. Fr. ripe Mch. Apl.

MYRSINE, Linn.

Trees or shrubs. Leaves coriaceous. Flowers polygamous, in dense clusters, from the axils of the fallen leaves. Calyx 4-5-lobed. Corolla deeply 4-5-lobed. Stamens 4-5. Ovary free, globose. Fruit small, globose, red or purple. Seed solitary, albumen pitted.

M. capitellata, Wall. in Roxb. Fl. Ind. ed. Carey & Wall. II. 295 ; Fl. Br. I. 3. 512 ; Brandis For. Fl. 286; Bedd. Fl. Sylv. t. 234. In the evergreen forests of North Kánara from Ainshi southwards, not common. Fl. Oct., Nov. Fr., Feb., Mch.

EMBELIA, Burm.

Shrubs or woody climbers. Leaves entire or toothed, petiole often glandular. Flowers small, in simple or branched racemes. Calyx free, deeply 5-lobed Petals usually 5, distinct, spreading. Stamens 5, filaments more or less adnate to the petals. Ovary superior. Fruit small, globose, 1-2-seeded, albumen pitted.

Inflorescence in terminal panicles. Lateral nerves of the leaves indistinct	*E. Ribes.*
Inflorescence in axillary racemes. Lateral nerves distinct	*E. robusta.*

E. Rib es, Burm. Fl. Ind. 62. t. 23 ; Fl. Br. I. 3. 513 ; Dalz. & Gibs. Bomb. Fl. 137 ; Brandis For. Fl. 284 *E. glandulifera*, Dalz. & Gibs. Bomb. Fl. 137. *Waiwarung, kárkannie*, Vern. Throughout the Konkan and North Kánara ; it is rare in the latter district and only found by me in the evergreen forests, near the Falls of Gairsoppah. Fl. C.S. Fr. May. Scandent.

E. robusta, Roxb. Fl. Ind. 1. 587 ; Fl. Br. I. 3. 515 ; Bedd. Fl. Sylv. 137 ; Brandis For. Fl. 284 ; *E. Basaal*, Dalz. & Gibs. Bomb. Fl. 136. *Kokla, carbati*, M.; *byebering*, Vern.; *ambati*, M. Throughout the presidency, common in the moist forests of the Konkan and North Kánara ; in the latter district it has often a scandent habit. Fl. Fr. Apl., June.

ARDISIA, Swartz.

Shrubs or small trees. Leaves coriaceous. Flowers small, hermaphrodite, in compound or simple umbels or racemes. Calyx 4-5-lobed. Corolla deeply 4-5-lobed. Stamens 4-5. Ovary free, globose. Fruit globose. Seed solitary, albumen pitted.

A. humilis, Vahl. Symb. III. 40; Fl. Br. I. III. 529; Dalz. & Gibs.
Bomb. Fl. 137; Brandis For Fl. 287. *Dikna,* Vern. In the evergreen
forests, throughout the Konkan and North Kánara, usually along the
banks of nálás and streams, common. Fl. Apl., May. Fr. ripe Aug.
There is a variety of *A. humilis,* a shrub with crenate, pellucid-punctato
leaves and small, white flowers in simple umbels, common near Kárwár in
North Kánara. This was declared to be "*exactly like A. polycephala,*
Wall., in Hb. Wight," at Kew. Fl. July, Aug. Fruit ripe Jany.

ÆGICERAS, Gærtn.

A shrub or small tree. Leaves coriaceous, 1-nerved. Flowers
in sessile umbels. Flowers hermaphrodite, white. Calyx-lobes 5,
imbricate. Corolla segments 5, acute, twisted to the right in bud.
Stamens 5. Filaments hairy at the base. Ovary oblong, ovules
many. Fruit cylindric, curved 1-seeded, dehiscent.

Æ. majus, Gærtn. Fruct. 1. 216, t. 46, fig. 1; Fl. Br. I. 3. 533; Dalz
& Gibs. Bomb. Fl. 137; Bedd. Fl. Sylv. 139. *Kanjala,* M. Common
near the sea-coast on marshy ground and along creeks. Fl. C.S.
Fr. R.S.

ORDER 53. SAPOTACEÆ.

Trees or shrubs. Leaves alternate, simple. Stipules 0. Flowers
axillary, solitary or clustered. Bracts small or 0. Calyx-lobes
4-8, 2-3-seriate, the outer series valvate, when 4, decussate, imbricate, when 5; 2-3 exterior, imbricate. Corolla-lobes as many or
twice as many as calyx-segments, imbricate in bud. Stamens on
the corolla-tube as many as the corolla-lobes and opposite to them
or 2-3-times as many, 1-3-seriate, filaments short, connective often
produced, staminodes when present, alternating with the stamens.
Ovary superior, 2-8-celled, cells 1-ovuled; ovule usually attached
to the inner angle. Fruit a drupe or berry, usually indehiscent,
1-8-seeded. Seeds ellipsoid, testa bony or crustaceous. Albumen
fleshy, or oily, or 0.

Calyx-lobes in one series, imbricated.	
Staminodes 0.	CHRYSOPHYLLUM.
Staminodes present	SIDEROXYLON.
Calyx-lobes 2-seriate, with the outer series val-	
vate. Calyx-segments 4.	
Stamens 8	ISONANDRA.
Stamens 12-10. Corolla deeply lobed	BASSIA.
Calyx-segments 6-8.	
Staminodes 0.	DICHOPSIS.
Staminodes present	MIMUSOPS.

CHRYSOPHYLLUM, Linn.

Trees. Leaves coriaceous, exstipulate. Flowers in axillary
fascicles. Calyx-lobes 5-6, entire, imbricate. Corolla-lobes 5-6,
Stamens 5-6; staminodes 0. Ovary 5-6-celled, villous. Berryfleshy
globose. Seeds 5-6.

C. Roxburghii, G. Don Gen. Syst. IV. 33 ; Fl. Br. I. 3. 535 ; Dalz.
& Gibs. Bomb. Fl. 139 ; Bedd. Fl. Sylv. t. 236. *Hale*, K. ; *tursiphal, dongrimaphal*, Vern. Star apple. In the evergreen forests of the
Konkan and North Kánara, common. Fl. Apl., May. Fr. Nov., Dec.

SIDEROXYLON, Linn.

Trees. Leaves alternate, exstipulate. Flowers small, in axillary
fascicles, shortly pedicelled ; pedicels pubescent or tomentose.
Calyx-segments 5. Corolla-tube campanulate, lobes 5. Stamens 5,
staminodes 5, lanceolate. Ovary villous, usually 5-celled ; style
cylindric. Berry 1-5-seeded. Seeds albuminous.

S. tomentosum, Roxb. Fl. Ind. I. 602. Fl. Br. 1. 3. 538. *Sapota
tomentosa*, Dalz. & Gibs. Bomb. Fl. 139 ; *Achras elengoides*, Bedd. Fl.
Sylv. t. 235. *Kumpoli*, K. ; *kumbul, kantakumla*, Vern. Common on the
evergreen forests of the North Kánara and Konkan gháts. Fl. C. S.
Fr. Oct., Jany.

ISONANDRA, Wight.

Trees. Leaves alternate, coriaceous, glabrous. Flowers small,
villous, in axillary clusters, shortly pedicelled. Calyx-segments 4.
Corolla deeply 4-lobed, longer than calyx. Stamens 8, all perfect.
staminodes 0. Ovary villous, 4-celled ; style linear. Berry 1-seeded,
flattened, smooth ; albumen fleshy.

I. Stocksii, Clarke ; Fl. Br. I. 3. 539. Konkan, Stocks.

DICHOPSIS, Thw.

Trees ; shoots rusty-tomentose. Leaves petioled, coriaceous,
Flowers fascicled, axillary. Calyx-lobes 6, 2-seriate, outer 3, valvate
inner 3, imbricate. Corolla-lobes 6. Stamens 12-18, 1-seriate or the
alternate a little higher up ; anthers lanceolate, connective produced, acute or bifid ; staminodes 0. Ovary villous, usually 6-celled.
Berry 1-2-seeded. Seed exalbuminous.

D. elliptica, Benth. in Gen. Plant. II. 658 ; Fl Br. I. 3. 542 ; *Bassia
elliptica*, Dalz. & Gibs. Bomb. Fl. 139 ; Bedd. Fl. Sylv. t. 43. Indian
gutta-percha tree. *Panchoti, palla*, M. In the evergreen forests of North
Kánara, Fl. H. S.

BASSIA, Linn.

Trees. Leaves coriaceous ; stipules caducous. Flowers on axillary,
generally fasciculate pedicels. Calyx deeply 4-lobed, lobes biseriate,
outer subvalvate, inner subimbricate in bud. Corolla-tube fleshy or
not ; limb of 5-14 divisions. Stamens numerous, fertile, in 1-3
series ; anthers erect, cordate, cuspidate or aristate, 2-celled. Ovary
villous, 4-12-celled. Berry globose or ellipsoid, 1-3-seeded ; albumen 0.

Leaves elliptic. Anthers 20-30, 3-seriate, hairy
acuminate... ... ... *B. latifolia.*
Leaves lanceolate. Anthers 16, 2-seriate ; tips 3-
toothed. Young fruit densely hirsute ... ... *B. longifolia.*
Leaves oblong-obtuse. Stamens 16, 2-seriate ; connective lanceolate-linear. Young fruit glabrous ... *B. malabarica.*

B. latifolia, Roxb. Fl. Ind. II. 526; Fl. Br. I. 3. 544; Brandis For Fl. 289; Bedd. Fl. Sylv. t. 41; Dalz. & Gibs. Bomb. Fl. 139. *Mohwa*, M. ; *ippi*, K. Common in the dry forests throughout the presidency. also in the Konkan and North Kánara forests, but nowhere abundent in them. Fl. Mch., Apl. Fr. June, July.

B. longifolia, Linn. Mant. 563; Fl. Br. I. 3. 544; Dalz. & Gibs. Bomb. Fl. 139; Bedd. Fl. Sylv. t. 42; Brandis For. Fl. 290. *Ippi*, K. ; *moha*, M. Common in the moist forests of the Konkan and North Kánara ; often along the banks of rivers and nálás; takes the place of *B. latifolia* in the moist forests of the southern parts of the presidency. Fl. Nov. Dec. Fr. ripe May June.

B. malabarica, Bedd. Fl. Sylv. 140; Fl. Br. I. 3. 544. Common in the southern parts of North Kánara along the banks of rivers and in moist forests at or near the sea-level. Fl. Nov.-Dec. Fr. ripe May. June.

MIMUSOPS, Linn.

Trees. Leaves coriaceous ; nervures numerous, parallel. Flowers axillary, clustered or solitary. Calyx-segments 6-10, in 2 series, Corolla-tube short; lobes 18-24, 2-3-seriate. Stamens 6-8; connective excurrent ; staminodes as many as the stamens, bifid or lacinate. Ovary 6-8-celled. Seeds compressed, albuminous.

Stamens 8, staminodes, lanceolate, hairy   ...   ... *M. Elengi.*
Stamens 6, staminodes bifid, glabrous   ...   . *M. hexandra.*

M. Elengi, Linn. ; Roxb. Fl. Ind. II. 236 ; Fl. Br. I. 3. 548 ; Dalz. & Gibs. Bomb. Fl. 140; Bedd. Fl. Sylv. t. 40. Brandis For. Fl. 293. *Buckhul*, K. ; *owli*, *rovali*, M. Common in the evergreen forests of North Kánara, also in the Konkan, a very large tree. Fl. Jan.-Mch. Fr. ripe R.S.

M. hexandra, Roxb. Fl. Ind. II. 238 ; Dalz. & Gibs. Bomb. Fl. 140; Bedd. Fl. Sylv. 142; *M. indica*, Brandis For. Fl. 291. *Khirni*, H. ; *ranjana*, *ráini*, M. In the dry forests of the Deccan, Khándesh and Gujárát, not observed by me in North Kánara or the Konkan. Fl. Nov.-Dec. Fr. ripe March.

ORDER 54. EBENACEÆ.

Trees or shrubs. Leaves alternate, entire. Flowers usually diœcious, regular, axillary ; pedicels articulated under the flower. Calyx inferior, gamosepalous, often accrescent in fruit. Corolla gamopetalous. Stamens as many or 2-3-times as many as the corolla-lobes; staminodes in the female flower resembling stamens or 0. Ovary superior ; styles 2-8; cells as many or twice as many as the styles, imperfectly septate ; ovules usually 1 in each cell; attached to the inner angle of the cells, pendulous. Fruit coriaceous or fleshy, indehiscent, several or few seeded. Seeds pendulous, longitudinally furrowed, albumen equable or ruminated.

Flowers 3-merous. Ovary 3-6-celled ...   . MABA.
Flowers 4-5-merous. Ovary 4-5 or 8-10-celled   ... DIOSPYROS.

●

MABA, J. R. & G. Forst.

Trees or shrubs. Leaves alternate. Flowers diœcious, axillary
Calyx 3-partite or 3-fid, enlarged in fruit. Corolla-tube longer than
calyx, lobes 3, twisted to the right. Male fl.: stamens 3-22; filaments distinct or united; ovary rudimentary. Female fl: staminodes 0-12; ovary 3-celled; ovules 6. Fruit 1-6-celled, 1-6-seeded.
Albumen not ruminated.

Ovary densely hairy   ... ... *M. nigrescens.*
Ovary glabrous   ... ... *M. micrantha.*

M. nigrescons, Dalz. & Gibs. Bomb. Fl. 142; Fl. Br. I. 3. 551.
Raktarohida, raktarora, M. In the evergreen forests of the Konкan
and North Kánara, common on the North Kánara ghats near Nilkund
and Gairsoppah. Fl. Dec., Feb. Fr. Apl.-May.

M. micrantha, Hiern. in Trans. Camb. Phil. Soc. XII. 133; Fl. Br. I.
3. 552; *Holochilus micranthus*, Dalz. & Gibs. Bomb. Fl. 142; Bedd. Fl.
Sylv. 147. On the southern ghats. Fl. Feb., Mch. Dalz.

DIOSPYROS, L.

Trees. Leaves entire, alternate. Flowers diœcious, axillary,
usually 4-5-merous. Calyx lobed, lobes often accrescent in fruit.
Corolla tubular or urceolate, lobed, lobes twisted to the right. Male
fl.: stamens 4-64 often 16; ovary rudimentary. Female fl.: staminodes 0-16: ovary 4-5-celled, cells 1-ovuled; stigmas 1-4. Fruit
1-10-seeded. Albumen equable or ruminated.

Corolla tubular in bud.
Corolla yellow, fulvous-tomentose without. Sta-
  mens 16. Fruit clothed with stinging hairs.
  Albumen equable...   ...   . *D. pruriens.*
Corolla glabrous without.
Stamens 24-32; filaments glabrous   ...   .*D. assimilis.*
Corolla white-tomentose without. Stamens 9-12.
  Calyx truncate. Albumen equable   ...   .*D. oocarpa.*
Corolla yellow-woolly without. Calyx triangular-
  toothed or lobed. Stamens 12. Albumen ru-
  minated   ...   .*D. Tupru.*
Corolla yellow-lanate without. Calyx ovate-tooth-
  ed. Stamens 12-16 Albumen ruminated   ...   ...*D. melanoxylon.*
Corolla yellow-tomentose without. Male flowers
  fascicled; calyx deeply lobed. Albumen rumi-
  nated   ...   .*D. Candolleana.*
Corolla black-velvetty without. Calyx-lobes foli-
  aceous, auricled in fruit. Stamens 20. Albumen
  equable   ...   .*D. paniculata.*
Corolla urceolate.
Tree often spinous. Corolla glabrous without.
  Stamens 16, glabrous. Albumen equable   ...   ...*D. montana.*
Corolla-lobes tomentose without. Stamens 16,
  pilose. Albumen equable   ...   .*D. Kaki.*
Corolla nearly glabrous without. Stamens 40,
  pilose. Albumen equable...   ...   .*D. Embryopteris.*
Corolla glabrate without. Stamens 13-22, glab-
  rous or sparsely pilose. Albumen ruminated ...   ...*D. sylvatica.*

Corolla very small, lobes pilose near the margin.
Stamens 16.   Albumen equable ...   .*D. microphylla.*
Corolla glabrous without. Stamens 16, glabrous.
Fruit small, globose. Albumen uniform   ...   ...*D. Chloroxylon.*

D. pruriens, Dalz. in Hook. Kew Jonr. IV. 110; Fl. Br. I. 3. 553; Dalz. & Gibs. Bomb. Fl. 141; Bedd. Fl. Sylv. 144. In the evergreen forests of the southern gháts of North Kánara, common. Fl. Nov., Jan. Fr. R. S.

D. montana, Roxb. Fl. Ind. II. 538; Dalz. & Gibs. Bomb. Fl. 142. Brandis For. Fl. 296; Bedd. Fl. Sylv. 143; *D. cordifolia,* Bedd. Fl. Sylv. 143; *D. Goinda.* Dalz. & Gibs. Bomb. Fl. 141. *Tendu,* K, ; *tembhurni. govindu, lohari,* M. ; *hadru,* Vern. Throughont the deciduous forests of the presidency, common on the North Kánara gháts. Fl. Mch., Apl. Fr. ripe R. S.   Calyx 4-lobed enlarged, foliaceous in fruit.

D. Kaki, Linn. f. Suppl. 434; Fl. Br. I. 3. 555; Grah. Cat. Bomb. Pl. 107. Cultivated near Bombay. Fruit yellow, globose, edible.

D. Embryopteris, Pers. Syn. II. 624; Fl. Br. I. 3. 556; Brandis For. Fl. 298; Bedd. Fl. Sylv. t. 69. *Timburi,* M. Near creeks and back-waters along the coast of North Kánara and the Konkan, also in the ever-green forests of the gháts. Fl. Mch., May. Fr. Dec. Fruit globose, with decidnous, rusty tomentum seeds 8, in viscid pulp. Calyx-lobes large accrescent.

D. assimilis, Bedd. in Mad. For. Rep. 1866-67, 20, t. 1; Fl. Br. I. 3. 558; *D. nigricans,* Dalz. & Gibs. Bomb. Fl. 141. *Kare,* K. ; *abnus, malia,* M. In the evergreen forests of the Konkan and North Kánara. Fl. Feb., Mch. This tree yields the ebony used for wood carving in Kumta and Honávar.

D. sylvatica, Roxb. Fl. Ind. II. 537; Fl. Br. I. 3. 559; Bedd. Fl, Sylv. 143. In the evergreen forests of the Konkan and North Kánara. common in the forests near Yellápur. Fl. Jany., Feb. Fr. ripe Sept., Oct. Fruit globose ¾ in. in diameter. Fruiting calyx with 4 spreading lobes, small tree.

D. microphylla, Bedd. Fl. Sylv. 145; Fl. Br. I. 3. 559. In the evergreen forests of North Kánara; common on the Yellápur gháts. Large white barked tree. Leaves resembling those of the boxwood, wood white infirm. Fl. H. S. Fr. C. S.

D. Chloroxylon, Roxb. Fl. Ind. II. 538; Fl. Br. I. 3. 560; Dalz. & Gibs. Bomb. Fl. 140; Brandis For. Fl. 297; Bedd. Fl. Sylv. 143. *Nensi,* Vern (Surat, Násik).; *ninai,* M. In the dry deciduous forests of the Deccan, not observed in the Konkan or North Kánara. Fl. June. Fr. Jan.-Feb. Fruit small, globose, ½ in. in diameter, a small tree sometimes spinescent.

D. oocarpa, Thwaites Enum. 180; Fl. Br. I. 3. 560. In the evergreen forests of North Kánara and Konkan; common on the gháts from Kárwár to Ainshi. Fl. Feb., Mch. Fr. ripe June. Calyx-tube nearly truncate.

D. Tupru, Buch.-Ham. Jour. 1. 183 ; Fl. Br. I. 3. 563 ; *D. exsculpta*, Dalz. & Gibs. Bomb. Fl. 142 (syn. excl.) *Turtar*, M. From the Konkan to Mysore. Fl. Br. 1.

D. melanoxylon, Roxb. Fl. Ind. II. 530 ; Fl. Br. I. 3. 564; Brandis For. Fl. 294 ; Bedd. Fl. Sylv. t. 67 ; *D. exsculpta*, Bedd. Fl. Sylv. t. 66, not of Ham. *Balai*, K. ; *timburni, tendu*, M. ; *temru*, Vern. Common in the dry deciduous forests of the presidency. Fl. May. Fr. ripe December. Fruit when quite ripe with an edible pulp. This is no doubt *D. exsculpta*, Ham., described in Dalz. & Gibs. Bomb. Fl. 142. I have, however, included *D. Tupru*, Buch.-Ham., of which *D. exsculpta*, Ham., is made a synonym, by Clarke, in this list, and which I am unacquainted with if different from *D. melanoxylon*. Roxb. Brandis unites *D. Tupru*, Buchanan. *D. exsculpta*, Ham., and *D. melanoxylon*. Roxb. *Vide* For. Flora p. 295.

D. Candolleana, Wight Ic. tt. 1221-2 ; Fl. Br. I. 3. 566 ; Dalz. & Gibs. Bomb. Fl. 132 ; Bedd. Fl. Sylv. 144 ; *D. Canarica*, Bedd. Fl. Sylv. 145. Throughout the Konkan and North Kánara, in evergreen forests, common near the coast at Kárwár. Fl. Apl., June. Fruit ripe Nov. with calyx-lobes reflexed, coriaceous.

D. paniculata, Dalz. & Gibs. Bomb. Fl. 141 ; Fl. Br. 1. 3. 570 ; Bedd. Fl. Sylv. 144. In the evergreen forests of the Konkan and North Kánara, frequent in the forests near the Nilkund and Gairsoppah gháts. Fl. Nov.-Dec. Fr. Apl.-June. Seeds ovoid ; albumen equable, horny : cotyledons foliaceous. A large tree.

ORDER 55. **STYRACEÆ.**

Trees or shrubs. Leaves alternate. Flowers hermaphrodite, regular. Calyx-tube superior or inferior ; limb 4-5-toothed or truncate, persistent. Petals 4-5, free or united into a short tube. Stamens attached to the corolla-tube, 8-10 or many. Ovary inferior or superior, 2-5-celled with two or more ovules in each cell, pendulous or erect. Fruit drupaceous, usually 1-seeded. Seeds albuminous ; embryo straight or curved.

SYMPLOCOS, Linn.

Trees or shrubs. Leaves alternate, toothed or entire ; often drying yellow. Flowers in simple or compound racemes or spikes, usually yellow or white. Calyx 5-lobed, lobes imbricate. Petals 5, imbricate free or connate. Stamens numerous in several series adnate to the corolla-tube. Ovary inferior 2-4-celled ; ovules 2 in each cell, pendulous. Fruit a berry, crowned with the calyx-lobes, 1-3-seeded, seeds albuminous.

> Fruit urceolate, seeds fluted. Embryo curved   ...*S. spicata.*
> Fruit ovoid, smooth, slightly curved. Embryo straight...*S. Beddomei.*

S. spicata, Roxb. Fl. Ind. II. 541 ; Fl. Br. I. 3. 573 ; Brandis For. Fl. 300 ; Bedd. Fl. Sylv. 149 ; *Hopea spicata*, Dalz. & Gibs. Bomb. Fl.

140. In the evergreen forests of the Konkan and North Kánara ghâts; common. Fl. Aug.-Dec. Fr. Mch.-Apl. Corolla and attached stamens caducous possess a sweet scent.

S. Beddomei, C. B. C. Fl. Br. I. 3. 582; *Hopea racemosa*, Dalz. & Gibs. Bomb. Fl. 140. *Lodhra, hura,* M. Along the borders of evergreen forests and in moist places near nálás, throughout the Konkan and North Kánara. Fl. and Fr. at different times throughout the year.

ORDER 56. OLEACEÆ.

Erect or scandent shrubs or trees. Leave sopposite, trifoliate, pinnate or simple. Inflorescence in terminal or cymose panicles. Flowers regular. Calyx free, limb 4-5-toothed; lobed or truncate. Corolla gamopetalous, 4-5, or more lobed. Stamens 2, inserted on the corolla, alternating with the carpels; filaments usually short. Ovary free, 2-celled; stigma simple or 2-lobed; ovules 1-2 (rarely 3-4) in each cell. Fruit succulent or dry. Seeds solitary or 2 in each cell, albumen fleshy, horny or 0.

Corolla-lobes imbricate.
Scaudent shrubs; fruit a berry ... ... JASMINUM.
Erect tree. Fruit a 2-partite capsule ... ... NYCTANTHES.
Tree. Fruit a woody, loculicidally 2-valved capsule... SCHREBERA.
Corolla-lobes valvate or petals distinct in pairs or 0,
Inflorescence axillary.
Petals 4, in distinct pairs ... ... LINOCIERA.
Corolla tubular or 0. ... ... OLEA.
Panicles terminal ... ... LIGUSTRUM.

JASMINUM, L.

Shrubs usually scandent. Leaves simple, 3-foliate or unequally pinnate; petiole usually articulated. Inflorescence cymose; flowers bracteate. Calyx 4-9-fid, tube funnel-shaped, teeth linear long or short. Corolla-tube narrow, lobes 4-10, spreading. Stamens 2, included, filaments short, anther oblong, connective usually shortly produced. Ovary 2-celled; style cylindric, stigmas linear, ovules 2 in each cell, basal. Berry 2-lobed, or entire by the failure of 1 carpel. Seed usually 1 in each lobe, exalbuminous.

Leaves simple, ovate.
Calyx pubescent. Bracts small.
Erect or subscandent shrub, Cymes lax, few flowered... *J. Sambac.*
Scandent hairy shrub. Cymes dense ... ... *J. pubescens.*
Bracts leafy, white ... ... *J. Rottlerianum.*
Leaves mostly subcordate, broad at the base.
Scandent shrub. Calyx-teeth acute, hairy, reflexed ... *J. malabaricum.*
Erect or scandent shrub or tree. Calyx-teeth linear,
subclavate ... ... *J. arborescens.*
Hairy, scandent shrub. Calyx-teeth minute ... ... *J. Roxburghianum.*
Calyx glabrous, ribbed ... ... *J. Ritchiei.*
Leaves compound, opposite.
Leaves simple or trifoliate, lateral leaflets very small ... *J. auriculatum.*
Leaves trifoliate, lateral leaflets large ... ... *J. flexile,*
Leaves alternate, trifoliate or pinnate ... ... *J. humile,*

J. Sambac, Ait. Hort. Kew 1. 8; Fl. Br. I. 3. 591 ; Dalz. & Gibs. Bomb.
Fl. 137 ; Brandis For. Fl. 311. *Mugra, bhat-mogra,* M. Throughout
the presidency, wild or cultivated, often in waste places near villages.
in North Kánara and the Konkan. Fl. throughout the year.

J. arborescens, Roxb. Fl. Ind. ed. Carey & Wall. 1. 94 ; Fl. Br. I. 3.
594 ; Brandis For. Fl. 311 ; Var. *latifolia,* Roxb. ; Dalz. & Gibs. Bomb.
Fl. 311. *Kundi,* M.; *ran jai, kusur, jungly chumbeli,* Vern. Throughout
the presidency, very common in the moist forests of the Konkan and North
Kánara. Fl. Feb., Mch. Fr. May, June. A scandent shrub with linear,
pubescent, divaricated calyx-teeth.

J. Roxburghianum, Wall. Cat. 2870 ; Fl. Br. I. 3. 595. In the deciduous forests of North Kánara near Sambrani, Haliyál taluka. Fl. Fr. H.S.
Probably only a villous variety of *J. arborescens,* Roxb.

J. Ritchiei, Clarke Fl. Br. I. 598. In the evergreen forests of the
Konkan and North Kánara gháts, common on the gháts, from Ainshi
southwards. Fl. Aug., Sept. Fr. Dec., Jan. Calyx-teeth short. Fruit-
carpels accurately spherical on slender pedicels, thickened at the apex.

J. pubescens, Willd. Sp. Pl. I. 37 ; Fl. Br. I. 3. 592 ; Dalz. & Gibs
Bom. Fl. 138 ; *J. hirsutum,* Brandis For. Fl. 312 ; *J. bracteatum,* Wgt.;
Dalz. & Gibs. Bomb. Fl. 138. *Kunda,* Vern. Throughout the presidency ;
in the moist forests of North Kánara and the Konkan. Fl. Dec., Feb.
Branchlets and calyx-teeth densely hairy.

J. Rottlerianum, Wall. Cat. 2865 ; Fl. Br. I. 3. 593 ; Dalz. & Gibs.
Bomb. Fl. 138. Common in the evergreen forests of North Kánara and
the Konkan. Fl. Jan., Mch. Fr. June, Aug. The terminal flower heads
have characteristic white prominent bracts.

J. malabaricum, Wgt. Ic. t. 1250 ; Fl. Br. I. 3. 594. On the gháts
from the Konkan southwards. Calyx-teeth acute, hairy, reflexed.

J. auriculatum, Vahl. Symb. III. 1 ; Fl. Br. I. 3. 600. *J. affine* and
ovalifolium, Wgt. Ic. tt. 1255, 1256. *Jai,* Vern. Throughout the dry
forests of the presidency, often in hedges, absent from the Konkan and
North Kánara, usually scandent, sometimes a bush. Fl. Aug., Sept. Fr.
Dec., Feb. Lateral leaflets very small, when present.

J. humile, Linn. ; DC. Prod. VIII. 313 ; Fl. Br. I. 3. 602 ; *J. revo-
lutum,* Brandis For. Fl. 313. Commonly cultivated in gardens.

J. flexile, Vahl. Symb. III. 1 ; Fl. Br. I. 3. 601. In the evergreen
forests of the southern gháts of North Kánara, common. Fl. Nov., Juny.
A scandent trifoliate shrub, calyx-tube nearly truncate.

NYCTANTHES, Linn.

A small tree. Leaves opposite, scabrous. Flowers bracteate,
in fascicles of three. Calyx sub-truncate, teeth inconspicuous.
Corolla-tube clyindrical, orange ; limb of 5-8-lobes, white. Anthers
2. Ovary 2-celled ; ovule 1 in each cell, erect. Capsule flat, splitting into 2 subdiscoid carpels.

N. Arbor-tristis, Linn.; Roxb. Fl. Ind. I. 86; Fl. Br. I. 3. 603; Bedd. Fl. Sylv. t. 240; Brandis For. Fl. 314; Dalz. & Gibs. Bomb. Fl. Suppl. 51. *Sephali, Sans.; hursing,* K.; *parijtak,* M.; *har, singahar, shiuli,* Vern. The Arabian jasmine. Cultivated in gardens throughout the presidency, indigenous in Assam and Northern India. Fl. throughout the year.

SCHREBERA, Roxb.

A tree. Leaves imparipinnate; leaflets 3-4-pairs. Flowers in terminal cymes. Calyx 4-7-lobed. Corolla hypocrateriform; limb 5-7-lobed, lobes patent. Stamens 2. Ovary 2-celled, 4 ovules in each cell, pendulous from its apex. Fruit a pear-shaped, 2-valved capsule, thick, woody, loculicidally dehiscent. Seeds pendulous, winged, albumen 0.

- **S. swietenioides,** Roxb. Fl. Ind. ed. Carey & Wall. 109; Fl. Br. I. 3. 604; Bedd. Fl. Sylv. t. 248; Brandis For. Fl. 305; Dalz. & Gibs. Bomb. Fl. 138. *Moka, mokari,* Vern. Throughout the presidency in deciduous forests, common in North Kánara. Fl. Apl.-May. Fr. ripe next cold season.

LINOCIERA, Swartz.

Shrubs or trees. Leaves opposite, entire. Flowers often in small axillary or terminal fascicles. Calyx small, 4-fid. Petals 4, often cohering in pairs. Stamens 2. Ovary 2-celled; ovules 2 in each cell, pendulous from its apex. Fruit an ellipsoid drupe, usually 1-seeded.

Seeds albuminous. Petals ¼ inch   ...   . *L. malabarica.*
Seeds exalbuminous. Petals ⅛ inch ...   . *L. intermedia.*

L. malabarica, Wall. Cat. 2328; Fl. Br. I. 3. 607; Dalz. & Gibs. Bom. Fl. 159; *Chionanthus malabarica,* Bedd. Fl. Sylv. t. 239. Throughout the evergreen forests of the Konkan and North Kánara. Fl. Nov.-Jan. Fr. Feb.-Mch. Flowers have the odour of ripe apples in the early morning.

L. intermedia Wgt. Ic. t. 1245; Fl. Br. I. 3. 609; Var. *Roxburghii, Olea Roxburghiana* Dalz. & Gibs. Bomb. Fl. 159; *Chionanthus intermedia* Bedd. Fl. Sylv. t. 237. Common on the Konkan gháts. Fl. H.S.

OLEA, Linn.

Trees or shrubs. Leaves opposite, entire or toothed. Flowers in axillary or terminal panicles, often diœcious. Calyx 4-toothed or lobed. Corolla-tube short, lobes 4 or 0. Stamens 2. Ovary 2-celled; ovules 2 in each cell. Drupe ellipsoid or globose, 1-seeded.

O. dioica, Roxb. Fl. Ind. ed. Carey & Wall. 1. 105; Fl. Br. I. 3. 612; Dalz. & Gibs. Bomb. Fl. 159; Bedd. Fl. Sylv. 154. Indian olive. *Parr jamb, karamba,* Vern. In the evergreen ghát forests of the Konkan and North Kanara, common. Fl. Feb.-Apl. Fr. R.S. Flowers diœcious.

LIGUSTRUM, Linn.

Shrubs or small trees. Leaves opposite, entire. Flowers bisexual, in terminal trichotomous panicles. Calyx small, 4-toothed. Corolla 4-lobed. Ovary 2-celled with 2 ovules in each cell. Fruit a berry. Seeds 4, albuminous.

L. neilgherrense, Wgt. Ic. t. 1243 ; Fl. Br. I. 3. 615 ; Dalz. & Gibs. Bomb. Fl. 159. *L. robustum*, Bedd. Fl. Slyv. 153. *Kangin*, M. On the higher gháts of the Konkan and North Kánara, common in the moist forests near the Tinai ghát. Fl. Aug., Sept. Fr. ripe January.

ORDER 57. SALVADORACEÆ.

Trees or shrubs, unarmed or spinous. Leaves opposite, entire. Flowers clustered or panicled, diœcious or polygamo-dimorphic. Calyx free, 3-5-fid. Corolla shortly campanulate or petals free, 4-merous, imbricate in bud. Stamens 4, filaments free or connate into a tube. Ovary free, 1-2 or imperfectly 4-celled ; style short, stigma 2-fid or subentire ; ovules 1-2 in each cell, erect from the base. Berry or drupe mostly 1-seeded ; seed erect, exalbuminous.

Petals free. Stamens monadelphous. Ovary 1-celled, 1-ovuled ... ... DOBERA.
Corolla gamopetalous. Stamens on corolla-tube. Ovary 1-celled, 1-ovuled ... ... SALVADORA.
Petals free. Stamens free. Ovary 2 (or falsely 4-celled.) AZIMA.

DOBERA, Juss.

A glabrous tree. Leaves elliptic-acute, coriaceous. Flowers polygamo-diœcious, clustered on the branches of an axillary inflorescence, trichotomous in the male flowers, subsessile in the female. Calyx 3-5-toothed. Petals 4-5, free, imbricate. Stamens 4-5, filaments connate into a tube. Scales 4-5 outside the staminal-tube, alternate with the stamens. Ovary 1-celled, ovule solitary, erect. Fruit subglobose.

D. Roxburghii, Planch. in Ann. Sc. Nat. 3, X. 191. Fl. Br. I. 3. 619 ; *Blackburnia monadelpha*, Roxb. Fl. Ind. I. 415. A native of the mountains of the Circars, Roxb.: also found in Bombay. Fl. Br. I. I do not know this tree ; it is said to flower at the beginning of the hot season.

SALVADORA, Linn.

Shrubs or trees. Leaves opposite entire. Flowers hermaphrodite or (mostly) functionally unisexual, in terminal or axillary panicles ; bracts minute. Calyx 4-lobed, lobes imbricate. Corolla 4-lobed. Stamens 4, alternating with the corolla-lobes. Ovary 1-celled, ovule 1, erect, basal. Drupe globose, supported by the slightly enlarged calyx. Seed erect exalbuminous.

Leaves ovate. Panicles compound. Drupes scattered... *S. persica.*
Leaves lanceolate acute. Panicles reduced to axillary fascicles of short spikes. Drupes clustered · ... ... *S. oleoides.*

S. persica, Linn.; A. DC. Prod. XVII. 28; Fl. Br. I. 3. 619; Dalz. & Gibs. Bomb. Fl. 312; Brandis For. Fl. 315; *S. Wightiana*, Bedd. Fl. Sylv. t. 247. Mustard tree of Scripture. *Khabbar, pilu*, Sind.; *khákhin*, M. In the dry districts of the presidency and Sind; in the Dhárwár district, often on the bunds of tanks and in open places near villages of the black soil country, also found near the coasts of Gujarát, the Konkan and North Kánara. Fl. Nov.-May. Fr. Jan., June, often on saline soils.

S. oleoides, Dene. in Jacq. Voy. Bot. 140 t. 144; Fl. Br. I. 3. 620; Brandis For. Fl. 316. t. 39; *S. Stocksii*, Wgt. Ill. II. 229. *Kabbar jhár, diar*, Sind. In the dry deserts of Sind on arid and saline soils, often associated with *S. persica*. Fl. Mch.-April. Fr. June.

AZIMA, Lamk.

Spinous rambling shrubs. Leaves opposite, entire. Flowers diœcious, axillary; bracts 0 or leaflike; bracteoles linear, small. Calyx 2-4-divided. Petals 4, imbricate in bud. Stamens 4. Ovary 2-celled; stigma large 2-fid; ovules 2-1 in each cell, erect. Berry globose, 2-1-seeded. Seed globose, exalbuminous.

A. tetracantha, Lamk. Dict. I. 343; Fl. Br. I. 3. 620; Dalz. & Gibs. Bomb. Fl. 143. Common throughout the dry districts of the presidency; often in hedges. Fl. Dec., Mch. Fr. H. S. Fruit white.

ORDER 58. APOCYNACEÆ.

Trees or shrubs, often climbing, rarely herbs. Leaves opposite or whorled, entire. Flowers regular, hermaphrodite. Calyx free, 5-divided, often glandular at base. Corolla 5-lobed, lobes spreading, often twisted-imbricate in bud, rarely valvate. Stamens 5, on throat or mouth of corolla, filaments short, anthers conniving in a cone round the stigma or free; pollen granular. Disc annular or lobed or 0. Ovary 1-celled with two parietal placentas or 2-celled with axile placentas or of two distinct or connate carpels. Ovules in each cell 2-many, rarely solitary. Fruit a drupe or berry or of two follicles opening along the inner edge. Seeds pendulous, rarely ascending or peltately attached, often winged or with a terminal coma of long silky hairs; albumen present or 0.

Anthers free.
Ovary of two combined carpels. Seeds without a coma.
  Spinous shrubs. Leaves opposite ... ... CARISSA.
Ovary of two distinct carpels. Calyx eglandular. Fruit of
  2, 1-seeded drupes ... ... RAUWOLFIA.
Leaves alternate. Fruit of 1 carpel, fibrous woody ... CERBERA.
Fruit of two follicles.
  Ovules 2-seriate.
  Seeds winged. Leaves alternate ... ... RHAZYA.
  Ovules many seriate.
  Climber. Seeds winged. Leaves whorled ... ... ELLERTONIA.
  Trees or shrubs. Leaves whorled. Seeds with a coma,
    ciliate along the edges ... ... ALSTONIA.

Trees or shrubs. Seeds with a coma at apex. Leaves
 opposite ... ... HOLARRHENA.
Trees or shrubs. Fruit more or less fleshy, dehiscent.
 Seeds without a coma, immersed in pulp ... ... TABERNÆMONTANA
Anthers conniving in a cone round the stigma.
Fruit of two follicles. Seeds with a coma.
Anthers exserted.
Filaments twisted. Carpels connate in flower ... PARSONSIA.
Filaments not twisted.
Climbers.—Corolla rotate, throat naked. Connective
 dorsally glandular ... ... VALLARIS.
Trees. Corolla salver-shaped, throat with fimbriato
 scales ... ... WRIGHTIA.
Anthers included.
Shrub. Corolla-lobes short. Leaves whorled ... NERIUM.
Corolla large, lobes overlapping to the right.
Lofty climbers.
Corolla bell-shaped ... ... BEAUMONTIA.
Corolla salver-shaped ... ... CHONEMORPHA.
Corolla small, lobes nearly straight. Seeds glabrous ... AGANOSMA.
Corolla small, lobes sharply twisted to the left, tips not
 deflected. Seeds beaked, coma white ... ... ANODENDRON.
Corolla-lobes twisted to left, tips deflected. Follicles
 slender... ... ... ICHNOCARPUS.

CARISSA, L.

Shrubs or trees, armed with axillary spines. Leaves opposite,
coriaceous. Flowers in peduncled trichotomous cymes. Calyx 5-
partite. Corolla-tube cylindrical, lobes spreading. Stamens included. Ovary 2-celled; ovules several in each cell; style filiform;
stigma minutely 2-fid. Fruit a globose or ovoid berry. Seeds
usually 2, peltately attached to the septum; albumen fleshy.

Spines straight.
Leaves glabrous. Berry size of cherry ... ...*C. Carandas.*
Leaves small, densely and finely pubescent. Fruit size of a pea.
A small tree or spreading shrub ... ...*C. spinarum.*
Spines decurved.
An erect shrub. Cymes puberulous ... ...*C. macrophylla.*
A climbing shrub. Cymes quite glabrous ... ...*C. suavissima.*

C. **Carandas**, Linn.; A. DC. Prod. VIII. 332; Fl. Br. I. 3. 630;
Brandis For. Fl. 320; Dalz. & Gibs. Bomb. Fl. 143; Bedd. Fl. Sylv. 156.
Karunda, corinda, carwand, hartundi, Vern.   Throughout the presi-
dency, common in the Konkan and North Kánara. Fl. Jan.-Apl.
Fr. ripe May-June.

C. **spinarum**, A. DC. Prod. VIII. 332; Fl. Br. I. 3. 631; *C.
diffusa,* Roxb.; Brandis For. Fl. 321; Bedd. Fl. Sylv. t. 157; *C. hirsuta,*
Roth., Dalz. & Gibs. Bomb. Fl. 143.   On the dry hills of the Deccan,
common in the Dhárwár and Belgaum districts, on dry stony ground.
Fl. C.S.   Fr. Aug., Sept; sometimes a prostrate shrub. Fruit
sweet to the taste, black when ripe.

C. **macrophylla**, Wall. Cat. 1679; Fl. Br. I. 3. 631; *C. lanceolata,*
Dalz. & Gibs. Bomb. Fl. 143; *C. Dalzellii,* Bedd. Fl. Sylv. 156.
Common in the evergreen forests along the Konkan and North Kánara

gháts, abundant near tho coast on the hills about Kárwár. Fl. Jany., Feb. - Fr. ripe June.

. **C. suavissima**, Bedd. MSS. Fl. Br. I. 3. 632. North Kánara and Konkan gháts, in evergreen forests, a fine lofty climber. I am of opinion that this species is the same as *C. macrophylla*, and differs only in the glabrous cymes and climbing habit. Fl. and Fr. at the same times as *C. macrophylla*, Wall.

RAUWOLFIA, Linn.

Glabrous shrubs. Leaves whorled, rarely opposite. Calyx 5-divided. Corolla salver-shaped, tube inflated above the middle, pilose within. Stamens included in the inflated part of the corolla-tube. Disc large, cup-shaped or annular. Ovary 2-celled, ovules 2, collateral in each cell. Fruit a didymous drupe, usually 1-seeded.

A small shrub, corolla-tube slender ... ...*R. serpentina.*
Large shrub. Corolla-tube broad, inflated at the top.
Peduncles erect or spreading in fruit ... ...*R. densiflora*

R. serpentina, Benth. in Gen. Plant. II. 695; Fl. Br. I. 3. 632; *Ophioxylon serpentinum*, Linn.; Dalz. & Gibs. Bomb. Fl. 143; Bedd. Fl. Sylv. 156. *Harki*, M. Throughout the moist forests of the Konkan and North Kánara. Fl. May. Fr. June. An undershrub, drupes often didymous, endocarp rugose.

R. densiflora, Benth. in Gen. Pl. II. 697; Fl. Br. 1. 3. 633; *R. decurva*, Hook. f. Fl. Br. I. 3. 633; *Ophioxylon neilgherrense*, Dalz. & Gibs. Bomb. Fl. 144; *O. densiflorum*, Bedd. Fl. Sylv. 156. In the moist forests of the gháts from the Konkan southwards, common in the forests of the Supa sub-division of North Kánara. *R. decurva*, Hook. f., is, in my opinion, only a variety of *R. densiflora*, Benth. Fruiting specimens of both being exactly similar. The decurved, short peduncles seem, however, to be a constant character of the form *decurva*.

CERBERA, Linn.

Tree. Leaves alternate. Flowers large, white or red, in terminal cymes. Calyx 5-partite. Corolla funnel-shaped, tube short, throat ribbed or scaly; lobes broad. Stamens inserted at the middle of the tube. Ovaries 2, distinct, style filiform, stigma 2-lobed; ovules 4 in each ovary or carpel. Fruit an ellipsoid drupe with a woody fibrous mesocarp. Seeds broad, compressed.

C. Odollam, Gærtn. Fruct. II. 193, t. 124; Fl. Br. I. 3. 638; Brandis For. Fl. 322; Bedd. Fl. Sylv. 157. *Sukinu*, M. Salt swamps of the South Konkan and North Kánara. Fl. at different times throughout the year.

RHAZYA, Dene.

Shrubs. Leaves alternate, thick, glabrous. Flowers in large terminal or axillary cymes. Calyx short, 5-divided. Corolla-tube

long, throat constricted, hairy; lobes 5, overlapping to the left
Stamens included. Disc obscure. Carpels 2, distinct; ovules many.·
Follicles 2, erect, slender. Seeds winged at both extremities;
albumen ruminate.

R. stricta, Decaisne in Ann. Nat. Sc. Ser. 2 IV. 81 ; Fl. Br. I. 3.
640 ; Brandis For. Fl. 322. *Sewar, sihar, ishawarg*, Sind. On the dry
plains of Sind and Affghanistan. Fl. April.

PLUMERIA, Linn.

Shrubs or trees. Leaves elongate, crowded at the ends of the
branches. Flowers large, in terminal cymes. Calyx 5-fid, eglandular
within. Corolla with spreading lobes, contorted in bud. Anthers
at the base of the corolla-tube near the ovary. Carpels 2, distinct.
Fruit of 2, linear oblong follicles. Seeds winged, albumen fleshy.

P. acutifolia, Poir ; A. DC. Prod. VIII. 392 ; Fl. Br. I. 3. 641 ;
Brandis For. Fl. 323 ; Dalz. & Gibs. Bomb. Fl. Suppl. 52. *Khair
champa, son champa,* Vern. Cultivated and naturalized throughout
the presidency near temples and villages. Fl. hot and rainy seasons.

ELLERTONIA, Wight.

Climbing shrubs. Leaves opposite or ternately whorled.
Flowers in terminal cymes. Calyx 5-fid. Corolla-tube cylindric ;
lobes 5, overlapping to the left. Stamens with ciliate filaments.
Carpels 2, distinct; ovules many. Follicles 2, linear, spreading
Seeds 2-seriate, winged above and below.

E. Rheedii, Wgt. Ic. t. 1295 ; Fl. Br. I. 3. 641. On the ghâts of the
Konkan and North Kánara in evergreen forests, in the forests near the
Ainshi ghât at about 1,000 feet elevation. Fl. C. S. Fr. H. S.

ALSTONIA, R. Brown.

Trees or shrubs. Leaves whorled or opposite. Flowers in
corymbose cymes. Calyx short, 5-lobed. Corolla-tube cylindrical,
lobes spreading. Stamens included. Ovary of 2, distinct carpels.
Fruit of 2, long follicles. Seeds peltately attached, densely ciliate
on the margins. Albumen scanty.

A large tree. Leaves obtuse. Disc 0 ... ... *A. scholaris.*
A shrub. Leaves lanceolate, acuminate. Disc of 2 ligulate
　glands ... ... *A. venenatus.*

A. scholaris, Brown in Mem. Wern. Soc. 1. 75; Fl. Br. I. 3. 642 ;
Dalz. & Gibs. Bom. Fl. 145 ; Bedd. Fl. Sylv. t. 242 ; Brandis For.
Fl. 325. *Mudhol, kodale,* K.; *satwin, saptaparni,* M. Throughout the
presidency, usually in deciduous forests common in the evergreen forests
of North Kánara. Fl. Dec.-Mch. Fr. June.

A. venenatus, Brown in Mem. Wern. Soc. 1. 75 ; Fl. Br. I. 3. 642 ;
Dalz. & Gibs. Bomb. Fl. Suppl. 52 ; Bedd. Fl. Sylv. 160. I have only
found this shrub in the evergreen forests near the Nilkund ghât

of North Kánara at about 1,800 feet elevation. Fl. R.S. Fruit Dec.-
Jany. Fruit long beaked.

HOLARRHENA, Br.

Trees or shrubs. Leaves opposite, entire. Flowers usually in
terminal cymes. Calyx 5-lobed ; lobes small, glandular within.
Corolla-tube swollen at its base round the anthers, lobes spreading
twisted to the left. Stamens included. Carpels 2-distinct ; ovules
numerous. Fruit of 2 long slender follicles. Seeds with a deciduous
coma, albumen 0.

H. antidysenterica, Wall. Cat. 1672 ; Fl. Br. I. 3. 644 ; Brandis For.
Fl. 326 ; Bedd. Fl. Sylv. 160; Dalz. & Gibbs. Bomb. Fl. 145. *Dowla,
kura, indrajar,* Vern. Throughout the presidency, common in deciduous
forests. Fl. Apl.-June. Fr. next. March.

TABERNÆMONTANA, Linn.

Trees or shrubs. Leaves opposite. Flowers white. Calyx
glandular within. Corolla-tube cylindrical, slender ; lobes overlapping to the left. Stamens inserted about the middle of the tube.
Carpels two, distinct. Fruit of two smooth or ribbed 1-many-
seeded dry or fleshy follicles. Seeds in pulp, albumen fleshy.

T. Heyneana, Wall. in Bot. Reg. t. 1273 ; Fl. Br. I. 3. 646 ; T.
crispa. Roxb. ; Dalz. & Gibs. Bomb. Fl. 144. *Naglkudó, pandra-kura,*
M. Throughout the Konkan and North Kánara, in evergreen forests,
common. Fl. Mch., Apl. Fr. July-Aug. Corolla deciduous. Seeds in
red pulp.

T. coronaria, Br. (*taggar,* Vern.) is a shrub commonly cultivated in
gardens throughout the presidency, not indigenous in Western India.
Flowers rainy season.

PARSONSIA, Br.

Twining shrubs. Leaves opposite. Flowers small, white.
Calyx 5-partite. Glandular or 0. Corolla-tube short ; lobes overlapping to the right. Stamens with twisted filaments ; anthers conniving over the stigma. Disc of 5 lobes or scales. Ovary of
2 cells. Fruit cylindric, carpels at length separating from the
placentas. Seeds tufted, albumen scanty.

P. spiralis, Wall. Cat. 1631 ; Fl. Br. I. 3. 650 ; *Heligme javanica,* A.
DC. ; Dalz. & Gibs. Bomb. Fl. 146. In evergreen forests, from the
Konkan southwards. Common on the southern gháts of North Kánara.
Fl. at different times throughout the year.

VALLARIS, Burm.

Climbers. Leaves dotted. Flowers white. Calyx 5-partite.
Corolla-tube short ; lobes overlapping to the right. Stamens at the

top of the tube; anthers exserted, conniving in a cone, connective
with a dorsal gland, cells spurred at the base. Carpels 2, connate,
many ovuled. Fruit oblong. Seeds compressed, tip comose, albumen scanty.

V. Heynei, Spreng. Syst. I. 635; Fl. Br. I. 3. 650; Dalz. & Gibs.
Bomb. Fl. 144; *V. dichotoma*, Wall.; Brandis For. Fl. 327. Through-
out the dry districts of the presidency, also in the Konkan and North
Kánara often; in hedges. Fl. Dec., Apl.

WRIGHTIA, Br.

Trees or shrubs. Leaves opposite. Flowers in corymbose cymes.
Calyx with 5-10 scales, inside at base. Corolla-tube short, throat
with 1-2 series of fimbriate scales, distinct or united in a ring;
lobes overlapping to the left. Stamens with 5 exserted, conniving
anthers, cells spurred at the base. Carpels 2, many ovuled, ovules
on axile placentas. Fruit of 2 linear follicles; seeds numerous,
tufted at the lower end, albumen 0.

Leaves glabrous. Coronal scales fimbriate ... ... *W. tinctoria.*
Leaves tomentose. Coronal scales short, obtuse ... *W. tomentosa.*

W. tinctoria, Br. in Mem. Wern. Soc. 1. 73; Fl. Br. I. 3. 653; Dalz.
& Gibs. Bomb. Fl. 145; Brandis For. Fl. 324; Bedd. Fl. Sylv. t. 241.
Kad murki, K.; *kala-kura,* M.; *bhurcuri,* Vern. Throughout the presidency
in deciduous forests. Fl. Mch.-Apl. Fr. Jany., Feb. The leaves yield
indigo.

W. tomentosa, Roem & Schultes Syst. IV. 414; Fl. Br. I. 3. 653;
Dalz. & Gibs. Bomb. Fl. 145; Brandis For. Fl. 323; Bedd. Fl. Sylv.
159; *W. Wallichii,* Dalz. & Gibs. Bomb. Fl. 145 Bedd. Fl. Sylv. 160
Kala inderjow, tambara-kura, Vern. Throughout the presidency, common
in the moist forests of the Konkan, rare in North Kánara, on the Supa
ghâts. Fl. Apl., June. Fr. Jan., Feb.

NERIUM, Linn.

Shrubs. Leaves narrow, opposite or whorled. Flowers in terminal cymes. Calyx-lobes with fleshy glands at the base inside.
Corolla funnel-shaped, lobes spreading, unequal-sided; throat with
5-toothed scales. Anthers included, conniving round the stigma;
tips filiform; cells with long twisted, hairy appendages. Ovary
of 2 carpels, many ovuled. Follicles adpressed separating when
ripe. Seeds villous, coma terminal, albumen fleshy.

N. odorum, Soland. in Hort. Kew ed. 1. V. 1. 297; Fl. Br. I. 3. 655;
Brandis For. Fl. 328; Grah. Cat. Bomb. Pl. 114. *Kunher,* Vern.
Sind. Graham says "grows wild by the banks of Deccan rivers," perhaps
only a variety of *N. oleander,* L., which is cultivated in gardens through-
out the presidency. Fl. Apl., June, and nearly throughout the year,
Brandis.

BEAUMONTIA, Wall.

Climbers. Leaves opposite. Flowers large white, in terminal cymes; bracts leafy. Calyx 5-divided, glandular inside at the base. Corolla bell-shaped with a short tube, lobes overlapping to the right. Stamens included, anthers horny, sagittate, cells spurred at the base. Ovary 2-celled; cells many ovuled. Fruit of 2 connate follicles, separating when ripe. Seeds with a crown of silky hairs at the hilum.

B. **Jerdoniana**, Wgt. Icones. t. 1314, 5; Fl. Br. I. 3. 661. In the evergreen forests of the Konkan and North Kánara gháts. Sir J. Hooker says, "probably will have to be ranked as a variety of *B. grandiflora*, Wall." Fl. Nov., Dec. Fr. ripe following Nov., Dec.

CHONEMORPHA, G. Don.

Climbing shrubs. Leaves broad, opposite. Flowers large, in terminal, peduncled cymes. Calyx 5-cleft, with a ring of glands at the bottom of the tube. Corolla salver-shaped, lobes twisted in bud, throat naked. Stamens at the top of the tube, filaments short, broad, villous; anthers sagittate, connate, pungent; cells spurred. Disc thick, annular. Ovaries 2, oblong, many ovuled. Follicles straight, hard, trigonous. Seeds with a crown of long silky hairs.

C. **macrophylla**, G. Don Gen. Syst. IV. 76; Fl. Br. I. 3. 661; Brandis For. Fl. 328. *Echites grandis*, Wall.; Dalz. & Gibs. Bomb. Fl. 147. Throughout the evergreen forests of the Konkan and North Kánara. Common in the Supa forests. Fl. Apl.-Sept. Fr. C.S.

AGANOSMA, G. Don.

Climbers. Leaves opposite. Flowers in terminal cymes. Calyx 5-divided, glandular at the base. Corolla salver-shaped with bearded, longitudinal bands behind the anthers; lobes overlapping to the right. Stamens included: anthers sagittate, rigid, connate. Disc 5-lobed, cupular. Carpels 2, many ovuled. Follicles woody, coriaceous. Seeds glabrous, albumen scanty.

A. **cymosa**, G. Don. Gen. Syst. IV. 77: Fl. Br. I. 3. 665; var. 4, *elegans*; *Aganosma Doniana*, Wight, Dalz. & Gibs. Bomb. Fl. 146. Throughout the evergreen forests of the Konkan and North Kánara, Fl. Apl.-June. Fr. next C. S. *A. caryophyllata*, G. Don. Fl. Br. I. 3. 664, is cultivated in gardens. The lax cymes and linear calyx-lobes distinguish it from the indigenous species.

ANODENDRON, A. DC.

Lofty climbers. Leaves opposite. Flowers small in corymbose panicles or cymes. Calyx 5-cleft, eglandular. Corolla salver-shaped, mouth contracted, lobes narrow, twisted to the left. Stamens with connate, sagittate anthers, cells shortly spurred at the base. Disc cupular. Obscurely 5-crenate. Ovaries 2, many ovuled. Follicles divaricate, woody. Seeds with a crown of long silky hairs.

A. paniculatum A. DC. Prod. VIII. 444 ; Fl. Br. I. 3. 668 ; Dalz. & Gibs. Bomb. Fl. 147. *Lamtani,* Vern. Common in the evergreen forests of the Konkan and North Kánara. Fl. Jany., Mch. Fr. May.

ICHNOCARPUS, Br.

Climbers. Leaves opposite. Calyx 5-divided, glandular within Corolla hypocrateriform ; lobes hairy inside. Fruit of 2, very slender terete follicles. Seeds with a slender coma at the hilum.

I. frutescens, Br. in Hort. Kew ed. 2. II. 69 ; Fl. Br. I. 3. 669 ; Brandis For. Fl. 327 ; Dalz. & Gibs. Bomb. Fl. 147. *Krishnasarwa, kantebhouri,* M. Throughout the presidency in deciduous forests ; common in North Kánara and the Konkan. Fl. Nov., Dec. Fr. Mch., Apl.

ORDER 59. ASCLEPIADEÆ.

Herbs or shrubs, often twining. Leaves opposite. Flowers regular, pentamerous. Calyx 5-divided, segments imbricate in bud. Corolla-tube short, often with a corona of hairs, scales, or processes. Stamens at the base of the corolla, filaments connate or rarely free. The fleshy staminal column usually bears attached to the filaments or back of the anthers a ring or series of scales or processes, (inner or staminal corona) ; anthers crowning the column, connate or free, adnate to the stigma by the connective ; tip often produced into an inflexed membrane ; pollen in 1-2 granular or waxy masses in each cell, masses united in pairs or fours to a gland (the corpuscle) which lies on the stigma. Ovary of two distinct carpels within the staminal column ; styles 2, short, united by the stigma, which is 5-angled. Fruit of 2 follicles. Seeds usually winged and plumose ; albumen copious, dense.

Filaments usually free. Pollen-masses granular, in pairs in each cell.
Corona of 5, free, short, thick scales.
Corolla-lobes valvate ...   . HEMIDESMUS.
Corolla-lobes imbricate.   ...   . CRYPTOLEPIS.
Coronal-scales connate into a lobed ring ...   . PERIPLOCA.
Filaments connate. Pollen-masses waxy.
Anthers with a membranous inflexed tip.
Pollen-masses in pairs in each cell, in 4s, on the corpuscle, pendulous.
Corolla-lobes imbricate   ...   . TOXOCARPUS.
Corolla-lobes valvate ...   . GENIANTHUS.
Pollen-masses solitary in each cell, in pairs on the corpuscle, pendulous.
Erect shrubs. Corolla-lobes valvate   ...   . CALOTROPIS.
Twining leafy shrubs. Corolla-lobes imbricate.
Corolla funnel-shaped...   ...   . DÆMIA.
Corolla rotate.

Corona staminal with five, horny processes behind the
anthers ...   . Holostemma.
Corona annular. staminal.
Leafy twining shrubs   ...   . Cynanchum.
Leafless, jointed shrubs with pendulous branches...   ... Sarcostemma.
Pollen-masses solitary in each cell, in pairs on the
corpuscle, erect.
Corolla-lobes imbricate.
Corona coralline   ...   . Gymnema.
Corona staminal.   Corolla urceolate, campanulate or
salver-shaped.
Coronal-scales simple on the backs of the anthers   ... Marsdenia.
Coronal-scales notched on the backs of the anthers   ... Pergularia.
Corolla rotate.
Cymes racemose.   Coronal-processes 2-fid...   ...   ... Cosmostigma.
Cymes in umbels   Coronal-scales spreading, cuspidate... Dregea.
Corolla-lobes valvate.
Corolla small, rotate ...   . Heterostemma.
Corolla urceolate or disciform, Corona cupular   ...   ... Oianthus.
Corolla rotate, Corona large, stellate   ...   . Hoya.
Anthers without a membranous inflexed tip.   Pollen-
masses 1 in each cell, in pairs on the corpuscle, erect.
Corolla-lobes valvate.
Twining shrubs   Corona double   ...   . Leptadenia.
Erect fleshy shrub.   Corona staminal, simple   ...   ... Frerea.

HEMIDESMUS, Br.

Twiners. Leaves elliptic-linear. Flowers in axillary, subsessile
cymes, small. Sepals glandular within. Corolla rotate, reddish.
Coronal-scales 5, on the corolla-throat, thick. Stamens with the
filaments distinct ; tips of the anthers connate, inflexed, membranous.
Appendages of the corpuscles dilated. Follicles slender, divaricate,
smooth. Seeds tufted.

H. indicus, Br. in Mem. Wern. Soc. 1. 57 ; Fl. Br. I. 4. 5 ; Dalz. &
Gibs. Bomb. Fl. 147. *Nannari*, Vern.; *uparsal, sariva, anantamul*, M.
Indian sarsaparilla. Throughout the presidency, common in hedges.
Fl. Throughout the year.

CRYPTOLEPIS, Br.

Twiners. Leaves pale beneath. Corolla-lobes linear, throat
naked, tube inflated at the middle, lobes twisted to the left. Corona
of five, fleshy scales, included within the tube. Stamens included,
sagittate, with a tuft of hairs at the back. Pollen granular ;
appendages oblong. Follicles lanceolate divaricate, smooth. Seeds
comose.

C. Buchanani, Rœm. & Sch. Syst. IV. 409 ; Fl. Br. I. 4. 5 ; Brandis
For. Fl. 330 ; Dalz. & Gibs. Bomb. Fl. 148. *Karanta*, Vern. Common.
in hedges throughout the presidency. Fl. June, Aug. Fr. C. S. Bracts
small, persistent. Milky juice abundant.

PERIPLOCA, Linn.

Erect or twining shrubs. Leaves lanceolate, minute or wanting.
Flowers in lax cymes. Corolla rotate ; lobes bearded within, longer

than the tube. Corona of five thick scales, connate into a 10-lobed ring, often produced in a long filiform arista. Filaments short, free ; anthers with a tuft of hairs at the back. Pollen grains in 4s ; appendages dilated. Follicles cylindric, smooth.

P. aphylla. Dcne. in Jacq. Voy. Bot. 109, t. 116; Fl. Br. I. 4. 12 ; Brandis For. Fl. 330. *Barrara, ransher, barai,* Vern. Common in Sind. Fl. Mch.-Apl.

TOXOCARPUS, W. & A.

Woody climbers. Leaves opposite. Flowers small, in axillary cymes. Corolla-lobes overlapping to the left, tube short. Coronal-scales minute, adnate to the column. Anthers small, often inappendiculate ; pollen-masses minute, waxy. Stigma produced in a long beak. Follicles spreading, smooth. Seeds comose.

T. Kleinii, Wgt. & Arn. Contrib. 61; Fl. Br. I. 4. 14. From the Konkan, southwards. Fl. Br. I.

GENIANTHUS, Hook. f.

Climbers. Leaves opposite. Flowers small, in axillary cymes. Corolla-lobes valvate, spreading and recurved, villous on the inner face. Corona of five scales, adnate to the column. Pollen-masses minute, waxy. Follicles terete, slender.

G. laurifolius, Hook. f. Fl. Br. I. 4. 16 ; *Toxocarpus crassifolius,* Dalz. & Gibs. Bomb. Fl. 148. On the Konkan and North Kánara gháts, in evergreen forests, common in the forests near the Nilkund and Gairsoppah gháts. Fl. Oct., Dec. Fr. H. S.

CALOTROPIS, Br.

Erect shrubs. Leaves broad, subsessile. Flowers medium-sized, cymose. Corolla-tube broad, lobes ovate, valvate. Corona of five, compressed, fleshy appendages. Anther-tips membranous, inflexed. Pollen-masses in pairs, waxy, pendulous. Follicles broad, ovoid.

Coronal-scales truncate, hairy	*C. gigantea.*
Coronal-scales acute, glabrous or pubescent.	*C. procera.*

C. gigantea, Br. in Hort. Kew Ed. 2, II. 78 ; Fl. Br. I. 4. 17 : Dalz. & Gibs. Bomb. Fl. 149 ; Brandis For. Fl. 331, *Madar, mudar,* M. ; *arka, akari, rowi,* Vern. Throughout the presidency in dry, waste places ; very common. Fl. throughout the year. Full of acrid, white juice.

C. procera, R. Br. in Ait. Hort. Kew Ed. 2. II. 78 ; Fl. Br. I. 4. 18 ; Dalz. & Gibs. Bomb. Fl. 149 ; Brandis For. Fl. 331. *Lalmandar, tambara, ak,* M. In the dry Deccan districts and Sind. Fl. Feb., May. Fr. C. S. Yields a valuable fibre.

DÆMIA, Br.

Climbers. Leaves cordate. Flowers in axillary cymes, pedicels filiform. Corolla green, tube short, lobes large, broad, overlapping to the right. Corona of an outer 5-10-crenate membrane and 5, compressed scales, adnate to the anthers and spurred behind with long subulate tips. Pollen masses in pairs, pendulous, waxy. Follicles echinate.

D. extensa, Br. in Mem. Wern. Soc. 1. 50 ; Fl. Br. I. 4. 20 ; Dalz. & Gibs. Bomb. Fl. 150. *Utaran,* M. Throughout the dry districts of the Deccan, also in hedges in the Konkan, common in the Dhárwár district. Fl. apparently throughout the year. Follicles softly echinate.

HOLOSTEMMA, Br.

Twining shrubs. Leaves membranous, cordate. Cymes axillary, of large purple flowers. Corolla-lobes thick, overlapping to the right. Corona 10-lobed, fleshy. Anthers large, horny, shining, cohering in a 10-winged column. Pollen-masses linear, falcate, compressed, pendulous ; pedicels long, black. Stigma 5-winged. Follicles short, thick, smooth.

H. Rheedei, Wall. Pl. As. Rar. II. 51; Fl. Br. I. 4. 21 ; Dalz. & Gibs. Bomb. Fl. 148. *Tultuli, shidori, dudurli,* M. In the moist forests of the Konkan and North Kánara, often in hedges. Fl. July, Aug. Fr. C. S. Flowers edible.

CYNANCHUM, Linn.

Shrubs or herbs, erect or climbing. Leaves usually broad, cordate. Flowers small, green, in axillary cymes. Corolla rotate, lobes valvate or overlapping to the right. Corona adnate to the base of the column, cupular, with scales or tubercles on the inner face. Pollen-masses not compressed, pendulous, waxy. Follicles terete or 2-winged.

Leaves not glaucous beneath. Follicles terete   ...   ... *C. pauciflorum.*
Leaves glaucous beneath. Follicles winged...   ...   ... *C. Callialata.*

C. pauciflorum. Br. in Mem. Wern. Soc. 1. 45 ; Fl. Br. I. 4. 23 ; Dalz. & Gibs. Bomb. Fl. 148. From the Konkan southwards, also in the Poona district. Fl. C. S.

C. Callialata, Ham. in Wgt. Contrib. 56 ; Fl. Br. I. 4. 24. From the Konkan southwards common in the evergreen forests near the Nilkund ghát of North Kánara. Fl. Oct., Dec. Fr. Jan., Feb.

SARCOSTEMMA, Br.

Leafless, jointed shrubs with pendulous branches. Cymes in terminal umbels. Corolla rotate, lobes overlapping to the right. Corona cup-shaped, lobed, alternate lobes with large fleshy processes, adnate to the back of the anthers. Pollen-masses pendulous, waxy. Follicles smooth.

Follicles thinly coriaceous. Seeds ⅓ inch, ovate   ... *S. brevistigma.*
Follicles thickly coriaceous. Seeds ⅓ inch, linear   ... *S. Stocksii.*

S. brevistigma, Wight and Arnott's Contrib. 59 ; Fl. Br. 1. 4. 26 ; Dalz. & Gibs. Bomb. Fl. 149. *Somvel,* M. Throughout the Deccan in stony places. Fl. June.

S. Stocksii, Hook. f. Fl. Br. I. 4. 27. Sind. *S. intermedium,* Dene. Fl. Br. I. 4. 27 ; Dalz. & Gibs. Bomb. Fl. 27. *Phok,* M. Common in the dry districts of the Deccan. Abundant in the Gadag táluka of the Dhárwár district. Fl. July, Sept.

GYMNEMA, R. Br.

Twining shrubs. Leaves opposite. Flowers small, umbellate. Corolla rotate or urceolate, throat with 5 scales, alternating with the lobes. Anthers terminating with a membrane. Pollen-masses in pairs, erect. Stigma exserted. Follicles slender or turgid, glabrous.

Pubescent. Leaves small, obovate, acute ... ... *G. sylvestre.*
Glabrous or nearly so. Leaves large, coriaceous, ovate-
oblong ... ... *G. montanum.*

G. sylvestre, Br. in Mem. Wern. Soc. 1. 33 ; Fl. Br. I. 4. 29 ; Dalz. & Gibs. Bomb. Fl. 151. *Kuwali, káli-kardori, vakhande,* M. Throughout the presidency, common in hedges in the Dhárwár district, and near the coast at Kárwár. Fl. June, Aug. Fr. Oct., Nov.

G. montanum, Hook. f. Fl. Br. I. 4. 31. *Bidaria elegans,* Dalz. & Gibs. Bom. Fl. 151. Higher ghats of the Konkan.

MARSDENIA, R. Brown.

Twining shrubs. Leaves often cordate. Flowers in umbelliform cymes. Corolla campanulate, limb spreading, lobes overlapping to the right. Coronal-scales 5, sometimes auriculate, adnate to the anthers dorsally, erect. Pollen-masses in stipitate pairs, erect. Stigma obtuse or beaked. Follicles lanceolate or poniard-shaped.

M. tinctoria, Br. in Mem. Wern. Soc. 1. 30 ; Fl. Br. I. 4. 34 ; Brandis For. Fl. 332. Cultivated in the Deccan. Fl. hot and rainy seasons.

PERGULARIA, Linn.

Woody climbers. Leaves membranous. Flowers greenish, in interpetiolar cymes. Corolla salver-shaped, lobes oblong, overlapping to the right. Coronal-scales large, membranous, adnate to the back of the anthers, erect, simple or transversely bifid. Pollen-masses erect, clavate. Follicles ventricose-lanceolate, smooth. Seeds ovate, concave.

P. pallida, Wight & Arn. Contrib. 42 ; Fl. Br. I. 4. 38 ; Brandis For Fl. 334. In the plains of the Deccan, in the scrub jungles, on the stony hills near Dhárwár. Fl. July, Aug.

P. odoratissima, L. West Coast or Primrose-Creeper is cultivated in gardens.

COSMOSTIGMA, Br.

A twining shrub. Leaves ovate or cordate. Flowers small, greenish, in axillary, racemose cymes. Corolla rotate. Coronal-scales adnate to the base of the anthers, erect broad, membranous, truncate or 2-fid. Pollen-masses obovoid, waxy, erect, pedicels long, flexuous. Follicles large, linear-oblong.

C. racemosum, Wight Contrib. 42; Fl. Br. I. 4. 46; Dalz. & Gibs. Bomb. Fl. 151. *Jati*, M. Throughout the Konkan and North Kánara; often in hedges. Fl. June, Aug. Fr. C.S. Stigma pentagonal, exserted.

DREGEA, E. Meyer.

Twining shrubs. Leaves ovate or cordate, acuminate. Flowers green in axillary umbelliform cymes. Corolla rotate. Coronal-scales 5, hemispheric, fleshy, adnate to the column, inner angle cuspidate, the tooth incumbent on the anthers. Pollen-masses erect, oblong, shortly pedicelled. Follicles thick, winged or ribbed, often yellow mealy.

D. volubilis, Benth. Gen. Pl. 775; Fl. Br. I. 4. 46; *Hoya viridiflora*, Br. Dalz. & Gibs. Bomb. Fl. 153. *Dori, ambri, herandori*, M. Common in hedges throughout the presidency. Fl. Mch., May. Fr. C. S.

HETEROSTEMMA, W. & A.

Twining shrubs. Leaves usually cordate, 3-5-nerved at the base. Flowers small, in umbelled cymes. Corolla rotate, lobes triangular, valvate. Corona of 5 large lobes, spreading from the column and lying flat on the corolla. Pollen-masses minute, broad, sessile, erect. Follicles slender, straight, terete.

H. Dalzellii, Hook. f. Fl. Br. I. 4. 48; *H. Wallichii*, Dalz. & Gibs. Bomb. Fl. 152. The Konkan; at Vengurla, and Málvan, also in the Deccan near Poona. Fl. Aug.-Sept.

Oianthus urceolatus, Benth.; *Heterostemma urceolatum*, Dalz. & Gibs. Bomb. Fl. 153, and O. disciflorus, Hook. f. Fl. Br. I. 4. 49, are twining undershrubs of the Konkan ghâts which Dr. Hooker thinks are possibly abnormal forms of *Heterostemma*.

HOYA, Br.

Twining shrubs. Leaves thickly fleshy or coriaceous. Flowers in axillary or terminal umbels. Corolla rotate, waxy, lobes 5, valvate in bud. Coronal-scales 5, stellately spreading. Pollen-masses pedicelled, erect. Follicles usually slender, acuminate. Seeds with a long coma.

Glabrous.
Leaves narrow, linear.	Corona pink ...	... *H. retusa.*
Leaves elliptic, broad.	Corolla cream-coloured	... *H. Wightii.*
Sparsely hairy.	Corolla white, lobes villous ...	... *H. pendula.*

H. retusa, Dalz. in Hook. Kew Jour. Bot. IV. 294; Fl. Br. I. 4. 56; Dalz. & Gibs. Bomb. Fl. 153. On trees in the moist forests of the Konkan and North Kánara gháts, common in the forests near Yellápur; appears to flower rarely and during the R.S.

H. Wightii, Hook. f. Fl. Br. I. 4. 59; *H. pallida*, Dalz. & Gibs. Bomb. Fl. 152. On trees throughout the Konkan and North Kánara ghúts, very common on the Supa ghúts. Fl. R. S., also in Jan. rarely. Fr. C. & H. seasons. Leaves usually short petioled. Petals reflexed.

H. pendula, Wgt. Icones t. 474; Fl. Br. I. 4, 61; Dalz. & Gibs. Bom. Fl. 152. Southern Konkan, hills of the Kolába district. Fl. R. & C. S.

LEPTADENIA, Br.

Erect or climbing shrubs. Leaves elliptic or ovate-cordate. Flowers small, in axillary umbelliform cymes. Corolla rotate, lobes bearded, valvate in bud. Corona double. Pollen-masses globose, sessile, erect. Stigma included, 2-cuspidate. Follicles acuminate, smooth.

> Twining. Leaves broad, coriaceous ... ... *L. reticulata.*
> Erect. Leaves linear, leathery or 0. ... ... *L. Spartium.*

L. reticulata, Wight. & Arn. Contrib. 47; Fl. Br. I. 4. 63; Dalz. & Gibs. Bom. Fl. 152. *Khar-kodi*, M. Common near the coast of the Konkan. Dalz. Throughout the dry Deccan districts, common near Dambal in the Dhárwár district. Fl. July Aug. Fr. Sept., Oct.

L. Spartium, Wight Contrib. 48; Fl. Br. I. 4. 64. *L. Jacquemontiana*, Dene. Dalz. & Gibs. Bom. Fl. 152. *Kip*, Vern. On the seacoast south of Gogha, very common in Sind, Dalz. Fl. Oct.

FREREA, Dalz.

A fleshy, erect shrub. Leaves oblong. Flowers large, solitary or in pairs, axillary. Corolla rotate, lobes acute, valvate. Corona annular round the column with five lobes alternating with 5 long, narrow processes inflexed over the anthers. Pollen-masses short, erect. Follicles terete, smooth.

F. indica, Dalz. in Jour. Linn. Soc. VIII. 10, t. 3; Fl. Br. I. 4. 76. The Konkan, near Hewra, 3,000 feet. alt., Dalz. I have not seen this shrub.

ORDER 60. LONGANIACEÆ.

Trees, shrubs or herbs. Leaves opposite, simple. Inflorescence cymose. Flowers regular. Calyx inferior, 4-5-toothed. Corolla 4-5-lobed. Stamens 4-5, on corolla tube and alternate with lobes. Ovary free 2-celled, ovules 1 or more in each cell. Fruit a capsule or berry, 1 to many seeded. Albumen copious.

> Corolla imbricate. Capsule 2 valved ... ... BUDDLEIA.
> Corolla contorted. Fruit indehiscent ... ... FAGRÆA.
> Corolla valvate. Fruit indehiscent ... ... STRYCHNOS.

BUDDLEIA, Linn.

Small trees or shrubs, often mealy or woolly tomentose. Leaves opposite, simple, united by a stipulary line. Flowers axillary or in terminal panicles, tetramerous. Corolla-lobes imbricate in bud. Stamens 4, anthers nearly sessile. Ovary 2-celled; style linear; ovules many in each cell. Capsule septicidally 2-valved, leaving a free dissepiment in the centre. Seeds small, many.

B. **asiatica**, Lour. Fl. Cochinch. 72 ; Dalz. & Gibs. Bom. Fl. 180 ; Fl. Br. I. 4. 82 ; Brandis For. Fl. 318 ; Bedd. Fl. Sylv. 163. Throughout the presidency ; common along the banks of nálás and ravines, also in deciduous forests. Fl. Jany. Apl. Fr. June.

FAGRÆA, Thunb.

Trees or scandent shrubs. Leaves entire ; petioles dilated at the base. Flowers bracteate, large or small, in axillary or terminal cymes. Calyx-lobes thick, broad, imbricate. Corolla with a long tube ; lobes 5, twisted to the left. Stamens 5. Ovary 2-celled ovules many in each cell. Berry with many seeds immersed in pulp. Albumen horny.

F. **obovata**, Wall. Cat. 1595 ; Fl. Br. I. 4. 83 ; Bedd. Fl. Sylv. 164. F. *coromandeliana*, Bedd. Fl. Sylv. t. 244. In the evergreen forests of the Kumta táluka of North Kánara. Scandent. Fruit smooth, size of a pigeon's egg. Fl. R.S. Fr. C.S.

STRYCHNOS, Linn.

Trees or scandent shrubs with short tendrils. Leaves entire, 3-5-nerved. Flowers axillary or terminal, pentamerous. Corolla-lobes valvate. Stamens in the throat of the corolla. Style filiform, stigma capitate ; ovules many on fleshy placentas. Fruit a berry, seeds compressed, immersed in pulp. Embryo eccentric, albumen cartilaginous.

Scandent, climbing shrubs.
Ovary hairy above. Fruit ovoid, small ...   ... *S. colubrina.*
Ovary glabrous. Fruit large ...   . *S. Dalzellii.*
Trees.
Fruit black, 1-seeded, ¾ in. diam.   ...   ... *S. potatorum.*
Fruit orange, many-seeded, 2-4-in. in diam.   ... *S. Nux-vomica.*

S. **colubrina**, Linn. Sp. Pl. 271 ; Fl. Br. I. 4. 87 ; Dalz. & Gibs. Bomb. Fl. 155 ; Bedd. Fl. Sylv. 163. *Kanal, kajer bel*, M. In the evergreen forests of the Konkan and North Kánara gháts, common in the forests near the Tinai ghát. Fl. Oct. Fr. Jan., Feb.

S. **Dalzellii**, C. B. C. Fl. Br. I. 4·87 ; *S. axillaris*, Dalz. & Gibs. Bomb. Fl. 155. Konkan gháts, Dalz.

' S. **Nux-vomica**, Linn. Roxb. Fl. Ind. ed. Carey & Wall. II. 261. Fl. Br. I. 4. 90 ; Dalz. & Gibs. Bomb. Fl. 155 ; Bedd. Fl. Sylv. t. 243 ; Brandis For. Fl. 317. *Kasarkana*, K.; *karo*, M.; *kajra*, Vern. Throughout the presidency in moist forests, very common in the Konkan and North Kánara. Fl. Mch. Apl. Fr. ripe at various times throughout the year.

S. potatorum, Linn. f. Suppl. 148; Fl. Br. I. 4. 90; Dalz. & Gibs. Bomb. Fl. 156; Bedd. Fl. Sylv. 163; Brandis For. Fl. 317. Clearing nut tree. *Nermali,* Vern. In the Southern Marátha Country, Belgaum district and in other parts of the presidency in dry forests. Fl. hot season. Fr. C.S.

Order 61. BORAGINEÆ.

Herbs, shrubs or trees. Leaves usually alternate, entire. Inflorescence definite; flowers in scorpioid cymes, rarely solitary and axillary. Calyx inferior, teeth 4-6, valvate in bud, usually persistent in fruit. Corolla gamopetalous, often with scales in the throat, lobes as many as those of the calyx. Stamens inserted in the corolla-tube, as many as the corolla-lobes and alternate with them. Ovary free, of 2 carpels, 2-ovuled, or 4-1-ovuled; style terminal or from between the ovary-lobes. Fruit drupaceous or dividing into 2-4-nutlets. Albumen 0 or scanty.

Style twice bifurcate. Drupe 4-1-seeded ... CORDIA.
Style 2-fid. ...   . EHRETIA.
Style 1   ...   . RHABDIA.

CORDIA, L.

Trees or shrubs. Leaves alternate, petioled. Flowers often polygamous in terminal or leaf opposed cymes, ebracteate. Calyx tubular or campanulate, teeth short, in fruit accrescent. Corolla funnel-shaped; lobes 4-8. Stamens 4-8, filaments often hairy at the base. Ovary 4-celled, glabrous; style terminal, twice bipartite. Drupe ovoid, endocarp bony, often perforated at the apex, cells 4 or fewer by abortion. Albumen 0.

Leaves alternate, 3-5-nerved at the base, without white discs.
Leaves ovate, rough, glabrous, not tomentose beneath. Cymes lax, glabrous. Calyx not ribbed ... *C Myxa.*
Leaves white tomentose beneath, glabrous above. Cymes lax, glabrous. Calyx not ribbed ... ... *C. Wallichii.*
Leaves with white discs above, scabrous, beneath villous or glabrescent. Cymes small, tomentose ... *C. monoica.*
Leaves feather veined, small, narrow, scabrous or glabrescent above, glabrescent beneath. Berry small ... ... *C. Rothii.*
Leaves broad, cordate, ovate, densely tomentose beneath, feather veined or faintly 3-5 nerved at the base. Cymes short tomentose. Calyx ribbed. Berry large ... ... *C, Macleodii.*
Leaves scabrid above with white discs, hard tomentose beneath, margin repand, base cuneate. Cymes tomentose ... . *C. fulrosa.*

I have followed Brandis and others in uniting *C. Myxa,* L., and *C. obliqua,* Willd. *C. Wallichii,* G. Don, I have kept distinct from *C. obliqua,* Willd., of which species it is made a variety in the Fl. Br. I.

C. **Myxa**, Linn. Fl. Br. I. 4. 136; Dalz. & Gibs. Bomb. Fl. 173; Brandis For. Fl. 336; Bedd. Fl. Sylv. 165. *C. obliqua*, Willd. Fl. Br. I. 4. 137; Dalz. & Gibs. Bomb. Fl. 173; *C. latifolia*, Roxb. Dalz. & Gibs. Bomb. Fl. 173. *Chella*, K; *shelu*, M.; *bhoknr*, *vargund*, Vern.; *lessuri*, *giduri*, Sind. Throughout the presidency in deciduous forests, also in Sind. Fl. Mch.-Apl. Fr. June, Sept.

C. **Wallichii**, G. Don.; Brandis For. Fl. 337; Dalz. & Gibs. Bomb. Fl. 174; Bedd. Fl. Sylv. t. 245. *Burgund*, *duhiwun*, Vern.; *geduri*, Sind; *sepistan*, *pistan*, Guj.; *buralessura* Hind. Common in Gujarát and in the drier districts of the presidency, also in the Mundgod subdivision of North Kánara in deciduous forests. Fl. Dec.-Jany. Fr. Apl. May. Leaves tomentose beneath. Calyx ribbed, fruit long acuminate,

C. **monoica**, Roxb. Fl. Ind. ed. Carey &· Wall. II. 334; Fl. Br. I. 4. 137; Bedd. Fl. Sylv. 166; *C. polygama*, Bedd. For. Man. 166. Grah. Cat. Bomb. Pl. 136. In the dry districts of the Deccan near Bádámi, Bijápur Collectorate, on sandstone. Fl. Aug., Sept.

C. **Rothii.**, Roem & Schul. Syst. IV. 798; Fl. Br. I. 4. 138; Dalz. & Gibs. Bomb. Fl. 174; Bedd. Fl. Sylv. 166. Brandis For. Fl. 338; *C. angustifolia*, Roxb. (not of Roem & Schul.) *Gund*, *gundni*, Vern.; *lijar*, *liai*, Sind. Throughout the dry districts of the presidency; often along the bunds of tanks and near villages. Fl. Apl., June. Fr. C.S.

C. **Macleodii**, Hook. f. & T. in Jour. Linn. Soc. II. 128; Fl. Br. I. 4. 139; Brandis For. Fl. 337. *Dhaican*, Sátára. Commonly planted along road-sides in the southern drier districts of the presidency, also in the moist forests of North Kánara and the Konkan. Fl. Mch.-Apl.-May. Fr. cold season. Leaves and inflorescence densely tawny tomentose.

C. **fulvosa**, Wight Ic. t. 1380; Fl. Br. I. 4. 140; Bedd. Fl. Sylv. 166. Konkan and Belgaum in dry forests. Fl. July. Aug. Leaves very scabrous above with white, small discs.

EHRETIA, Linn.

Trees or shrubs. Leaves alternate, entire or toothed. Flowers small in axillary or terminal panicles, rarely solitary. Calyx deeply 5-divided. Corolla rotate, lobes imbricate in bud. Stamens 5, on the corolla-tube; anthers usually exserted. Ovary 2-celled; cells 2-ovuled. Fruit a 1-4-seeded drupe.

Style bifid. Drupe small, 1-4-seeded.
Style simple, bifid.
Leaves elliptic, glabrous. Flowers sessile　...　　... *E. lævis.*
Leaves obovate, obtuse, hairy. Flowers pedicelled and sessile　...　...　...　...　...　...　... *E. obtusifolia.*
Styles 2. Leaves small, fascicled ...　...　...　... *E. buxifolia.*

E. **lævis**, Roxb. Cor. Pl. t. 55; Fl. Br. I. 4. 141. Dalz. & Gibs. Bomb. Fl. 170; Bedd. Fl. Sylv. t. 246; Brandis For. Fl. 340. *Adak*, K.; *datrang*, M.; *tamboli*, Vern. Throughout the presidency in deciduous forests. The variety *canarensis* is common in the Konkan and North Kánara in the ghát forests. Fl. Jany. to July, Fr. ripe March-June.

E. obtusifolia, Hochst. ; A. DC. Prodr. IX. 507 ; Fl. Br. 1. 4. 142 ; Brandis For. Fl. 340. Sind. Fl. Mch.-Apl.

E. buxifolia, Roxb. Cor. Pl. 142, t. 57 ; Fl. Br. I. 4. 144 ; Bedd. Fl. Sylv. 167. *Pala,* M. In the dry forests of the Deccan ; common on the stony hills near Dhárwár. Fl. Mch. Fr. ripe June, July.

RHABDIA, Mart.

Shrubs. Leaves alternate or clustered. Flowers small in few flowered racemes. Calyx 5-divided. Corolla pink, tube short, limb 5-lobed. Stamens 5. Ovary 2-celled ; style undivided. Fruit a drupe with 4 pyrenes. Seeds albuminous.

R. lycioides, Mart. Nov. Gen. & Sp. II. 137, t. 195 ; *R. viminea,* Dalz. & Gibs. Bomb. Fl. 170 ; Brandis For. Fl. 341. Common in the beds of the Konkan and North Kánara rivers. Fl. Oct., Dec. Fr. Jany.

ORDER 62. CONVOLVULACEÆ.

Shrubs or herbs, usually twining. Leaves alternate. Flowers cymose, bracteate, 1 or more together, regular, hermaphrodite, often showy. Calyx deeply 5-lobed, persistent, sometimes enlarged in fruit. Corolla usually campanulate or funnel-shaped ; limb shortly or deeply lobed, often 5-plaited in bud. Stamens 5, on the corolla-tube. Ovary superior, 2 carpels, 2-celled or by false dissepiments 4-celled ; style 1, rarely 2, stigma capitate 2-lobed, or 2-branched ; ovules 2 in each carpel. Fruit indehiscent, 2-4-valved or breaking up irregularly, 2-4, rarely 1-seeded. Seeds erect, albumen 0.

Fruit indehiscent.
Ovary 1-celled. Stigma large, globose, subsessile ... ERYCIBE.
Ovary 4-celled.
Stigmas 2, linear-oblong. Fruit dehiscent or indehiscent ... ... RIVEA.
Stigmas 2, globose ... ... ARGYREIA.
Ovary 2-celled ... ... LETTSOMIA.
Capsule 2-4-valved or fragile and breaking up.
Bracts and bractlets deciduous, often small.
Styles 2 ... ... BREWERIA.
Style simple ; stigmas capitate or filiform.
Sepals not or little enlarged in fruit. Capsule 2-4-celled ... ... IPOMŒA.
Sepals enlarged in fruit. Capsule 1-celled, 1-seeded. PORANA.
Bract orbicular, leafy with the 4-8-valved, small capsule, adnate to its centre ... ... NEUROPELTIS.

ERYCIBE, Roxb.

Climbing shrubs. Leaves coriaceous, evergreen. Corolla with a short tube, lobes bifid, plaited in bud. Stamens at the base of the corolla-tube. Ovary 1-celled with 4, erect ovules. Fruit a fleshy, 1-seeded berry ; cotyledons plicate.

E. paniculata, Roxb. Fl. Ind. I. 585; Br. Fl. I. 4. 180; Dalz. & Gibs. Bomb. Fl. 169; Brandis For. Fl. 344. Var. *Wightiana*; *E. Wightiana*, Dalz. & Gibs. Bomb. Fl. 170; Brandis For. Fl. 344. Throughout the presidency. The variety *Wightiana* is common in the moist forests of the Konkan and North Kánara. Fl. June, Dec. Fr. ripe May onwards.

RIVEA, Chois.

Climbers. Leaves cordate, silky beneath, petiole long. Flowers few on axillary peduncles. Corolla large, white, with a long cylindrical tube and a wide-mouthed, plaited limb. Stamens included. Ovary 4-celled, 4-ovuled; disc annular; stigmas linear, oblong. Fruit indehiscent, sub-globose. Seeds 1-4, in mealy pulp.

Leaves white-tomentose beneath. Corolla silky outside. *R. ornata.*
Leaves silky beneath. Corolla glabrescent without ... *R. hypocrateriformis.*

R. ornata, Chois. Convolv. Or. 27, t. 3; Fl. Br. I. 4. 183; *Argyreia ornata,* Brandis For. Fl. 343. Dalz. & Gibs. Bomb. Fl. 168; *Lettsomia ornata,* Roxb. Fl. Ind. 1. 496. *Phand,* M. Throughout the presidency in deciduous forests, not common. **Fl. R.S.**

R. hypocrateriformis, Chois. Convolv. Or. 26; Fl. Br. I. 4. 184; Dalz. & Gibs. 168; *R. bona-nox,* Roxb.; Dalz. & Gibs. Bomb. Fl. 168. *R. fragrans,* Nimmo. Grah. Cat. Bomb. Pl. 127. Clove-scented creeper. *Kulmiluta,* Vern. Common in hedges and in dry forests throughout the presidency. Fl. R.S. Fr. ripe. Sept. Oct. Flowers sweet-scented.

ARGYREIA, Lour.

Large twining shrubs. Leaves silky or pubescent beneath. Flowers large, in axillary cymes. Corolla usually campanulate. Stamens included. Ovary 4-celled, 4-ovuled; disc annular; stigmas 2, globose. Fruit indehiscent.

Climbing, scandent or prostrate, creeping shrubs.
Leaves broad, cordate, glabrous above, white tomentose beneath. Bracts deciduous, softly woolly ... ... *A. speciosa.*
Leaves elongate, usually rounded at the base, glabrescent beneath. Bracts large, persistent, unequal, glabrescent or sparsely hairy beneath. *A. involucrata.*
Leaves broad, ovate, cordate, hispid above, densely white silky beneath. Bracts with 3-5 basal nerves, densely white silky beneath, persistent ... ... *A. sericea.*
Leaves acute, fulvous strigose beneath, scabrous above. Bracts linear, persistent, bristly ... *A. pilosa.*
Leaves elliptic, acute, base rhomboid, softly strigose on both surfaces. Bracts narrowly oblong ... ... *A. Lawii.*
Suberect. Leaves obovate, villous beneath. Bracts deciduous, small. Flowers bright purple ... *A. cuneata.*

A. speciosa, Sweet. Hort. Brit. ed. 2, 373; Fl. Br. I. 4. 185; Dalz. & Gibs. Bomb. Fl. 168; Brandis For. Fl. 343. Elephant creeper.

Samudra-shoka, guguli, Vern. Throughout the presidency, common near the coasts of the Konkan and North Kánara, in moist forests and open places near villages. Fl. Aug., Sept. Fr. Dec.

A. **involucrata**, Clarke ; Fl. Br. I. 4. 187 ; var.*inequalis.* Near the coasts of the Konkan and North Kánara ; very common near Kárwár in open places. Fl. July, Aug. Fr. Nov., Dec. Leaves often purple-coloured.

A. **sericea**, Dalz. & Gibs. Bomb. Fl. 169 ; Fl. Br. I. 4. 188 ; *Ipomœa bracteata,* Grah. Cat. Bomb. Pl. 131. *Gavel,* M. Konkan and North Kánara, in moist forests. Fl. Sept., Oct. Fr. C.S. The specimens of my collection have the bracts with 3-5 basal parallel nerves.

A. **pilosa**, Arn. Pugill. Pl. Ind. Or. 38 ; Fl. Br. I. 4. 189. Throughout the Konkan and North Kánara, very common in the forests near Yellápur, in moist shady places, also near the banks of nálás. Fl. Sept., Oct. Fr. Oct., Nov. ; often procumbent. Corolla dark red or purple.

A. **Lawii**, C.B.C. Fl. Br. I. 4. 190. The Konkan, Fl. Br. I.

A. **cuneata**, Ker in Bot. Reg. 1. 661 ; Fl. Br. I. 4. 191 ; Dalz. & Gibs. Bomb. Fl. 169 ; Brandis For. Fl. 344. Throughout the dry districts of the presidency, common in open situations, twining when it meets with support, but usually an erect shrub. Fl. Aug., Sept. Fr. C.S.

LETTSOMIA, Roxb.

Scandent shrubs. Leaves with a rounded or cordate base. Flowers in peduncled, axillary, dense cymes. Corolla funnel-shaped, limb plaited in bud. Stamens included or exserted. Ovary 2-celled, 4-ovuled. Fruit a dry indehiscent berry, 1-4-seeded.

Corolla small. Stamens exserted. Fruiting calyx within and fruit red	*L. aggregata.*
Corolla comparatively large. Anthers included. Leaves elliptic. Corolla glabrous. Fruit dark orange	*L. elliptica.*
Leaves ovate, cordate. Corolla densely hairy. Fruit red	*L. setosa.*

L. **aggregata**, Roxb. Fl. Ind. 1. 488 ; Fl. Br. I. 4. 191 ; *Argyreia aggregata,* Choisy ; Dalz. & Gibs. Bomb. Fl. 169. Dry districts of the presidency, very common in hedges in the Belgaum and Dhárwár districts and in the drier parts of North Kánara. Fl. Aug., Sept. Fr. Jany., Feb.

L. **elliptica**, Wight, note to Icones t. 1356 ; Fl. Br. I. 4. 192 ; *Argyreia elliptica,* Choisy ; Dalz. & Gibs. Bomb. Fl. 169. *Bondwail,* Vern. ; *sonariel, khedari,* M. On the Konkan and North Kánara gháts in moist forests, common in the Supa ghát jungles. Fl. Oct., Nov. Fr. Dec., Feb.

L. **setosa**, Roxb. Fl. Ind. 1. 490 ; Fl. Br. I. 4. 194 ; *Argyreia setosa,* Choisy ; Brandis For. Fl. 343 ; Dalz. & Gibs. Bomb. Fl. 168. Throughout the presidency, common in the Konkan and North Kanara from the coast inland. Fl. Oct., Jany. Fr. Feb., Mch.

IPOMŒA, Linn.

Herbs, rarely shrubs. Leaves entire lobed or digitate. Corolla bell or funnel-shaped, limb 5-plaited. Stamens unequal, included or exserted, not connivent over the ovary. Ovary 2-celled, 4-ovuled, rarely 4-celled, 4-ovuled or 3-celled, 6-ovuled. Fruit a dry dehiscent capsule.

Seeds villous, furred or woolly,
Leaves large cordate. Corolla large, salver-shaped
  with a long tube, white.
Seeds softly villous ... ... *I. grandiflora,*
Corolla medium-sized,
Leaves cordate, 5-7 palmately lobed. Corolla red,
  Seeds with long wool ... ... *I. digitata.*
Leaves cordate, entire or sinuate. Corolla pink,
  tube dark purple within. Seeds furred ... *I. sepiaria.*
Leaves shallowly cordate. Panicles many flowered. Corolla purplish-white, Seeds with short
  silky hairs ... ... *I. staphylina.*
Leaves cordate, acute. Corolla large, campanulate, white with a deep purple centre. Panicles large, many flowered. Seeds with long
  silky hairs... ... ... *I. campanulata.*
Leaves cordate or hastate, elliptic, acute. Cymes few flowered, compact. Corolla tubular, pure
  white. Seeds with stiff patent hairs ... ... *A. cymosa.*
Seeds glabrous,
Branchlets winged, leaves ovate. Cymes few flowered. Corolla large, white, tubular-campanulate ... ... *I. Turpethum.*
Leaves hairy, cordate, palmately lobed, corolla
  middle-sized, yellow. ... ... *I. vitifolia.*

I. grandiflora, Lamk. Ill. 1. 467 ; Fl. Br. I. 4. 198.   North Kánara, in moist forests and along the banks of nálás, very rare. Fl. R.S. Oct. Stems muricate.

I. digitata, Linn. ; Meissn. in Mart. Brasil. VII. 278 ; *Batatus paniculata,* Dalz. & Gibs. Bomb. Fl. 167. *Bhuikohola, vidarikand,* M. Throughout the Konkan and North Kánara, common near the sea-coast in moist forests. Fl. R.S., Aug., Sept.

I. sepiaria, Koen. ; Roxb. Fl. I. 1. 500 ; Dalz. & Gibs. Bomb. Fl. 166. *Amti,* M. Throughout the presidency, common in the Konkan and North Kánara, in hedges near the sea-coast. Fl. R.S.

I. staphylina, Roem. & Sch. Syst. IV. 249 ; Fl. Br. I. 4. 210. In hedges and along the banks of nálás in the southern parts of the Dhárwár district. Fl. Dec. Jany. Fr. Mch. Flowers rather small.

I. campanulata, Linn. ; Chois. Convolv. Or. 69 ; Dalz. & Gibs. Bomb. Fl. 165 ; Fl. Br. I. 4. 211. Common throughout the Konkan and North Kánara in deciduous forests, a very beautiful climber when in full bloom during Dec. and Jany. Fr. ripe May.

I. cymosa, Roem. & Sch. Syst. IV. 241 ; Fl. Br. I. 4. 211 ; *Convolvulus blandus,* Roxb. Fl. Ind. 1. 470. Throughout the presidency, very common in the forests of the Supa sub-division of North Kánara. Fl. Feb., Mch. Cymes usually recurved, corolla in bud hairy at apex.

I. Turpethum, Br. Prod. 485 ; Fl. Br. I. 4. 212 ; Dalz. & Gibs. Bomb. Fl. 165. *Nishottar, ter, shetar, phutkari,* M. Throughout the presidency in the dry districts, also in North Kánara, common. Fl. Oct., Nov. Fr. Dec., Jany.

I. vitifolia, Sweet. Hort. Brit. ed. 2. 372 ; Fl. Br. I. 4. 213 ; Dalz. & Gibs. Bomb. Fl. 165. _Navalichável,_ M. Throughout the moist forests of the Konkan and North Kanara, common on the Supa gháts. It is either a large climbing shrub or a creeper along road sides. Fl. Jany., Feb. Corolla sulphur yellow-coloured, capsule membranous. *I. bona-nox,* L. (*chandrakant,* M.), *I. muricata,* Jacq., (*bhauri,* M.) and other species are commonly cultivated. *I. biloba,* Forsk., (*maryadvel, marjad-vel,* M.) is common along the sea-coast.

PORANA, Burm.

Large climbers. Leaves entire. Flowers usually in terminal panicles, often small. Sepals much enlarged, wing like in fruit. Ovary 2-celled 4-ovuled. Capsule usually 1-seeded by abortion.

P. malabarica, C. B. C. Fl. Br. I. 4. 223 ; *P. racemosa,* Dalz. & Gibs. Bomb. Fl. 162. *Bowri, gariya,* Vern. Mahábaleshvar and the higher gháts, not seen in the southern parts of the presidency. Fl. Sept., Oct. Fr. C.S.

BREWERIA, R. Brown.

Herbs or climbing shrubs. Leaves simple. Flowers terminal or axillary ; bracts small. Corolla funnel-shaped, limb plaited. Stamens included, bases dilated. Ovary 2-celled, 4 ovuled ; styles 2, nearly distinct ; stigmas capitate. Capsule globose or ovoid.

B. cordata, Blume Bijd. 722 ; Fl. Br. I. 4. 223 ; *B. Roxburghii,* Dalz. & Gibs. Bomb. Fl. 162. Throughout the Konkan and North Kanara, rare ; usually near the coast. Fl. Oct., Dec. Fr. Jany.

NEUROPELTIS, Wall.

A lofty climber with tubercled stems. Leaves alternate. Flowers small, bracteate, in short villous racemes ; bract inconspicuous in flower. Corolla funnel-shaped, lobes entire. Stamens 5 ; filaments villous. Ovary 2-celled 4-ovuled ; styles 2, separate from the base. Fruit a small capsule sessile in the centre of a large, scarious, veined, orbicular bract.

N. racemosa, Wall. Cat. 1322. Fl. Br. I. 255 ; Bedd. Ic. Pl. Ind. Or. t. 291. In the evergreen forests of the Supa sub-division North Kánara, near Potolli. This remarkable plant has not been hitherto observed in the Bombay Presidency. Fl. Jany., Feb. Fr. May.

ORDER 63. **SOLANACEÆ.**

Herbs, shrubs or small trees.　Leaves alternate.　Flowers regular bisexual.　Calyx 5-lobed.　Corolla campannlate or rotate, lobes 5. Stamens 5, on corolla-tube.　Ovary 2-celled.　Seeds indefinite discoid.

Corolla-lobes plaited in bud　　...　　...　　...SOLANUM.
Corolla-lobes imbricate　...　　...　　...　　...LYCIUM.

SOLANUM, Linn.

Herbs, shrubs or small trees prickly or unarmed.　Leaves simple. Corolla rotate.　Stamens 5, anthers opening by terminal pores. Fruit a berry.

Small tree, unarmed.　Berry yellow ...　　...*S. verbascifolium.*
Small tree, prickly.　Berry red...　　...　　.. *S. giganteum.*

S. giganteum, Jacq.　Coll. IV. 125; Fl. Br. I. 4. 233; Dalz. & Gibs. Bomb. Fl. 175.　*Katri*, M.　*Cheena* or *chunna jhár* of Mahábaleshvar, Graham.　Throughout the North Konkan and Kánara, usually in evergreen forests or in the vicinity of evergreen forests.　Cymes stellately white woolly, corolla blue.　Fl. R.S.　Fr. Oct., Nov.

S. verbascifolium, Linn.　Dunal in DC. Prod. XIII. pt. 1. 114; Fl. Br. I. 4. 230; Dalz. & Gibs. Bomb. Fl. 175.　Throughout the presidency. In the Supa sub-division of North Kánara in deciduous forests, common. Fl. Oct., Nov.　Fr. Dec., Jany.　Corolla white, pubescence brownish yellow.

LYCIUM, Linn.

Spinous shrubs.　Leaves small, often fascicled.　Flowers small Stamens usually unequal, anther cells dehiscing longitudinally. Berry small, seeds few.

Stamens not exserted ...　　...　　...　　...*L. europæum*
Stamens exserted　　...　　...　　...　　...*L. barbarum.*

(These 2 species are doubtfully distinct. Fl. Br. I. 1. 241.)

I. europæum L.; Sibth. Fl. Græc. t. 236; Fl. Br. I. 4. 240; Brandis For. Fl. 345.　*Gangro, chirchitta*, M.　In the dry Deccan districts and Sind, in the plains.　Fl. Oct., Mch.

L. barbarum L.; Berss. Fl., Orient. IV. 289; Fl. Br. I. 4. 241; Brandis For. Fl. 345. Sind.　Fl. Oct., Mch.

ORDER 64. **GESNERACEÆ.**

Herbs or shrubs.　Leaves opposite or alternate.　Flowers herma-phrodite, irregular, bracteate, axillary or terminal.　Calyx 5-divided. Corolla gamopetalous; tube long or short; segments 5, imbricate in bud.　Stamens on the corolla-tube 5, didynamous or 2 only fertile. Disc various.　Ovary 1-celled or sub-2-celled; style linear; stigma capitate or 2-lobed; placentæ parietal, bifid; ovules many.　Fruit a capsule or berry.　Seeds many, minute; albumen 0 or scanty.

ÆSCHYNANTHUS, Jack.

Epiphytic shrubs. Leaves opposite, fleshy, nerves obscure. Calyx truncate, 5-fid. Corolla tubular-ventricose; limb 2-lipped. Stamens 4, didynamous. Disc annular. Ovary superior; stigma peltate; placentæ deeply inflexed, then recurved, bearing the ovules. Capsule long linear, loculicidally 2-valved. Seeds many, small, sessile, pendulous, rugose, hairy at the end next the hilum, 1 hair at the apex.

Æ. Perrottetii, A. DC. Prod. IX. 261; Fl. Br. I. 4. 339; Dalz. & Gibs. Bomb. Fl. 135. On the Konkan and North Kánara ghâts on trees in the evergreen forests, common on the Nilkund ghât. Fl. Dec., Jany. Fr. ripe next cold season.

Order 65. BIGNONIACEÆ.

Trees. Leaves opposite, pinnate. Flowers showy, hermaphrodite, irregular, racemose or panicled. Calyx campanulate, 2-5 lobed or spathaceous. Corolla 2-lipped. Stamens 4, didynamous or 5. Disc cushion-like or annular. Ovary 2-celled, ovules numerous. Fruit a capsule usually elongated. Seeds discoid, broadly winged, exalbuminous.

Stamens 5 perfect	...OROXYLUM.
Stamens 4 perfect.	
Leaves undivided, (compound in extra Indian species)	...TECOMA.
Leaves 1-2 pinnate.	
Calyx large, Bell-shaped	...PAJANELIA.
Calyx spathaceous	...DOLICHANDRONE.
Calyx irregularly 2-3 lobed	...HETEROPHRAGMA.
Calyx shortly lobed or truncate	...STEREOSPERMUM.
Leaves 2-3 pinnate	..MILLINGTONIA.

OROXYLUM, Vent.

A tree. Leaves bipinnate, large. Flowers large, in erect terminal racemes. Calyx persistent. Leathery, truncate. Corolla campanulate, oblique, 5-lobed. Stamens 5, all fertile; anthers 2-celled. Capsule large flat, linear, the dissepiment parallel to the valves, septicidally dehiscent. Seeds with a broad transparent wing.

O. indicum, Vent. Dec. Gen. Nov. 8; Fl. Br. I. 4. 378; Bedd. Fl. Syv. 148; *Spathodea indica,* Pers. *Calosanthes indica,* Brandis For. Fl. 347; Dalz. & Gibs. Bomb. Fl. 161. *Tetu,* K. Throughout the Konkan and North Kánara ghâts in moist forests, very common in the evergreen forests near the Nilkund ghât. Fl. May, July. Fr. ripe Jany.; remains long on tree. Flowers fœtid, fleshy.

MILLINGTONIA, Linn, f.

A large tree. Leaves opposite, bi-tri-pinnate. Flowers in numerous flowered panicles. Calyx 5-toothed, teeth recurved, truncate. Corolla with a long slender tube and an oblique mouth, 5-lobed.

Stamens 4; anthers 1-celled, the other cell reduced to a small hook. Capsule flat, linear, dehiscing at the edges. Seeds with a broad hyaline wing.

M. hortensis, Linn. f. Suppl. 291 ; Fl. Br. I. 4. 377 ; Bedd. Fl. Sylv. t. 249 ; Brandis For. Fl. 347 ; Wight. Ill. 101. *Nimi-chambeli, akas-nimb,* M. Cultivated along road sides and in parks and gardens throughout the presidency. Fl. Aug., Sept., also during the cold weather in the Konkan. Indigenous in the Malay Archipelago.

TECOMA, Juss.

Shrubs or small trees. Leaves simple or pinnate. Flowers in terminal racemes or panicles. Calyx campanulate, 5-toothed. Corolla tubular-ventricose, lobes 5. Stamens glabrous, inserted in the lower part of the corolla-tube ; anthers 2-celled. Ovary 2-celled on the annular disc. Capsule linear, loculicidally 2-valved. Seeds flat, imbricate with a thin, membranous wing.   •

Leaves simple   ..   ..*T. undulata.*
Leaves compound ...   .*T. Stans.*

T. undulata, G. Don. Gen. Syst. IV. 223 ; Fl. Br. I. 4. 378 ; Dalz. & Gibs. Bomb. Fl. 161 ; Brandis For. Fl. 352. *Lohuri, lohero,* Sind ; *rakta rohida,* M. In Sind and Gujarát, planted in gardens at Bombay, Fl. Mch.-Apl. Fr. May-July.

T. Stans, Spreng. Syst. 2. 834 ; Dalz. & Gibs. Bomb. Fl. Suppl. 55. In gardens, and run wild in the Konkan and North Kánara. Fl. R. and C. seasons. Corolla yellow. A native of South America.

DOLICHANDRONE, Seem.

Trees. Leaves 1-pinnate. Flowers in racemes or terminal panicles. Calyx spathaceous, circumciss-deciduous. Corolla tubular, 5-lobed ; lobes imbricate in bud. Stamens didynamous 4, with a rudimentary 5th. Anther-cells diverging. Ovary 2-celled. Capsule cylindrical or compressed, loculicidally dehiscent, septum free. Seeds discoid, winged.

Leaflets glabrous, acuminate, petiolule ½ inch. Capsule
   slightly compressed...   ...   .*D. crispa.*
Leaflets pubescent, obtuse, petiolule 0 or very short.
Capsule much compressed, falcate ...   .*D. falcata.*

D. crispa, Seem. in Seem. Jour. Bot. VIII. 381 ; Fl. Br. I. 4. 379 *Spathodea crispa,* Brandis For. Fl. 350 ; Bedd. Fl. Sylv. 168. Dalz. & Gibs. Bomb. Fl. 160. *Mushwal,* K. Common in the dry forests of the Kuputgudda range in the Dhárwár district. Fl. Aug., Sept. Fr. ripe next R.S.

D. falcata, Seem. in Seem. Jour. Bot. VIII. 381 ; Fl. Br. I. 4. 380 ; *Spathodea falcata,* Wall. ; Dalz. & Gibs. Bomb. Fl. 160 ; Bedd. Fl. Sylv. t. 71 ; Brandis For. Fl. 350 ; *Dolichandrone Lawii,* Seem. Fl. Br. I. 4.

380. *Mersingi, medasinghi,* M. Throughout the Konkan and North Kánara in deciduous forests. Fl. Mch., May. Fr. ripe Jany.

HETEROPHRAGMA, DC.

Trees. Leaves large, 1-pinnate. Flowers in terminal, woolly panicles. Calyx irregularly lobed or lipped. Corolla bell-shaped, 5-lobed, tomentose or glabrous. Capsule elongate, spuriously 4-celled. Seeds winged on 2 sides.

H. Roxburghii, DC. Prod. IX. 210 ; Fl. Br. I. 4. 381 ; Dalz. & Gibs. Bomb. Fl. 160 ; Bedd. Fl. Sylv. 169 ; *Spathodea Roxburghii,* Brandis For. Fl. 350. *Wurus, panlag,* M. Throughout the presidency in deciduous forests. Fl. Feb., Apl. Fr. R.S.

STEREOSPERMUM, Cham.

Trees. Leaves 1-2-pinnate. Flowers in large terminal panicles. Corolla bell or funnel-shaped, yellow or rose, 5-lobed. Stamens 4 with a rudimentary 5th. Disc fleshy, annular. Capsule elongate-linear or compressed quadrangular, loculicidally 2-valved. Seeds winged on 2 sides.

Leaves 1-pinnate.
Glabrous. Flowers yellow. Capsule obscurely quadrangular...*S. chelonoides.*
Inflorescence viscous pubescent. Corolla lilac or purple.
Capsule elongate-linear with 4 prominent lines...   ...*S. suaveolens.*
Leaves 2-pinnate. Capsule cylindric, tubercled   ...*S. xylocarpum.*

S. chelonoides, DC. Prod. IX. 210 ; Fl. Br. I. 4. 382 ; Bedd. Fl. Sylv. t. 72 ; Brandis For. Fl. 352 ; *Heterophragma chelonoides,* Dalz. & Gibs. Bomb. Fl. 160. *Padel, paral,* Vern. Throughout the Konkan and North Kánara in evergreen forests, also in deciduous moist forests. Fl. Apl., June. Fr. Aug., Jan.

S. suaveolens, DC. Prod. IX. 211 ; Fl. Br. I. 4. 382 ; Bedd. Fl. Sylv. 169 ; Brandis For. Fl. 351 ; *Heterophragma suaveolens,* Dalz. & Gibs. Bomb. Fl. 161. *Purul,* Vern. *Kálgari, patala,* M. In the deciduous forests of the dry districts of the presidency. Common in the Dhárwár district forests. Fl. Apl.-May. Fr. Nov.-Dec. Flowers viscous tomentose, fragrant.

S. xylocarpum, Wgt. Ic. t. 1335-6 ; Fl. Br. I. 4. 383 ; Bedd. Fl. Sylv. t. 70 ; Dalz. & Gibs. Bomb. Fl. 159 ; *Spathodea xylocarpa,* Brandis For. Fl. 349. *Kursing,* M. ; *genasing,* K.: *bairsing,* Vern. Throughout the presidency, common in the Dhárwár district dry forests. Fl. Apl., May. Fr. next Apl., May. Capsule long, woody.

PAJANELIA, DC.

Trees. Leaves 1-pinnate. Flowers large, in terminal panicles. Calyx bell-shaped. Corolla large, ventricose, lobes unequal. Stamens 4 with a rudimentary 5th. Disc annular. Ovary 2-celled. Capsule large, oblong, winged. Seeds winged all round.

P. Rheedii, DC. Prod. IX. 227; Fl. Br. I. 4. 384; Dedd. Fl. Sylv.
169. In the moist forests of North Kánara. Along the banks of nálás
in the Yellápur táluka. Fl. C.S. Flowers very large, filled when young
with mucilage.

Order 66. ACANTHACEÆ.

Herbs, shrubs or trees. Leaves opposite, simple. Flowers bracteate
and 2 bracteolate. Calyx 5-partite. Corolla 2-lipped or subequally
5-lobed. Stamens 4-2, on corolla-tube. Ovary superior, 2-celled.
Ovules 1 or more in each cell. Capsule loculicidal, often elastically
dehiscent. Seeds generally seated on retinacula and often clothed
with elastic hairs, seen when wetted.

Stamens usually 4.
Climbing shrubs. Retinacula 0 ... ...THUNBERGIA.
Erect shrubs. Anther-cells parallel, not spurred.
Seeds on retinacula. Corolla 5-lobed, not lipped ...STROBILANTHES.
Corolla deeply 2-lipped ... ...CALACANTHUS.
Corolla 1-lipped, the other lip obsolete ... ...ACANTHUS.
Stamens 2.
Anther-cells one higher than the other, often spurred
  at the base.
Corolla-tube short, lips subequal.
Anther-cells obtuse with a white basal appendage ...JUSTICIA.
Anther-cells acute at apex, minutely tailed at base...ADHATODA.
Corolla white with a very long tube, 2-lipped, lips
  very unequal ... ...RHINACANTHUS.

The following undershrubs are found throughout the forests of the Deccan,
Konkan and North Kánara :—*Dœdalacanthus roseus*, T. Anders.; *D. pur-
purascens*, T. Anders. (*gulsham*, M.); *D. montanus*, T. Anders.; *Blepharis
asperrima*, Nees; *Barleria Prionitis*, Linn. (*kalsunda, pivala-koranta*, M.);
B. cuspidata, Heyne, (common in the dry districts of the presidency); *B.
Hochstetteri*, Nees (Sind); *B. mysorensis*, Roth. (Dhárwár district); *B.
acanthoides*, Vahl. (Sind); *B. involucrata*, Nees (var. *elata*); *B. Lawii*, T.
Anders.; *B. montana*, Nees (*kolista, ikhari*, M.); *B. Gibsoni*, Dalz.; *B.,
cristata*, Linn.; *B. courtallica*, Nees; *B. strigosa*, Willd. (*wihii, kala
koranta*, M.); *Neuracanthus sphærostachyus*, Dalz.; *N. trinervius*, Wight,
(moist forests of North Kánara); *Crossandra undulæfolia*, Salisb. (*Aboli,
M.*) (often cultivated); *Eranthemum malabaricum, C. B. C.* (in evergreen
forests); *Gymnostachyum latifolium*, T. Anders. (in evergreen forests);
Ecbolium Linneanum, Kz. (*ranaboli, dhakta-adalsa*, M.) (in the evergreen
forests of North Kánara).

THUNBERGIA, Linn. f.

Climbing shrubs. Leaves cordate, angular. Flowers in axillary,
bracteate racemes. Calyx an entire or toothed ring, 2 bracteolate.
Corolla curved, ventricose with 5 subequal lobes. Anther-cells
ciliate. Ovary 2-celled, 2-ovuled. Disc thick, surrounding the
ovary. Capsule globular, beaked, 2-4-seeded at the base.

T. mysorensis, T. Anders. in Jour. Linn. Soc. IX. 448; Fl. Br. I. 4. 393 ; *Hexacentris mysorensis*, Wight Ic. t. 871 ; Dalz. & Gibs. Bomb. Fl. 182. In the evergreen forests of North Kánara near the Falls of Gairsoppah. Fl. Dec., Jany. An ornamental plant.

STROBILANTHES, Bl.

Shrubs or herbs. Leaves entire or serrate, often with raphides. Flowers axillary or in terminal, interrupted, or strobiliform spikes or panicles. Bracts leafy, deciduous or persistent, bracteoles small or 0. Calyx 5-divided, segments equal or unequal. Corolla campanulate, lobes nearly equal. Stamens 4 or 2. Capsule 4-2-seeded. Seeds compressed, hairy or glabrous ; hairs elastic when wetted ; retinacula curved.

Seeds glabrous.
Stem 4-winged. Flowers in capitate, axillary heads.
  Corolla white, hairy within. Capsule 2-seeded...   ...*S. barbatus.*
Flowers spicate, spotted-purple. Calyx much enlarged
  in fruit. Capsule 4-seeded.
Spikes hairy   ...   .*S. warreensis.*
Spikes glabrous ...   .*S. ciliatus.*
Seeds elastically hairy when wetted.
Flowers in small dense cymes. Bracts coriaceous, softly
  hairy. Corolla white...   ...   .*S. lupulinus.*
Spikes subglobose. Bracts membranous, glabrous, con-
  cave, white. Corolla blue   ....*S. Heyncanus.*
Heads of flowers ovoid. Bracts viscous-hairy. Corolla pale
  blue or nearly colourless   ...   .*S. ixiocephalus.*
Spikes short, numerous, along the naked branches. Bracts
  narrow, hairy, persistent. Corolla deep blue ...   ...*S. scrobiculatus.*
Flowers strobilate.
Leaves petioled, glabrate, lineolate above. Bracts white,
  concave, glabrous in flower. Corolla blue, purple or rose. *S. callosus.*
Leaves petioled, softly villous. Bracts oblong, or ovate,
  softly hairy, viscous, bracteoles linear. Corolla pale blue.*S. asper.*
Leaves sessile. Bracts large, acute, hairy, bracteoles 0.
  Corolla pale purple ...   .*S. sessilis.*
Spikes lax, viscid. Leaves sessile, winged at base, not
  perfoliate, pale beneath. Bracts oblong, linear. Corolla
  blue   ...   .*S. perfoliatus.*

S. barbatus, Nees in Wall. Pl. As. Rar. III. 85 ; Fl. Br. I. 4. 437 ; *S. tetrapterus*, Dalz. & Gibs. Bomb. Fl. 187. In the evergreen forests of the Konkan and North Kánara. Common in the Yellápur Táluka. Flowers once in every seven years, a gregarious shrub with winged stems and white flowers, hairy within. Fl. Sept., Oct. Fr. Apl., May. The bracts and bracteoles become densely viscous hairy in fruit.

S. warreensis, Dalz. in Hook. Kew Jour. II. 341 ; Dalz. & Gibs. Bomb. Fl. 187 ; *S. parviflorus*, Bedd. Ic. Pl. Ind. Or. t. 197. Evergreen forests of the Konkan and North Kánara, common in the Nilkund ghát forests. Fl. Oct. Fr. ripe Feb. Leaves with 7-8 pairs of nerves. Peduncles jointed, softly woolly. Capsule ¾ inch, exserted from the hairy calyx. Seeds thin, flat, shining, on large retinacula.

S. ciliatus, Nees in Wall. Pl. As. Rar. III. 85 ; Fl. Br. I. 4. 439 ;
Bedd. Ic. Pl. Ind. Or. t. 211 ; Wight Ic. t. 1517. Evergreen forests of
the Yellápur táluka of North Kánara. Fl. Oct., Nov. Leaves with 4-5 pairs
of nerves. Peduncles jointed ; leafy, bracteate, glabrous. Stamens
exserted. My specimens agree well with the plate in Wight's Icones.

S. lupulinus, Nees in Wall. Pl. As. Rar. III. 85 ; Fl. Br. I. 4. 443 ; *S.
Dalzellii,* T. Anders. On the Konkan and North Kánara gháts in moist
forests, common near Yellápur, and in the forests near the Ram and Coessi
Gháts. Fl. R.S. Fr. C.S.

S. Heyneanus, Nees in Wall. Pl. As. Rar. III. 85 ; Fl. Br. I. 4.
443 ; Dalz. & Gibs. Bomb. Fl. 187 ; *S. rugosus,* Wight Ic. t. 1619 ;
S. asperrimus, Dalz. & Gibs. Bomb. Fl. 187. *Akra,* M. On the gháts of
the Konkan and North Kánara, common in the evergreen forests between
the Nilkund and Gairsoppah gháts. Flowers apparently every year during
Nov., Dec. The bracts are white, glabrous. Corolla blue. A large
shrub. I have a small shrub from the Bababuden hills of Mysore, 5,000
ft. alt, which has the red bracts and deep blue flowers of *S. rugosus,* Wgt.
Ic. t. 1619. Fl. during Aug., Sept. Fr. C.S.

S. ixiocephalus, Benth. in Flora 1849, p. 557; Fl. Br. I. 4. 444 ; Bedd.
Ic. Pl. Ind. Or. t. 203 ; *S. Neesianus,* Wight; Dalz. & Gibs. Bomb.
Fl. 188 ; *S. glutinosus,* Grah. Cat. Bomb. Pl. 162 ? Throughout the
Konkan gháts, common in North Kánara from the coast inland in both
evergreen and deciduous forests, often on laterite. Flowers annually
during the cold season. Fr. H.S. A large gregarious shrub, with pale
blue or almost white flowers in viscous hairy heads. The branchlets are
sometimes clothed with long woolly hairs.

S. scrobiculatus, Dalz. MS. Fl. Br. I. 4. 445. On the higher
gháts of the Konkan and North Kánara, common on the summit of
Dursing gudda in the Supa táluka of North Kánara, 3,400 feet elevation.
Fl. annually Dec., Jany. Fr. Jany., Feb. I have a shrub from a lower level
(2,000 ft.) in the same táluka, which differs from *S. scrobiculatus,* in having
longer spikes, larger persistent bracts and much smaller corollas. It
grows in the beds of streams or on the rocky banks. Fl. Fr. Feb., Mch.

S. callosus, Nees in Wall. Pl. As. Rar. III. 85 ; Dalz. & Gibs. Bomb.
Fl. 188 ; *S. Grahamianus,* Wight Ic. t. 1520 ; Dalz. & Gibs. Bomb. Fl.
187. *Karvi,* Vern. Covers large areas on the Konkan and North Kánara
gháts, and forms the undergrowth in many of the deciduous moist forests.
Sometimes a very large shrub (30 ft. in height and 2½ in. in diameter). A
general flowering takes place every seven years. The white, glabrous bracts
become covered after fertilization of the ovaries, with viscous, strongly
smelling hairs. The flowers vary in colour from purple-blue to pink.
Last general flowering in North Kánara took place in Sept., Oct., 1887.
The capsules ripen during the cold and hot seasons and are elastically
dehiscent, making a peculiar, almost continuous, noise during the shedding
of the seeds in a forest of this species. The viscous, fruiting bracts are no
doubt protective against boring insects.

S. asper, Wight Ic. t. 1518, not of Dene. Fl. Br. I. 4. 452. *S.* Konkan ghåts, 5-8,000 ft. Stocks. Bababuden hills, Mysore, 6,000 ft. alt. Fl. Aug. Sept. Fr. C.S.

S. sessilis, Nees in Wall. Pl. As. Rar. III. 85; Fl. Br. I. 4. 452; var. *Ritchiei, S. sessiloides,* Dalz. & Gibs. Bomb. Fl. 187. *Bukra,* M. Very common on the ghåts of the Belgaum district. In North Kánara common on laterite in the Supa, Siddápur and Sirsi tálukas, also on the ghåts of the Konkan near Bombay. A gregarious small species. General flowering once in fifteen years, last took place near the Rám ghát of the Belgaum district in Sept. 1889. Fr. ripe Nov., Dec. Leaves shortly petiolate, more glabrous than in *S. sessilis,* Nees.

S. perfoliatus, T. Anders. in Jour. Linn. Soc. IX. 471; Fl. Br. I. 4. 458; *Endopogon integrifolius,* Dalz. & Gibs. Bomb. Fl. 185. On the Konkan ghåts, near Panvel, Dalz.; common in North Kánara; usually in evergreen or near evergreen forests from Ainshi southwards, gregarious. General flowering once in seven years; last took place near Ainshi from Dec. till Mch. 1887. Fr. Mch.-May. Leaves light-coloured or steel-grey beneath. Bracts linear, viscous-ciliate. Flowers blue, very handsome.

CALACANTHUS, T. Anders.

A shrub. Leaves large, entire. Flowers in terminal, bracteate, hairy spikes. Calyx 5-divided. Corolla purple, deeply 2-lipped. Stamens didynamous. Ovary glabrous; style hairy. Seeds densely hairy.

C. Dalzelliana, T. Anders.; Benth. in Gen. Plant. II. 1088; Fl. Br. I. 4. 478; *Lepidagathis grandiflora,* Dalz. & Gibs. Bomb. Fl. 190. *Motayén,* M. On the Konkan ghåts, near Bombay. Dalz. Common on the Supa ghåts of North Kánara. Fl. Oct., Nov. Fr. January. Flowers annually.

ACANTHUS, Linn,

Shrubs or herbs. Leaves often spinous. Flowers in terminal, bracteate spikes. Calyx 4-partite. Corolla 1-lipped, lip 3-lobed. Stamens didynamous; anthers bearded. Capsule compressed, 4-seeded. Seeds large, flat; retinacula thick.

A. ilicifolius, Linn. Benth. Fl. Austral. IV. 548; Fl. Br. I. 4. 481; *Dilivaria ilicifolia,* Nees; Dalz. & Gibs. Bomb. Fl. 192. Sea holly. *Márándi, moránná,* M. Along the coasts of the Konkan and North Kánara. Common in tidal swamps and along the banks of creeks and tidal rivers. Fl. Apl., May. Fr. June, July.

JUSTICIA, L.

Herbs or shrubs. Leaves entire. Calyx 4-5-divided. Corolla 2-lipped. Stamens 2, anther-cells oblique, the lower one spurred. Capsule 2-4-seeded, base narrow, sterile. Seeds cordate, scaly or tubercular, without elastic hairs.

Bracts larger than calyx.
Spikes interrupted. Bracts green, herbaceous. Corolla
  white with purple spots in the throat ... ... *J. montana.*

Spikes continuous. Bracts white, green-nerved.
Corolla white, rose-spotted ... ... *J. Betonica.*
Bracts linear, shorter than the calyx. Spikes inter-
rupted.
Leaves lanceolate, glabrous. Flowers and capsule
glabrous ... ... *J. Gendarussa.*
Leaves elliptic, pubescent. Flowers and capsule pubes-
scent ... ... *J. wynaadensis.*

J. montana, Wall. Cat. 2471, not of Roxb.; Fl. Br. I. 4. 525; *Hemichoriste montana,* Dalz. & Gibs. Bomb. Fl. 194. Konkan and North Kánara gháts, in evergreen forests. Fl. Dec., Feb. Fr. Jany., Mch. Seeds very rugose when ripe.

· **J. Betonica,** Linn.; Roxb. Fl. Ind. 1. 128; Fl. Br. I. 4. 525; var. *ramosissima; Adhatoda ramosissima,* Nees; Dalz. & Gibs. Bomb. Fl. 193. *Sut,* M. Throughout the presidency, often in hedges and near villages. Fl. Fr. Dec., Feb.

J. Gendarussa, Linn. f. Suppl. 85; Fl. Br. I. 4. 532; Dalz. & Gibs. Bomb. Fl. Suppl. 71. *Bakas, teo,* M. Throughout the presidency, naturalized near villages and along the banks of nálás. Fl. & Fr. Jany., Mch.

J. wynaadensis, Wall. Cat. 2474; Fl. Br. I. 4. 533; *Adhatoda wynaadensis,* Nees; Dalz. & Gibs. Bomb. Fl. 194. Common in the evergreen forests of North Kánara, also in the forests of the Belgaum gháts. Fl. and Fr. Dec., April.

ADHATODA, Nees.

Large shrubs. Characters of *Justicia,* but anther-cells are acute at apex, scarcely spurred at base,

A. vasica, Nees in Wall. Pl. As. Rar. III. 103; Fl. Br. I. 4. 540; Dalz. & Gibs. Bomb. Fl. 194. *Adulsa, bakus, vasuka,* Vern. Throughout the presidency, common as a hedge plant. Flowers at different times throughout the year. Corolla white, throat with a few irregular rose-coloured bars.

RHINACANTHUS, Nees.

Shrubs. Leaves entire. Flowers in dense or spreading, lax panicles. Calyx small. Corolla white, limb 2-lipped, upper short, 2-lobed, lower broad, 3-lobed; tube long, slender. Stamens 2 at the top of the corolla-tube; anther-cells muticous. Capsule clavate, 4-seeded. Seeds compressed, tuberculate.

R. communis, Nees in Wall. Pl. As. Rar. III. 109; Fl. Br. I. 4. 541; Dalz. & Gibs. Bomb. Fl. 194. *Gájkarni.* Apparently wild in the deciduous forests of the Belgaum district, also on the gháts of North Kánara near Supa and generally throughout the presidency. Fl. Mch., Apl. Fr. May, June.

ORDER 67. VERBENACEÆ.

Herbs, shrubs or trees. Leaves opposite or whorled, simple or digitate. Inflorescence cymose. Cymes often panicled, bracteate.

Calyx 5-4-toothed, persistent. Corolla gamopetalous, 2-lipped or
subequally 5-lobed. Stamens generally 4, didynamous. Ovary
superior, 2-4-celled, 4-ovuled. Fruit a drupe, 1-4-celled; cells
1-seeded.

Seeds with integuments.
Inflorescence in many flowered spikes or bracteate heads.
Pyrenes of fruit 2-1, 1-seeded... ... ... LANTANA.
Inflorescence cymose. Cymes panicled.
Corolla regular.
Drupe of 4, 1-celled pyrenes ... ... CALLICARPA.
Drupe of 1, 4-celled pyrene, in accrescent calyx ... ... TECTONA.
Corolla 2-lipped.
Drupe with 1 pyrene.
Leaves simple. Flowers small ... ... PREMNA.
Leaves simple ; flowers large ... ... GMELINA.
Leaves digitate ... ... VITEX.
Drupe with four pyrenes. ... ... CLERODENDRON.
Bracteate capitate cymes, 3-9-flowered. Bracts 6, large,
    involuceriform ... ... SYMPHOREMA.
Seeds without integuments, germinating on plant... ... AVICENNIA.

LANTANA, Linn.

Shrubs. Branches 4-sided, sometimes prickly. Leaves simple,
crenate often rugose. Flowers in peduncled bracteate heads. Calyx
small, membranous. Corolla-tube slender, limb spreading, lobes
unequal. Ovary 2-celled, 2-ovuled. Drupe fleshy, containing two
bony pyrenes.

Branches hairy or strigose, not prickly. Bracts broad,
    veined ... ... *L. indica.*
Branches with recurved prickles. Bracts linear, small ... *L. Camara.*

L. indica, Roxb. Hort. Beng. 46 ; Fl. Ind. III. 89 ; Fl. Br. I. 4. 562 ;
L. alba. Dalz. & Gibs. Bomb. Fl. 198 ; Brandis For. Fl. 369. *Ghaneri*,
M. Thinly scattered throughout the presidency. Dalzell ; common in
the Kadur district of Mysore, near coffee plantations, (*coffee weed*). The
variety *albiflora*, Wight, is also abundant at the base of the Bababuden
hills, in dry forests. Flowers throughout the year.

L. Camara, Linn. Fl. Br. I. 4. 562; *L. aculeata*, Linn. ; Dalz. & Gibs.
Bomb. Fl. Suppl. 68. *Ghaneri*, M. Very common in waste places
near villages throughout the presidency, often in hedges ; sometimes
scandent ; an introduced shrub. Fl. and Fr. throughout the year.

CALLICARPA, Linn.

Stellately tomentose shrubs or trees. Leaves toothed or subentire. Cymes axillary. Calyx very small, limb minutely 4-lobed,
not accrescent. Corolla small, tubular. Stamens 4; anthers exserted. Drupe small ; pyrenes 4.

C. lanata, Linn. Mant. 331; Fl. Br. I. 4. 567 ; Brandis For. Fl. 368 ;
Bedd. Fl. Sylv. 173 : *C. cana*, Dalz. & Gibs. Bomb. Fl. 200, not of Linn.
Kan phulia, M. ; *eisur, eshwar*, Vern. In the evergreen forests of the
Konkan and North Kánara gháts, common, sometimes a small tree, but

usually a large shrub. Fl. Feb., May. Fr. Aug., Oct. Stamens exserted.

TECTONA, Linn. f.

Trees stellately tomentose. Leaves large, entire. Cymes in large terminal panicles. Bracts small, narrow. Calyx 5-6-lobed, in fruit enlarged, urceolate. Corolla equally 5-6-lobed. Ovary 4-celled. Drupe densely hairy, 4-celled.

T. grandis, Linn. f. Suppl. 151 ; Fl. Br. I. 4. 570 ; Dalz. & Gibs. Bomb. Fl. 199 ; Bedd. Fl. Sylv. t. 250 ; Brandis For. Fl. 354. *Ságwan, ság,* M. ; *tegina,* K. Teak tree. Throughout the presidency in deciduous forests, grows to a large .size along the well-drained slopes of the Kálá-nadi in North Kánara, the underlying rock being usually metamorphic schist. Fl. June, Aug. Fr. ripe Nov., Jan.

PREMNA, Linn.

Trees or shrubs, sometimes climbing. Leaves entire or toothed. Cymes terminal, panicled, or corymbose. Flowers greenish-white or purplish, often odorous. Calyx 2-lipped or cup-shaped. Corolla short, tubular ; limb 2-lipped. Drupe small, globose, 1-4-celled and seeded.

Climbing shrub. Corymbs large, purple. Calyx cup-
    shaped...   ...   .   *P. coriacea.*
Erect tree, sometimes spinous. Corymbs greenish-yellow.
Calyx 2-lipped ...   .   *P. integrifolia.*

P. coriacea, Clarke, Fl. Br. I. 4. 573 ; *P. scandens,* Dalz. & Gibs. Bomb. Fl. 199, not of Roxb. *P. cordifolia,* Grah. Cat. Pl. Bomb. 155, not of Roxb. *Chambari,* Vern. Konkan and North Kánara gháts, in moist forests. Common from Ainshi southwards. The large purple corymbs (not greenish-yellow) and young leaves appear in Apl., May. Fr. ripe June, an ornamental climber. Clarke no doubt follows Dalzell and Graham as to the colour of the flowers ; but my plant, determined at Kew and Calcutta, has the corymbs purplish and not greenish-yellow.

P. integrifolia, Linn. Mant. 252 ; Fl. Br. I. 4. 574 ; Brandis For. Fl. 366 ; Grah. Cat. Bomb. Pl. 155 ; *P. serratifolia,* Bedd. Fl. Sylv. 172 ; *P. latifolia,* Dalz. & Gibs. Bomb. Fl. 200. *Khárá-narvel, aran,* M. Throughout the Konkan and North Kánara, usually near the sea-coast. Fl. June, July. Fr. ripe July, Aug. The greenish-yellow flowers have a peculiar odour. The seeds are pear-shaped with a fluted top—one to each fruit, not 3-4, in the North Kánara plant. *P. cordifolia,* Dalz. & Gibs. Bomb. Fl. 199 ; Grah. Cat. Bomb. Pl. 155, is doubtfully referred to *P. corymbosa,* Rottl. in the Fl. of Br. I. by Clarke ; it is noted from Khandála on the Konkan gháts ; leaves long acuminate, seeds rough. Is this simply a form of *P. integrifolia,* Linn.? In Graham's Catalogue, p. 155, there is a species, called *P. Nimmoniana,* G. (*gura, kal gura,* Vern.), which is said to grow at Mahábaleshvar, also on the Rotunda ghát. This species is not quoted in the Fl. Br. I.

GMELINA, Linn.

Trees or shrubs. Leaves entire or lobed. Cymes panicled.
Flowers large, yellow. Calyx campanulate, 5-toothed. Corolla-
tube ventricose upwards, limb 4-5-lobed. Stamens didynamous.
Style slender, shortly bifid. Fruit a succulent drupe.

Unarmed tree	...	...	...	...	...	...	*G. arborea.*
Spinescent shrub		...	...	...	...	...	*G. asiatica.*

G. arborea, Linn.; Roxb. Fl. Ind. III. 84 ; Fl. Br. I. 4. 581 ; Dalz.
& Gibs. Bomb. Fl. 201 ; Bedd. Fl. Sylv. t. 253 ; Brandis For. Fl. 364 ;
Gamble Ind. Timbers 295. *Shivani,* K. ; *shiran,* M. Throughout the
presidency in deciduous forests, yields an excellent, strong, light timber,
useful for ornamental work and which does not warp or shrink. Fl.
March, April. Fr. May-June.

G. asiatica, Linn. ; Roxb. Fl. Ind. III. 87 ; Brandis For. Fl. 365 ;
Bedd. Fl. Sylv. 172 ; Grah. Cat. Bomb. Pl. 158. Common in gardens
Bombay, forms an excellent hedge plant. Fl. throughout the year.

VITEX, Linn.

Trees or shrubs. Leaves usually 3-5-foliate. Calyx truncate or
5-toothed. Corolla 2-lipped, 5-toothed. Central lobe of lower lip
largest. Stamens 4, in pairs, exserted. Ovary 2-4-celled, 4-ovuled.
Drupe globose, supported by the more or less accrescent calyx.

Flowers in terminal panicles.

Leaves 1-3-foliate, leaflets sessile, white mealy beneath	...					*V. trifolia.*
Leaves 3-5-foliate, leaflets petioluled, grey-pubescent beneath	...	...	...	...	...	*V. Negundo.*
Leaves 3-5-foliate ; petioles 0 or slightly winged	...	...				*V. altissima.*
Leaves 3-5-foliate ; petioles broadly winged, wing cordate at the base	...	...	...	...	...	*V. alata.*
Flowers in axillary corymbose cymes	...	...	...	...	...	*V. Leucoxylon.*

V. trifolia, Linn. f. Suppl. 293; Bedd. Fl. Sylv. 172 ; Brandis For.
Fl. 370. *Nirgunda,* Vern. : *indráni, lingur,* M. Scattered throughout
the presidency. Scarcely differs from the next species, has shorter calyx-
teeth than *V. Negundo,* Linn. The leaves are sometimes unifoliate.
Fl. Mch.-June and at other times during the year.

V. Negundo, Linn. Roxb. Fl. Ind. III. 70 ; Fl. Br. I. 4, 583 ; Bedd.
Fl. Sylv. 171 ; Brandis For. Fl. 369 ; *V. bicolor,* Willd.; Dalz. & Gibs.
Bomb. Fl. 201. *Lekkigidda, shurnboli,* K.; *nirgunda, nengar,* H. Indian
privet. Throughout the presidency and Sind, very common along
the banks of rivers and in moist situations in or near deciduous forests.
Fl. Mch.-May ; also throughout the year at other times.

V. altissima, Linn. f. Suppl. 294; Fl. Br. I. 4. 585 ; Dalz. & Gibs.
Bomb. Fl. 201 ; Bedd. Fl. Sylv. t. 252 ; Brandis For. Fl. 370. *Balgi*
nauladi-mara, K. In the evergreen forests of the Konkan and North
Kánara, abundant in the Kumta and Yellápur táluka forests. Yields
a valuable timber. Fl. May. Fr. June, July.

V. alata, Heyne; Roth. Nov. Sp. 316; Fl. Br. I. 4. 584; Dalz. & Gibs. Bomb. Fl. 201; Grah. Cat. Bomb. Pl. 155. In the moist forests of the Konkan and North Kánara; common in the Yellápur táluka in evergreen forests, scarcely separable from *V. altissima*, L. f. Fl. Apl.-May. Fr. June, July.

V. Leucoxylon, Linn. f. Suppl. 293; Fl. Br. I. 4. 587; Dalz. & Gibs. Bomb. Fl. 201; Bedd. Fl. Sylv. 171; Brandis For. Fl. 370. *Hola naki, senkani*, K.; *songarbi*, M.; *sherus*, Vern. Throughout the Konkan and North Kánara gháts, along the banks of rivers and nálás, in moist forests, attains a considerable size; stem thick and short. Fl. Feb., Apl. Fr. May, June. Leaves often 5-foliate, petiole not winged.

CLERODENDRON, Linn.

Shrubs or small trees. Leaves simple, opposite or whorled. Flowers axillary or terminal. Calyx 5-toothed or lobed, accrescent in fruit. Corolla-tube long, cylindric; lobes 5, equal. Stamens 4, exserted. Ovary 4-celled, cells 1-ovuled; style filiform, bifid. Fruit a drupe, dry or succulent.

Leaves opposite. Corolla white.
Leaves small, obovate. Cymes few flowered, axillary.
Calyx in fruit campanulate, slightly lobed or truncate ... *C. inerme.*
Leaves ovate, sinuate. Flowers in axillary or terminal panicles.
Calyx deeply cleft, lobes caudate-acuminate ... ... *C. phlomoides.*
Leaves large, ovate, cordate, tomentose. Flowers in terminal panicles.
Calyx in fruit much enlarged, red within ... ... *C. infortunatum.*
Leaves long, lanceolate. Flowers in lax, terminal panicles.
　Calyx ½ inch, deeply divided ... ... *C. Siphonanthus.*
Leaves in verticels of 3. Flowers blue. Calyx ¼ inch,
　cup-shaped ... ... *C. serratum.*

C. inerme, Gærtn. Fruct. 1. 271, t. 57, fig. 1. Fl. Br. I. 4. 589; Dalz. & Gibs. Bomb. Fl. 200; Bedd. For. Man. 174; Brandis For. Fl. 363. Common near the sea-coast of the Konkan and North Kánara. Flowers throughout the year, abundantly during the rainy season. Tube of corolla nearly 1 inch long.

C. phlomoides, Linn. f. Suppl. 292; Fl. Br. I. 4. 590; Dalz. & Gibs. Bomb. Fl. 200; Bedd. Fl. Sylv. 174; Brandis For. Fl. 363. *Irun*, Vern.; *arni*, Gujarát. In the dry districts of the presidency, usually in hedges, common in the south of Dhárwár. Fl. throughout the year.

C. serratum, Spreng. Syst. II. 758; Fl. Br. I. 4. 592; Dalz. & Gibs. Bomb. Fl. 200; Brandis For. Fl. 364. *Barungi*, Vern. Throughout the presidency, often in deciduous forests, a small shrub, scarcely woody. Fl. May-Aug. Fruit a drupe of four pyrenes.

C. infortunatum, Gærtn. Fruct. 1. 271, t. 57, fig. 1; Fl. Br. I. 4. 594; Bedd. Fl. Sylv. 173; Dalz. & Gibs. Bomb. Fl. 200. *Bhant, bhat*, Vern. Throughout the presidency in deciduous forests, very common in North Kánara. Fl. Dec.-Apl. Fr. R.S. The flowers are tinged with red. Calyx in fruit red, leathery.

C. Siphonanthus, Br. in Ait. Hort. Kew ed. 2. IV. 65; Fl. Br. I. 4. 595; Brandis For. Fl. 364; Dalz. & Gibs. Bomb. Fl. Suppl. 69. *Barangi*, Vern. Cultivated in gardens and near villages throughout the presidency, sometimes a garden escape. Fl. June-Oct. Tube of white corolla very long, 4 inches.

SYMPHOREMA, Roxb.

Scandent shrubs. Leaves entire or toothed. Flowers in 7-flowered capitate cymes. Involucre of 6, obovate bracts, accrescent in fruit. Calyx 4-8-toothed. Corolla small, white. Stamens 6-16; anthers exserted. Ovary 2-celled, cells 2-ovuled. Fruit dry, 1-seeded.

Leaves thin, entire or slightly toothed. Flowers small, ½ inch. Bracts linear, spathulate, membranous. Stamens few ... ... *S. involucratum.*
Leaves leathery, deeply repand or toothed. Flowers 1-inch. Bracts obovate, broad. Stamens many ... ... *S. polyandrum.*

S. involucratum, Roxb. Cor. Pl. II. 46. t. 186; Fl. Br. I. 4. 599; Dalz. & Gibs. Bomb. Fl. 199. Throughout the Konkan and North Kánara in moist forests, common in the forests near Yellápur. Fl. Mch., Apl. Fr. May, June.

S. polyandrum, Wight Ic. t. 363; Fl. Br. I. 4. 599. In the Belgaum and Dhárwár districts in open situations, also in dry deciduous forests. Bare of leaves when in flower. Flowers profusely during Feb., Mch. Fr. May, takes the place of *S. involucratum* in the dry zone.

AVICENNIA, Linn.

A small tree. Leaves coriaceous, entire. Flowers sessile, yellow in peduncled heads. Sepals 5. Bracts and bracteoles small. Corolla 4-divided; divisions equal. Stamens 4, on the throat of the corolla. Capsule compressed, ovate, mucronate, 1-seeded; cotyledons thick, fleshy, folded; radicle villous.

A. officinalis, Linn. Schauer in DC. Prod. XI. 700; Fl. Br. I. 4. 604; Bedd. Fl. Sylv. 174; Brandis For. Fl. 371. White mangrove. *Tivar*, M. Common in salt marshes and along the banks of tidal rivers and creeks, throughout the Konkan and North Kánara. Leaves glaucous beneath. Fruit tomentose without. Fl. Mch.-Apl., May. Fr. ripe July, Aug.

ORDER 68. **LABIATÆ.**

Herbs rarely shrubs, usually with many oil-glands. Stem quadrangular. Leaves opposite or whorled, stipules 0. Flowers irregular. Calyx 4-5-cleft or 2-lipped. Corolla monopetalous, limb 4-5-lobed or 2-lipped, lobes imbricate in bud. Stamens on the corolla-tube, 4, didynamous, or the 2 upper imperfect. Disc prominent. Ovary 4-lobed and celled, ovule solitary erect, 1 in each cell. Fruit of 4, small, 1-seeded nuts, enclosed in the calyx.

COLEBROOKIA, Smith.

A woolly shrub. Leaves opposite or in verticels of 3. Flowers small, in panicled spikes of dense whorls. Calyx with long feathery teeth. Corolla very small. Stamens 4. Nutlets hairy, obovoid.

C. oppositifolia, Smith. Exot. Bot. II. t. 111 ; Fl. Br. I. 4. 642 ; *C. ternifolia*, Roxb. ; Dalz. & Gibs. Bomb. Fl. 209. *Dussarika jhár*, *bhamini*, Vern. Throughout the moist forests of the Konkan and North Kánara ghâts, common. Fl. Dec., Jan. Fr. Mch., Apl. Graham says "spikes like a squirrel's tail." The feathery calyx divisions are accrescent in fruit.

Several species of *Ocimum, Orthosiphon, Plectranthus* and *Pogostemon* are common undershrubs throughout the presidency.

ORDER 69. NYCTAGINEÆ.

Herbs, shrubs or trees. Leaves usually opposite, simple. Inflorescence various, flowers often bracteate-involucrate, or bracts small, deciduous. Perianth petaloid ; monopetalous, tube persistent ; limb 3-5-lobed, plaited in bud. Stamens 4-5, rarely more (up to 20), sometimes unilateral, filaments unequal, inflexed in bud. Ovary free, 1-celled ; stigma simple or multifid ; ovule 1, erect. Fruit membranous, enclosed in the perianth-tube. Seed erect, albumen floury or soft. Embryo curved, folded or convolute.

PISONIA, Linn.

Trees or shrubs, often spinous. Leaves opposite or alternate. Flowers in corymbs, usually diœcious. Perianth 5-10-toothed. Stamens 6-10, exserted. Ovary oblique ; stigma capitate. Fruit clavate ; cotyledons crumpled.

P. aculeata, Linn. ; Fl. Br. I. 4. 711 ; Grah. Cat. Bomb. Pl. 167. South Konkan. Grah. Fl. Jan., Feb. Spines axillary, recurved.

ORDER 70. POLYGONACEÆ.

Herbs or shrubs. Leaves usually alternate. Stipules scarious, sheathing. Flowers bisexual, jointed on the pedicil. Perianth of 3-6, free or connate, persistent sepals. Stamens 5-8, opposite the sepals. Ovary free, 2-4-angled ; styles 1-3 ; ovule 1, basal. Fruit a hard nut, enclosed in the calyx. Albumen floury or hard horny.

Stamens 12-18. Styles 4	...	...	...	... CALLIGONUM.
Stamens 8. Styles 3	...	...	...	... PTEROPYRUM.

CALLIGONUM, Linn.

Shrubs. Leaves alternate, linear or 0. Sheathing stipule short. Flowers 2-sexual. Sepals 5. Ovary 4-angled. Fruit a 4-angled nut, winged, hairy or bristly.

C. polygonoides, Linn. Meissn. in DC. Prod. XIV. 1. 29 ; Fl. Br. I. 5. 22. Sind.

PTEROPYRUM, Jaub. & Spach.

Rigid shrubs. Leaves small, alternate or fascicled. Sheathing stipule short or 0. Flowers small. Sepals 5. Ovary 3-angled; styles 3. Fruit a 6-winged, beaked nut.

P. Oliveri, Jaub. & Spach. Ill. Pl. Orient. II. 9, t. 108; Fl. Br. I. 5. 23. Wight Ic. t. 1809. Sind. Stocks. Fl. Sept., Oct.

ORDER 71. MYRISTICEÆ.

Evergreen trees. Leaves alternate, entire, usually dotted. Flowers diœcious, regular, fascicled, umbelled or panicled. Perianth deciduous, usually 3-lobed, lobes valvate in bud. Male fl. Anthers 3 or more, connate in a sessile or stipitate column, head, ring or disc, 2-celled. Female fl, Ovary superior, free, sessile, 1-celled; style short or 0; stigma capitate, discoid or lobed; ovule 1, basal, erect. Fruit fleshy, opening in 2-valves. Seed more or less covered with a lobed or lacerate, often coloured arillus; testa thin, albumen hard, ruminate. Embryo at base of seed, very small, cotyledons divaricate.

MYRISTICA, Linn.

The only genus, same characters as the order.

Male flowers bracteolate.
Anthers elongate, connate, stipitate.
Male flowers axillary, crowded on short, stout peduncles. *M. laurifolia.*
Male flowers in axillary, branched, trichotomous cymes. *M. malabarica.*
Male flowers with the anthers on the edge of a peltate, toothed, stipitate disc. Cymes fascicled on a short peduncle. ... ... *M. attenuata.*
Male flowers in spreading panicles, ebracteolate.
Anthers elongate, connate, stipitate ... ... *M. Farquhariana.*

M. laurifolia, Hook. f & T. Fl. Ind. 163; Fl, Br. I. 5. 103; Bedd. Fl. Sylv. t. 267. *M. Beddomei,* G. King. Ann. Cal. Gard. III. 291. *Jajikai,* K. ; *jayaphul,* M. Very common in the evergreen forests of the southern ghâts of North Kânara, also on the Konkan ghâts. Fl.Nov., Dec. Fr. June, July. Fruit brown minutely pubescent, furrowed longitudinally. The nutmegs *(rámphul)* and mace *(rámpatre)* of this species are exported to Bombay from North Kânara.

M. malabarica, Lamk. in Act Par. 1788, 162 ; Fl. Br. I. 5. 103 ; Dalz. & Gibs. Bomb. Fl. 4; Bedd. Fl. Sylv. t. 269; King Ann. Cal. Bot. Gard. III 288. In the evergreen forests of the Konkan and North Kânara ghâts, common in the Kumta táluka forests. Fl. Nov., Mch. Fr. R.S.

M. Farquhariana, Wall. Cat. 6795 ; Fl. Br. I. 5. 108 ; Bedd. Fl. Sylv. t. 270; *M. canarica,* Bedd. MSS. King Ann. Cal. Bot. Gard. III. From the Konkan southwards. Fl. Br. I. Not observed by me in North Kânara.

M. attenuata, Wall. Cat. 6791 ; Fl. Br. I. 5. 110 ; Bedd. Fl. Sylv. 176 ; King Ann. Cal. Bot. Gard. III. 316 ; *M. corticosa,* Bedd. Fl. Sylv. t. 271 (not of Hook. f. & T.) *M. amygdalina,* Grah. Cat. Bomb. Pl. 175;

Dalz. & Gibs. Bomb. Fl. 4. *Rukt mara*, K. Very common in the evergreen forests of the Konkan and North Kánara gháts. Fl. Nov., Jan. Fr. R.S.

ORDER 72. LAURINEÆ.

Trees or shrubs except *Cassytha*. Leaves alternate, rarely opposite, gland-dotted. Flowers usually 1-2-sexual; bracts 0 or deciduous, often involucriform. Perianth regular, 6-4-cleft, tube sometimes enlarged in fruit. Stamens normally 12, biseriate; usually half or more of the inner or outer stamens are wanting or reduced to staminodia. Anthers erect, 2-4-celled, cells opening by upcurved deciduous lids or valves. Ovary sessile, 1-celled, ovule solitary, pendulous. Fruit a 1-seeded berry or drupe, peduncle often thickened.

Flowers usually bisexual. The anthers of the 3
  inner stamens extrorse. Anthers 2 celled.
Fruit enclosed in perianth-tube ... ... CRYPTOCARYA.
Fruit not wholly included in perianth-tube ... BEILSCHMIEDIA
Anthers 4-celled.
Perianth-tube only persistent in fruit ... ... CINNAMOMUM.
Perianth-lobes persistent, reflexed in fruit ... MACHILUS.
Perianth altogether deciduous ... ... ALSEODAPHNE.
Flowers diœcious. All anthers introrse 4-celled.
Leaves sub-verticillate. Bracts densely imbricat-
  ing, caducous not whorled ... ... ACTINODAPHNE.
Leaves alternate. Bracts whorled ... ... LITSÆA.

CRYPTOCARYA, Brown.

Evergreen trees. Leaves alternate, penninerved. Flowers hermaphrodite, in axillary and terminal panicles. Perianth-segments 6, equal. Stamens 12, 6 outer perfect; anthers introrse, 3 inner with extrorse anthers alternating with 3 staminodes. Ovary immersed in the perianth-tube. Fruit included in the perianth-tube.

Leaves large. Panicles spreading. Fruit globose. ... ... *C. Wightiana*.
Leaves small. Panicles dense, contracted. Fruit ovoid. ... *C. Stocksii*.

C. Wightiana, Thw. Enum. 254 ; Fl. Br. I. 5. 120 ; Bedd. Fl. Sylv. t. 299 ; *C. floribunda*, Wight, Dalz. & Gibs. Bomb. Fl. 222. *Gulmur*, K. In the evergreen forests of the Konkan and North Kánara, common on the Yellápur gháts. Fl. Nov., Jan. Fl. May, June.

C. Stocksii, Meissn. in DC. Prod. XV. 1. 71 ; Fl. Br. I. 5. 120. Kánara, probably from the South Kánara gháts; I have not seen it in any of the North Kánara forests. I have specimens in fruit collected in October on the Bababuden hills of Mysore at 5,000 feet altitude. Flowers during May, June, on the Sispára ghát.

Besides the above two species there is a large tree, common on the Nilkund ghát of North Kánara, flowering specimens of which were said to be near to *C. amygdalina*, Nees, at Kew. Flowers in long peduncled panicles Dec., Jany. It is probably, however, only *C. Wightiana*, Thw.

BEILSCHMIEDIA, Nees.

Evergreen trees. Leaves alternate, penninerved. Flowers in short panicles or racemes, in the axils of the upper leaves. Perianth deciduous, segments nearly equal. Perfect stamens 9, anthers 2-celled; inner 3 stamens with extrorse anthers; outer 6 with introrse anthers. Staminodes sessile or cordate. Berry ovate, on the short thick pedicel. Perianth deciduous.

B. **fagifolia**, Nees in Wall. Pl. As. Rar. II. 69 ; Fl. Br. I. 5. 122. Var. ? *Dalzellii*, Meissn ; *B. fagifolia*, Bedd. Fl. Sylv. t. 263 ; *B. Roxburghiana*, Dalz. & Gibs. Bomb. Fl. 222. Evergreen forests of the Konkan and North Kánara gháts ; common on the Ainshi ghát. Fl. C. S. Fr. H. S.

CINNAMOMUM, Blume.

Evergreen trees. Leaves opposite or alternate, usually 3-nerved at the base. Flowers in axillary or terminal panicles, often unisexual. Perianth of 6, sub-equal segments. Perfect stamens 9, 6 outer eglandular with introrse, 4-celled anthers, the inner 3 perfect stamens with 2 glands at their base, and extrorse anthers. 4th series of 3 short staminodia. Ovary free. Fruit supported by the thickened receptacle and perianth.

> Fruiting peduncle ⅛ in. in diameter   Fruit ⅔ in. long.   *C. zeylanicum.*
> Fruiting peduncle ⅜ in. in diameter   Fruit 1 in. long.   *C. macrocarpum.*

C. **zeylanicum**, Breyn in Ephem. Nat. Cur. Dec. 1. Ann. 4. 139 ; Fl. Br. 1. 5. 131 ; Bedd. Fl. Sylv. t. 262. *C. iners*, Grah. Cat. Bomb. Pl. 173. *Dalchini*, K.; *ohez, bojevar,* Vern. Wild cinnamon tree. In the evergreen forests of the Konkan and North Kánara, very abundant on the southern gháts in the Kumta and Siddápur talukas. Fl. Nov., Feb. Fr. June, July.

C. **macrocarpum**, Hook. f. Fl. Br. I. 5. 133. North Kánara in evergreen forests. Supa taluka. Fl. Jany.

MACHILUS, Nees.

Evergreen trees. Leaves alternate, penninerved. Flowers bisexual, in axillary panicles. Perianth of 6 segments, unchanged in fruit. Stamens as in *Cinnamomum.* Berry globose.

M. **macrantha**, Nees in Wall. Pl. As. Rar. II. 70 ; Fl. Br. I. 5. 140 ; Bedd. Fl. Sylv. t. 264 ; Dalz. & Gibs. Bomb. Fl. 221 ; *M. glaucescens,* Wight Ic. 1825 ; Dalz. & Gibs Bomb. Fl. 221. *Gulum*, M. Throughout the gháts of North Kánara and the Konkan, in moist forests. There may be two species here. *M. macrantha,* Nees, has a small black fruit size of a large currant, and flowers during Jany., Feb. Fruit ripe May, June. *M. glaucescens,* Wight, has a large fruit, depressed succulent, size of a small plum, green with white dots, common in the evergreen forests near Yellápur. Flowers Nov.-Dec. Fr. Feb.-Mch.

ALSEODAPHNE, Nees.

Evergreen trees. Leaves alternate, penninerved, subverticillate at the ends of the branchlets. Flowers in axillary, cymose, panicles. Stamens as in *Cinnamomum.* Fruit ellipsoid, seated on the swollen peduncle.

A. semicarpifolia, Nees in Wall. Pl. As. Rar. II. 72 ; Fl. Br. I. 5.ᵃ144 ; Dalz. & Gibs. Bomb. Fl. 222 ; Bedd. Fl. Sylv. t. 297. *Nelthare,* K.; *phulgus,* M. On the Konkan and North Kánara ghâts, usually in evergreen forests, from the coast inland. Fl. July, Aug., Sept. Fr. ripe Apl., May. The panicles often exceed the leaves in the North Kánara tree. Var. *angustifolia,* Meissn., is common near Yellápur.

ACTINODAPHNE, Nees.

Evergreen trees. Leaves subverticillate, penni or triplenerved. Flowers diœcious, in axillary, bracteate umbels or clusters ; bracts imbricating, caducuous. Perianth-segments 6. Stamens 6-9, anthers 4-celled all introrse, outer 6 stamens eglandular, inner 3, 2-glandular. Staminodes of female 9. Perianth-tube enlarged in fruit.

A. Hookeri, Meissn, in DC. Prod. XV. 1. 218 ; Bedd. Fl. Sylv. t. 296;Fl. Br. I. 5. 149 ; *A. lanceolata,* Dalz. & Gibs. Bomb.Fl. 312. *A angustifolia,* Nees, Wight Ic. 1811. *Pisha,* M. Evergreen forests of the Konkan and North Kánara. Dalzell says "very common at Mhábaleshvar," a small tree in the Yellápur forests. Fl. Oct., Nov. Fr. Jany.

LITSÆA, Lamk.

Usually evergreen trees or shrubs. Leaves alternate, penninerved, rarely triplenerved. Flowers diœcious, umbellate ; umbels few flowered ; involucral-bracts 4-6, concave. Perianth tubular 4-6-lobed. Stamens 6, 9, 12, outer 6 usually eglandular ; anthers introrse, 4-celled. Fruit a succulent drupe, seated on the much enlarged perianth-tube.

Perianth-segments incomplete, not or slightly accrescent in fruit. Stamens often more than 12.
Umbels solitary, many flowered .. .. *L. tomentosa.*
Umbels corymbose or racemose, few flowered ... ... *L. sebifera.*
Perianth-segments 6, enlarged, cupular in fruit. Stamens about 12
Leaves glaucous beneath. Umbels on short peduncles or in sessile clusters ... ... *L. Stocksii.*
Umbels in stout racemes. Leaves rusty-tomentose beneath .. .. *L. Wightiana.*
Perianth-segments 4, deciduous. Leaves strongly triple-nerved at the base, glaucous beneath. Stamens 6 ... ... *L. zeylanica.*

L. tomentosa, Herb. Heyne, ex Wall. Cat. 2550 ; Fl. Br. I. 5. 157 ; *Tetranthera apetala,* Dalz. & Gibs. Bomb. Fl. 222. *Chikna,* M.

Throughout the Konkan and North Kánara in evergreen forests, not so common as the next species, from which it scarcely differs. Fl. Dec. Fr. Feb. Mch. Stamens 18-20.

L. sebifora, Pers. Syn. II. 4 ; var. 3. *tomentosa*, Fl. Br. I. 5. 158 ; *Tetranthera laurifolia*, Jacq. Brandis For. Fl. 379. *Maidalakri*, M. Throughout the Konkan and North Kánara, in moist forests, common on the southern ghāts of the Kumta and Siddápur tálukas. Fl. May, July. Fr. Jany., Feb. Stamens 9-20.

L. Stocksii, Hook. f. Fl. Br. I. 5. 176 : *Tetranthera lanceæfolia*, Grah. Cat. Bomb. Pl. 171 (Var. *acutata*). Evergreen forests of the Konkan and North Kánara, common on the ghāts from Ainshi southwards. Fl. Aug., Nov. Fr. Mch., Apl. Doubtfully distinct from *Litsæa laeta*, Wall. Var. *glauca* ; a Ceylon species.

L. Wightiana, Wall. Cat. 2557 ; Fl. Br. 1. 5. 177 ; *Tetranthera Wightiana*, Bedd. Fl. Sylv. t. 293 ; *Cylicodaphne Wightiana*, Dalz. & Gibs. Bomb. Fl. 222. On the southern ghāts of North Kánara in evergreen forests, common in the forests near the Falls of Gairsoppah. Fl. Oct., Nov. Fr. Apl., May. A large tree.

L. zeylanica, C. & Fr. Nees in Amoen. Bot. Bonn. Fasc. 1. 58 Brandis For. Fl. 382 ; Dalz. & Gibs. Bomb. Fl. 223 ; Bedd. Fl. Sylv. t.ⁿ 294. *Kanvel, chirchira*, M. Throughout the Konkan and North Kánara moist forests, common. Fl. Oct., Jany. Fr. Jany., Feb.

Cassytha filiformis, Linn., a slender twining parasite, is common throughout the presidency, often growing on *Ipomœa biloba*, near the sea-coast.

ORDER 73. THYMELACEÆ.

Trees or shrubs. Leaves alternate or opposite, entire. Flowers 2-sexual, regular. Inflorescence various. Perianth tubular or campanulate, 4-5-lobed ; throat with or without scales. Stamens usually as many or twice as many as the perianth segments. Ovary free, 1-celled (in *Thymeleæ* proper), with a single pendulous evule ; style simple or 0, stigma capitate. Fruit indehiscent. Seed pendulous or lateral ; albumen fleshy or 0.

LASIOSIPHON, Fresen.

Small trees or shrubs. Leaves opposite or scattered. Flowers bisexual, in bracteate heads. Perianth-tube circumciss above the ovary ; lobes 5, spreading ; scales above the stamens, 5-10. Stamens 10. Ovary sessile, 1-celled ; style filiform, stigma capitate. Fruit small, dry ; pericarp membranous. Albumen scanty or 0.

L. eriocephalus, Dene. in Jacq. Voy. Bot. 148 ; Fl. Br. I. 5. 197 ; Bedd. Fl. Sylv. 179 ; *L. speciosus*, Dene. ; Dalz. & Gibs. Bomb. Fl. 221. *Rami, ramita*, M. Woolly-headed gnidia. Throughout the Konkan and North Kánara, common on the Supa ghāts, in deciduous forests. Fl. Nov., Feb. Fr. Mch., Apl.

ORDER 74. ELÆAGNACEÆ.

Shrubs or trees, silvery scaly. Leaves alternate or opposite,
entire. Flowers regular, 1-2-sexual, white or yellow. Perianth
tubular, 2-6-cleft, valvate in bud. Stamens usually 4, alternate
with the perianth lobes. Ovary free, 1-celled, with a solitary erect
ovule. Fruit indehiscent, enclosed within the accrescent perianth
tube.

ELÆAGNUS, Linn.

Trees or shrubs. Leaves alternate, entire, densely silvery scaly,
on the lower surface. Flowers hermaphrodite or unisexual by
abortion. Perianth-tube constricted above the ovary ; limb valvate,
4-cleft, deciduous. Stamens 4, on the mouth of the corolla. Fruit
with a bony or coriaceous endocarp.

E. latifolia, Linn. Sp. Pl. Ed. 2. 177 (excl. syn.) ; Fl. Br. I. 5. 202 ;
Bedd. Fl. Sylv. t. 180 ; Brandis For. Fl. 390 ; *E. Kologa,* Schl. ;
Dalz. & Gibs. Bomb. Fl. 224. *Nurgi, ambgool,* Vern. Throughout the
Konkan and North Kánara, usually in or near evergreen forests. Fl.
Nov., Feb. Fr. May-July. Branches often with curved spines. Fruit
acid, astringent.

ORDER 75. LORANTHACEÆ.

Parasitic evergreen shrubs. Leaves coriaceous, generally opposite.
Flowers regular, hermaphrodite or unisexual, usually bracteate.
Calyx adnate to the ovary, limb truncate or 0. Petals 4-8, free or
connate, valvate in bud. Stamens as many as corolla lobes, usually
inserted on them. Ovary inferior, 1-celled. Ovule 1, erect. Fruit a
1-seeded drupe. Albumen fleshy.

Flowers 2 sexual ...   . LORANTHUS.
Flowers 1 sexual ...   . VISCUM.

LORANTHUS.

Shrubs, often with stellate hairs. Flowers usually large, coloured.
Petals free or more or less connate into a tube. Stamens with
versatile anthers.

Flowers small in glabrous spikes. Corolla segments
    4-5, free. Bract scale like.
Leaves short petioled. Buds cylindric ...   . *L. Wallichianus.*
Leaves long petioled. Buds angled ...   . *L. obtusatus.*
Flowers in axillary fascicles or racemose, mealy
    tomentose. Bract scale like. Anther cells indistinct .
Flowers, &c., rusty villous. Corolla buff or pink   ... *L. scurrula.*
Flowers yellowish-white tomentose, scurfy. Corolla
    green...   ...   . *L. pulverulentus.*
Flowers racemose or in clusters. Corolla-tube gibbous-
    ly inflated or not. Bract scale like. Anther cells
    indistinct.

Flowers pubescent or tomentose.
Fruit ovoid, pale red. Flowers fascicled ... *L. tomentosus.*
Fruit pyriform. Flowers solitary ... .. *L. Stocksii.*
Flowers glabrous. Corolla lobes not elastic.
Leaves small, cuneate. Flowers fascicled ... ... *L. cuneatus.*
Leaves large, oblong, elliptic. Flowers racemose, yellow,
 with green lobes. Calyx-limb entire... ... ... *L. longiflorus.*
Corolla lobes elastic ... ... *L. elasticus.*
Flowers 4-5, enclosed in a large bell-shaped involucre,
 ebracteolate ... ... *B. lageniferus.*
Flowers in axillary clusters. Branches triquetrous.
 Leaves mostly in verticels of 3 ... ... *L. trigonus.*
Flowers in decussate spikes. Bracts and bracteoles
 large, orbicular, deciduous. Calyx-tube truncate.
 Anthers multilocellate.
Corolla curved, 1-2 inches ... ... *L. loniceroides.*
Corolla straight, ?-? inch ... ... *L. capitellatus.*

L. Wallichianus, Schultz. Syst. VII. 100 ; Fl. Br. I. 5. 205 ; Dalz.
& Gibs. Bomb. Fl. 109, *Banda,* M. Throughout the Konkan and
North Kánara, parasitic on *locundi* (*Memecylon amplexicaule,* Roxb.) and
other trees. Common in the forests near Kárwár. Fl. Apl., June.
Fr. Aug. Sept.

L. obtusatus, Wall. Cat. 526 ; Fl. Br. I. 5. 205 ; Dalz. & Gibs. Bomb.
Fl. 109. North Kánara and Konkan gháts. common on *jamba* (*Xylia
dolabriformis,* Benth.) and other trees on the Supa ghát deciduous forests.
Fl. Feb., Apl., June, July.

L. scurrula, Linn.; Fl. Br. I. 5. 208 ; *L. buddleioides,* Desrouss. ;
Dalz. & Gibs. Bomb. Fl. 110. Throughout the presidency, common in
North Kánara, often growing on *Leea sambucina, Dillenia pentagyna* and
Pterocarpus marsupium. Fl. Nov. Feb. Leaves opposite.

L. pulverulentus, Wall. in Roxb. Fl. Ind. ed. Carey & Wall. 221 ;
Fl. Br. I. 5. 211 ; Brandis For. Fl. 396. Konkan, Stocks. Fruit large
club-shaped, Kurz. In the Central Provinces on *Butea frondosa,* Brandis.

L. tomentosus, Heyne in Roth. Nov. Sp. 191 ; Fl. Br. I. 5. 212.
Common in the deciduous forests of the North Kánara gháts near
Yellápur, very common on *Phyllanthus emblica* and *Mussœnda frondosa.*
Fl. Nov., Feb. Fr. Apl., May. Leaves alternate, calyx toothed.

L. Stocksii, Hook. f. Fl. Br. I. 5. 213. Konkan, Stocks. Leaves sossile,
opposite. Fruit pyriform.

L. cuneatus, Heyne in Roth. Nov. Sp. 193. Throughout the Konkan
and North Kánara, common near Yellápur, a dense shrub on *Terminalia
paniculata.* Fl. Nov., Dec. Leaves small, cuneate.

L. longiflorus, Desrouss. in Lamk. Encycl. III. 498 ; Dalz. & Gibs.
Bomb. Fl. 110 ; Brandis For. Fl. 397 ; *L. bicolor,* Roxb. Fl. Ind. I. 548.
Var. *falcata.* Kurz. For. Fl. II. 321 ; Var. *amplexifolia.* Thw. Enum.
134 ; Grah. Cat. Bomb. Pl. 86. *Banda, cainguli,* Vern. Throughout the
presidency. Parasitic on many trees in North Kánara. The var. *amplexifolia* often grows on *Flacourtia montana* and the var. *falcata* on *Careya
arborea.*

Var. nov. coccinea. Branches stout, lenticellate. Leaves usually sub-opposite, broad ovate, obtuse, thickly coriaceous, shortly petioled ; lateral-nerves 5-6 pairs. Flowers in rather lax, axillary racemes. Calyx urn-shaped, limb often 5-1-notched, very rugose outside. Corolla-tube long, slender, 2½ inches, bright scarlet without; lobes short, light scarlet or greenish. Filaments scarlet. Fruit ovoid, crowned with the calyx-limb. This variety has the racemes with the flowers secund, but more lax than in *L. longiflorus, proper,* and the corolla is always bright red or scarlet. The leaves are shortly petioled and neither cordate or amplexicaul ; it is a stout parasite and grows nearly exclusively on *Terminalia bellerica* in North Kánara. Flowers during the cold season Jan., Feb. Fruit ripe May, June.

L. elasticus, Desrouss. in Lamk. Encycl III. 599 ; Fl. Br. I. 5. 216; Dalz. & Gibs. Bomb. Fl. 109. Throughout the Konkan and North Kánara, very common on mango trees, near the sea-coast. The corolla lobes coil elastically. Fl. July, Aug. Fr. April, May. A common species.

L. lageniferus, Wight Ic. t. 306 ; Fl. Br. I. 5. 218 ; Dalz. & Gibs. Bomb. Fl. 110. *Bandyali,* M. Hills of the Konkan, common in the deciduous forests of North Kánara, near Dandeli, growing on *Adina cordifolia.* Fl. Apl., May. Fr. July, Aug. The bell-shaped involucre distinguishes this species from all others growing in the Bombay Presidency.

L. trigonus, Wight. & Arn. Prod. 386; Fl. Br. I. 5. 219 ; on the gháts of the South Konkan and North Kánara, common on the Supa gháts, growing on *Eugenia jambolana, Dalbergia latifolia* and *Ziziphus xylopyra.* Flowers small, acicular. Fruit orange, shortly pedicelled, crowned with the calyx-limb Fl. Dec., Feb. Fr. ripe May.

L. loniceroides, Linn. Sp. Pl. Ed. 2. 473 ; Fl. Br. I. 5. 221; Dalz. & Gibs. Bomb. Fl. 110. Throughout the presidency, common in the Konkan and North Kánara growing on many different trees. Fl. Mch., May. Fr. R S.

L. capitellatus, W. & A. Prod. 382 ; Fl. Br. I. 5. 221 ; Dalz. & Gibs. Bomb. Fl. 109. From the Konkan, southwards. In North Kánara, on the Supa gháts, growing on *Artocarpus integrifolia* I have never seen it on any other tree; it may be however only a variety of *L. loniceroides,* Linn. Fl. Feb., May. Fr. R. S.

VISCUM.

Shrubs with opposite leaves or leafless. Flowers axillary, unisexual. Petals 3-4, sessile. Anthers on the petals, opening by many pores. Ovary inferior. Fruit succulent. Embryo 1-2 in each seed, in fleshy albumen.

Leafy shrubs.
Large shrub, branchlets angular...   ...   ...   *V. orientale.*
Dwarf tufted shrub, branches stout, short   ...   ...   *V. capitellatum.*

Branches terete, long, slender. Leaves 0 or few ... *V. ramosissimum.*
Leafless.
Branches angled, not contracted at nodes ... *V. angulatum.*
Branches flattened, contracted at nodes ... *V. articulatum.*

V. orientale, Willd. Sp. Pl. IV. 737; Fl. Br. I. 5. 224; Brandis For. Fl. 393. North Kánara and Konkan gháts, common near the Nilkund ghát in evergreen forests, growing on *Terminalia paniculata.* Fl. Sept., Jany. Leaves 3-5-longitudinal nerved.

V. capitellatum, Sm. in Rees' Cycl. XXXVII; Fl. Br. I. 5. 225. On the North Kánara gháts, commonly parasitic on other *Loranthaceœ,* also on *Terminalia paniculata.* Fl. Nov., Dec. Fr. Jan. Leaves thick spathulate. Flowers on fascicled peduncles. I have a variety of this or another species with much thinner ovate leaves and sessile, fascicled flowers, growing on *Loranthus trigonus,* Wgt. No. 1619, prop. herb.

V. ramosissimum, Wall. Cat. 6876; Fl. Br. I. 5. 225. From the Konkan southwards, Fl. Br. I; on the lower slopes of the Bababuden hills of Mysore on *Flacourtia Ramontchi.* Fl. H.S. Fr R S. Leaves linear, few.

V. angulatum, Heyne MSS. DC. Prod. IV. 283; Fl. Br. I. 5. 225; Dalz. & Gibs. Bomb. Fl. 110; *V. attenuatum,* Brandis For. Fl. 394. Very common on the gháts from the Konkan southwards on many different kinds of trees. Fl. apparently throughout the year.

V. articulatum, Burm. Fl. Ind. 311; Brandis For. Fl. 394; Fl. Br. I. 5. 226. Throughout the presidency, not common, growing on *Diospyros melanoxylon* in the Dhárwár district. Fl. June, Aug. Fr. Dec., Feb.

ORDER 76. SANTALACEÆ.

Trees, shrubs or herbs. Leaves entire, alternate or opposite, without stipules. Flowers regular, hermaphrodite or unisexual. Perianth superior or inferior, 3-8-toothed, lobed or partite. Stamens 3-6, opposite to perianth lobes. Disc various. Ovary inferior, 1-celled, with a free central placenta, bearing 2-5, pendulous ovules. Fruit a nut or drupe. Seed globose, albumen copious.

Tree. Leaves opposite. Disc of scales between the
 stamens ... .SANTALUM.
Shrub. Leaves alternate. Disc angled between
 the stamens ... ...OSYRIS.
Spinous tree. Leaves alternate. Filaments 2-fid.
 Disc annular ... ...SCLEROPYRUM.

SANTALUM, L.

Glabrous trees or shrubs. Leaves simple, coriaceous. Flowers in terminal and axillary cymes, bisexual. Perianth tube campanulate with a concave disc adhering to the base. Ovary at first free, at length ½ inferior, 1-celled with 2-4 pendulous ovules, attached near the base of the central placenta. Drupe globular, fleshy; endocarp ribbed.

S. album, Linn. Sp. Pl. 497; Fl. Br. I. 5. 231; Dalz. & Gibs. Bom. Fl. 224; Brandis For. Fl. 398; Bedd. Fl. Sylv. t. 256. Sandalwood tree. *Gundada*, K.; *chandan*, M. Indigenous throughout the dry districts of the presidency, also abundant in some of the dry deciduous forests of North Kánara; often along the bunds of tanks and in hedges. Fl. Mch., Aug. Fr. C.S. Yields the sandalwood of commerce.

OSYRIS, L.

Glabrous shrubs; branches angular. Leaves alternate. Flowers hermaphrodite or unisexual. Perianth of male fl. slender, of hermaphrodite fl. obconical, 3-4 lobed. Stamens 3-4-5. Ovary inferior, ovules at apex of short central placenta. Fruit a drupe. Seed globose, albuminous.

O. arborea, Wall. Cat. 4035; Fl. Br. I. 5. 232; Brandis For. Fl. 399; *O. Wightiana*, Dalz. & Gibs. Bomb. Fl. 223. *Popoli, lotal*, M. Throughout the Konkan and North Kánara, common in moist forests from the sea-coast inland. Fl. throughout the year.

SCLEROPYRUM, Arnott.

Trees often spiny. Leaves coriaceous. Flowers in short spikes at the leafless nodes, polygamous. Perianth-tube of male fl. solid; of female fl. adnate to the ovary; lobes 5. Stamens 5, filaments 2-fid. Anther cells dehiscing transversely. Ovary inferior; style short, stigma large, peltate; ovules 3, pendulous from the top of a central column. Drupe pyriform, pedicelled. Seed subglobose.

S. Wallichianum, Arn. in Jard. Mag. Zool. & Bot. II. (1858) 550; Fl. Br. I. 5. 234; *Pyrularia Wallichiana*, Bedd. Fl. Sylv. t. 304. In the evergreen forests of the Konkan and North Kánara gháts; a moderate-sized tree in the forests near Yellápur, common near the Nilkund ghát. Fl. Dec., Feb. Fr. ripe May, June.

ORDER 77. EUPHORBIACEÆ.

Trees or shrubs with milky juice. Leaves alternate or opposite. Flowers unisexual. Perianth calycine, rarely double. Stamens various. Ovary superior, 3-2 carpellary. Ovules 1-2, in each cell, pendulous from inner angle of cell. Funicle often thickened. Fruit a capsule or drupe. Seed |with or without an arillus. Albumen fleshy, rarely 0.

Cells of ovary 2-ovuled.
Flowers enclosed in an involucre or not
Flowers enclosed in an involucre .. ..EUPHORBIA.
Flowers not inclosed in an involucre.
Petals minute or 0.
Petals present. Calyx-lobes valvate. Filaments united.
Fruit a drupe... ... ...BRIDELIA.
Fruit a capsule ... ...CLEISTANTHUS,
Calyx-lobes imbricate. Filaments free ... ...ACTEPHILA.
Petals 0.

Stamens 3-6. Styles or their arms slender (except
　　Glochidion).
Perianth of male flowers 6-lobed, lobes spreading.
　　Flowers monoicous.
Disc present in both sexes ... ...PHYLLANTHUS.
Disc absent. Styles confluent ... ...GLOCHIDION.
Perianth of male flowers turbinate, lobes minute ...BREYNIA.
Perianth of male flowers rotate, 6-lobed, lobes spurred .SAUROPUS.
Sepals 5. Flowers diœcious .. ..FLUEGGIA.
Stamens few or many. Styles or stigmas 2-3, dilated.
Leaves entire or serrulate. Fruit a drupe.
Stamens 2-3. Ovary 2-3-celled ... ...PUTRANJIVA.
Stamens 8 or many. Ovary 1-celled ... ...HEMICYCLIA.
Stamens many. Ovary 2-4-celled... ... ...CYCLOSTEMON.
Leaves trifoliate. Stamens 5 ... ...BISCHOFIA.
Styles or stigmas very minute.
Male flowers in catkins. Fruit dehiscent... ...APOROSA.
Male flowers in spikes or racemose.
Fruit small, compressed, rugose, not arillate ... ...ANTIDESMA.
Fruit large, coriaceous. Seeds arillate ... ...BACCAUREA.
Cells of ovary 1 ovuled.
Flowers in terminal, unisexual cymes or with the cen-
　　tral flower female. Petals present.
Fruit capsular . ...JATROPHA,
Flowers in terminal, androgynous spikes. Males
　　petaliferous.
Females often apetalous. Filaments inflexed in bud ...CROTON.
Male flowers petaliferous. Females often apetalous.
　　Filaments straight.
Sepals imbricate. Petals united. Fruit a drupe ...GIVOTIA,
Sepals imbricate. Petals free. Fruit a capsule.
Sepals enlarged in fruit.
Petals present in female flower ... ...BLACHIA.
Petals absent in female flower ... ...DIMORPHOCALYX.
Sepals valvate. Petals 4-8... ... .AGROSTISTACHYS.
Flowers apetalous. Calyx of male closed in bud,
　　membranous, splitting into 3-5 concave sepals.
　　Styles long, bifid or multifid.
Filaments free.
Stamens 4. Anthers 2-celled.
Capsule drupaceous Anther cells parallel ... .. TREWIA.
Capsule dry. Anther cells globose ... ...MALLOTUS.
Stamens 1, few or many. Anther cells 3-4-locellate.
Styles very long 2-fid ... ...CLEIDION,
Styles entire ... ...MACARANGA.
Filaments variously connate in bundles.
Flowers dioecious in axillary spikes ... . HOMONOIA.
Flowers monoecious in terminal panicles ... ...RICINUS.
Calyx of male minute and open in bud or obsolete.
Male calyx terete, 2-3-lobed ... ...SAPIUM.
Calyx terete, 3-partite ... ...EXCŒCARIA.

EUPHORBIA, Linn.

Herbs or shrubs often with much milky juice. Leaves of the
stem alternate, with or without stipular spines or opposite and stipulate, leaves of the flowering branches opposite. Flowers monoicous, consisting of many male and 1 female flower in a small perianth like involucre; lobes with thick glands at the sinuses; glands
with often a petaloid wing. Male fl. : 1, 2-celled anther on an arti-

culated filament. Female fl. : a stipitate 3-celled ovary in the centre of the involucre ; styles 3, free or combined, simple or 2-fid. Capsule of 3, 2-valved cocci.

Unarmed tree ; branches cylindric   ...   .*E Tirucalli.*
Armed with stipular spines.
Branches angled and winged ; leaves minute or
  wanting ...   .*E. antiquorum.*
Branches jointed, cylindric or with 5 spirally twisted
  ribs...   ...   .*E. neriifolia.*
Branches sub-cylindric   ...   .   *E. Nivulia.*

E. Tirucalli, Linn. Hort. Cliff. 197 ; Brandis For. Fl. 439 ; Dalz. & Gibs. Bom. Fl. Suppl. 76. Bedd. Fl. Sylv. 217. Milk bush. *Nevli, thuvar, seyr,* M. : *yele gulla,* K. Naturalized throughout the dry districts of the presidency ; usually in hedges, planted. Fl. Aug., Sept.

E. antiquorum, Linn. Hort. Cliff. 196 ; Fl. Br. I. 5. 255 ; Dalz. & Gibs. Bom. Fl. 226 ; Brandis For. Fl. 438 ; Bedd. Fl. Sylv. 217. *Narsej,* M. Throughout the dry districts of the presidency, Fl. R.S. in the Belgaum and Bijápur districts.

E. neriifolia, Linn. Hort. Cliff. 196. in part ; Fl. Br. I. 5. 255. Dalz. & Gibs. Bomb. Fl. 226 ; Brandis For. Fl. 439 ; Bedd. Fl. Sylv. 216. *Thor, nivarung, seej, mingut,* Vern. ; *yellikalli,* K. Throughout the presidency ; often in waste, stony, dry situations, generally planted in hedges. Fl. Fr. Feb., Mch. New leaves in March, Apl.

E. Nivulia, Ham. in Trans. Linn. Soc. XIV. 286 ; Dalz. & Gibs. Bomb. Fl. 225 ; Brandis For. Fl. 439. Bedd. Fl. Sylv. 216 ; Fl. Br. I. 5. 255 ; Wight Ic. 1862. Dry rocky hills in Gujarát and Sind. Fl. Mch. Fr. Apl.-May.

BRIDELIA, Will.

Shrubs or trees. Leaves alternate, entire, with prominent parallel lateral nerves. Flowers monoicous or diœcious, bracteate, in axillary or spicate clusters. Calyx 5-cleft ; lobes valvate. Petals 5. Male fl. : stamens 5, on a central column, situated on a flat, sinuate disc. Female fl. : ovary 2-celled, surrounded by a membranous cup-shaped or tubular disc, variously lobed ; styles 2, bifid, connate at the base. Fruit a berry, with 1-2, 1-seeded cocci or pyrenes.

Tree. Lateral nerves 15-20-pairs   ...   .   *B. retusa.*
A climbing shrub. Lateral nerves 8-12-pairs ...   ...   *B. stipularis.*
A straggling shrub. Lateral nerves 6-9-pairs...   ...   *B. Hamiltoniana.*

B. retusa, Spreng. Syst. Veg. III. 48 ; Fl. Br. I. 5. 268 ; Brandis For. Fl. 449 ; Bedd. Fl. Sylv. t. 260 ; *B. montana,* Dalz. & Gibs. Bomb. Fl. 233. *Mulla honne,* K. ; *kanta kauchi,* M. ; *asana, asauna,* Vern. Throughout the presidency in deciduous forests. Fl. Sept., Oct. Fl. Oct.-Jan. A spinescent tree.

B. stipularis, Blume Bijd. 597 ; Fl. Br. I. 5. 270 ; Brandis For. Fl. II. 449 ; Bedd. Fl. Sylv. 201. Common in deciduous forests throughout the presidency. Fl. May, Nov.

B. Hamiltoniana, Wall. Cat. 7882 ; Fl. Br. I. 5. 271 ; Bedd. Fl. Sylv. 202. Higher ghâts of the Konkan, Law, Stocks ; not seen by me.

CLEISTANTHUS, Hook. f.

Trees or shrubs. Leaves alternate, penniveined, lateral nerves prominent. Flowers small, in axillary clusters or short spikes, monoicous. Calyx-segments 5, valvate in bud. Petals 5, minute. Disc of the male flat, of the female conical. Stamens 5, united in a column in the centre of the disc. Ovary 3-celled, long-hairy; styles 3 bifid. Fruit a 3-celled, 6-valved capsule. Seeds exarillate, albuminous.

C. malabaricus, Muell. Arg. in DC. Prod. XV. II. 508; Fl. Br. I. 5. 276; Bedd. Fl. Sylv. 203. In the evergreen forests of the Konkan and North Kanara, usually near streams and rivers, common in the forests near the Falls of Gairsoppah. Fl. Nov., Dec. Fr. Jany., Feb.

ACTEPHILA, Blume.

Trees or shrubs. Leaves alternate, large. Flowers in axillary clusters, petals present or 0. Male fl. Sepals 5-6, imbricate. Petals as many inserted under the 5-lobed disc or 0. Stamens 5, on the disc. Pistillode 3-cleft. Female fl. Ovary sessile on the 5-lobed disc; styles bifid. Capsule long peduncled 3-lobed, loculicidally dehiscent.

A. excelsa, Muell. Arg. in Linnæa XXXII. 78; Fl. Br. I. 5. 282; Bedd. Fl. Sylv. 189; *Anomospermum excelsum*, Dalz. & Gibs. Bomb. Fl. 233. Konkan gháts.

PHYLLANTHUS, Linn.

Trees, shrubs or herbs. Leaves entire, stipulate, usually distichous. Petiole short. Flowers in axillary clusters, mon or dioicous. Calyx-segments usually 5-6, imbricate, in 2 series. Stamens 3-5, free or connate. Disc of distinct glands alternating with the calyx-segments. Female fl. Sepals of the male. Ovary 3 or more celled; styles free or connate, usually 2-fid; ovules 2 in each cell. Fruit dehiscent, sometimes fleshy. Seeds without arillus or strophiole.

Shrubs with small leaves.
Scandent. Leaves 1 in. Fruit a small black, 4-8-celled
   berry ... ... *P. reticulatus.*
Erect. Leaves 1-5-in. Fruit a small 3-celled crustaceous
   capsule, cells 2-valved ... ... *P. Lawii.*
Trees. Leaves large or small.
Leaves ¼-½ in. Fruit large, fleshy, globose, with 3
   bony, 2-celled cocci ... ... *P. Emblica.*
Leaves 2-3-in. Fruit large, fleshy, globose, with a 3-4-
   celled bony endocarp ... ... *P. distichus.*
Leaves 3-6-in. by 1-3 in. Fruit dry globose, bursting
   irregularly, ½-in. ... ... *P. indicus.*

P. reticulatus, Poir. Encycl. V. 298; Fl. Br. I. 5. 288; Brandis For. Fl. 453; Bedd. Fl. Sylv. 190; *Anisonema reticulatum*, A. Juss. Dalz. & Gibs. Bomb. Fl. 234. *Pavan*, M.; *dalwan* Guj; *kale-madh-ka-jhar*, H. Throughout the presidency and Sind in deciduous forests, often in hedges, very common. Flowers throughout the year.

P. **Emblica**, Linn. Sp. Pl. 982 ; Fl. Br. I. 5. 289 ; Brandis For. Fl.
454; Bedd. Fl. Sylv. t. 258; *Emblica officinalis*, Gœrtn. Dalz. & Gibs.
Bomb. Fl. 235. *Nelli*, K. ; *awla*, M. Throughout the presidency in deciduous forests. Fl. Mch.-May. Fr. ripe Oct., Feb. A substance called
white catechu is extracted from the wood chips.

P. **Lawii**, Grah. Cat. Bomb. Pl. 181 ; Fl. Br. I. 5. 290 ; *P. spinulosus,*
Herb. Heyne Wall. Cat. 7897 ; *P. polyphyllus*, Dalz. & Gibs. Bomb.
Fl. 234; *P. juniperinoides*, Muell. Arg. Along the banks of streams and
rivers in the Konkan and North Kánara, very common along the Kálánadi,
forms dense thickets near the water's edge. Fl. Sept. Oct. Fr. Nov.

P. **indicus**, Muell. Arg. in Linnæa XXXII. 52; Fl. Br. I. 5. 305 ;
Bedd. Fl. Sylv. 191 ; *Prosorus indica*, Dalz. & Gibs. Bomb. Fl. 236.
On the Konkan and North Kánara ghâts in deciduous forests. Fl. Feb.,
Mch. Fr. May-June.

P. **distichus**, Muell. Arg. in DC. Prod. XV. II. 413; Bedd. Fl. Sylv.
191 ; *Cicca disticha*, Dalz. & Gibs. Bomb. Fl. Suppl. 78. Country goose-
berry. *Harparawri*, *raiavala*, Vern. In gardens of the Konkan and
Deccan, planted. Fl. May.-Nov. Fr. R.S., C.S. Male flowers red, minute
fascicled.

GLOCHIDION, Forst.

Trees or shrubs. Leaves alternate, bifarious. Flowers small,
usually monoicous, apetalous. No disc glands. Male fl. Sepals 6, in
2 series, imbricate. Anthers 3-8, connate in a column ; cells linear ;
connective produced. Female fl. Calyx of 6 short sepals. Ovary of
3-15-cells. Styles connate in a column ; ovules 2 in each cell. Capsule
of 3 or more 2-valved cocci. Seeds albuminous.

Female calyx with 5-6 sepals. Anthers 4. Style
conical ... ...
Leaves elliptic, base acute, glabrous. Style truncate ... *G. lanceolarium.*
Leaves softly pubescent, large, ovate or cordate. Style
4-5-cleft. .. .. *G. tomentosum.*
Style 6-8-toothed. Leaves glabrous, shining. Capsule
globose, not lobed ... ... *G. zeylanicum.*
Female calyx irregularly 4-6-lobed or toothed. Anthers
3. Style globose or subglobose... ... ...
Ovary glabrous, 8-lobed. Capsule much depressed,
6-8-lobed... ... ... *G. Hohenackeri.*
Ovary very short, villous, 3-6-celled. Capsule faintly
lobed ... ... *G. Ralphii.*
Ovary glabrous, 4-5-celled. Capsule obtusely 3-angu-
lar, 6-Lobed ... ... *G. Johnstonei.*
Female calyx of 6 free sepals. Anthers 3. Style minute.
Stipules subulate. Style very minute, conical ... ... *G. malabaricum.*
Stipules triangular, hastate, or falcate. Style short,
stout, conical ... ... *G. ellipticum.*
Leaves pubescent or tomentose. Style stout, exceeding
the sepals, obconic, tip lobed ... ... *G. velutinum.*

G. **lanceolarium**, Dalz. in Bomb. Fl. 235 ; Fl. Br. I. 5. 308 ; Bedd. Fl.
Sylv. 192 ; *Phyllanthus lanceolarius*, Muell. Arg. Brandis For. Fl. 453.
Bhoma, M. Throughout the presidency, in deciduous forests, common
on the Konkan and North Kánara ghâts. Fl. Dec., Apl.

G. tomentosum, Dalz. in Hook. Jour. Bot. III. (1853), 38 ; Fl. Br.
I. 5. 309 ; Bedd. Fl. Sylv. 192. Along nálás and water-courses in North
Kánara, in deciduous forests, common about Yellápur. Fl. Apl., May.
Fr. June, July. There is a small tree near the Falls of Gairsoppah with
the style stout conical in fruit (*G. tomentosum* proper). Fl. Nov., Dec.
Fr. Jany.

G. zeylanicum, A. Juss. Tent. Euphorb. 107, t. 3 ; Fl. Br. I. 5.
310 ; Bedd. Fl. Sylv. 192 ; *G. nitidum,* Dalz. & Gibs. Bomb. Fl.
235 ; *G. canarum,* Bedd. Fl. Sylv. 192. Common on the Konkan and
North Kánara gháts, along streams and water-courses, both in evergreen
and in deciduous forests. Fl. and Fr. at different times throughout the
year, usually Dec., June, often associated with *G. tomentosum,* , Dalz., to
which species it is closely allied.

G. Hohenackeri, Bedd. Fl. Sylv. 193 ; Fl. Br. I. 5. 314. On the
Konkan and North Kánara gháts, in evergreen forests ; common in
the forests near the Nilkund and Gairsoppah gháts. Fl. Jany., Mch.
Fr. Apl., May.

G. Ralphii, Hook. f. Fl. Br. I. 5. 314. Konkan and North Kánara
gháts, common in the evergreen forests near the Falls of Gairsoppah,
along water-courses. Fl. Feb., June. Fr. June-Nov.

G. Johnstonei, Hook. f. Fl. Br. I. 5. 314.

A small tree, common on the Supa gháts of North Kánara near Anmode.
Fruit small, much depressed, crowned by the 3-lobed stigma, usually
fascicled along leafless branches ; peduncles short, stout. Fl. Feb. Fr. Apl.

G. malabaricum. Bedd. Fl. Sylv. 194 ; Fl. Br. I. 5. 319. On the
Konkan and North Kánara gháts.

G. ellipticum, Wight Ic. t. 1906 ; Fl. Br. I. 5. 321 ; Bedd. Fl. Sylv.
193 ; var. *Wightiana* ; *G. diversifolium,* Bedd. Fl. Sylv. 193. South
Konkan, also in the forests near the Ainshi ghát of North Kánara ;
a small tree. Fl. Dec.-Feb. Fr. Feb., Mch. Flowers in dense fascicles. Capsule very small.

G. velutinum, Wight Ic. t. 1907-2 ; Fl. Br. I. 5. 322 ; Bedd. Fl.
Sylv. 195 ; *Phyllanthus nepalensis,* Brandis For. Fl. 453 Throughout
the deciduous forests of the Konkan and North Kánara gháts. Common in the Supa táluka dry forests. Fl. Dec., Mch. Fr. Mch., June. A
distinct species. Capsule depressed, lobed.

FLUEGGIA, Willd.

Armed or unarmed shrubs. Leaves small, alternate, stipulate.
Flowers small, axillary, monœcious. Petals 0. Male fl. Sepals 5,
imbricate. Stamens 5 with alternating disc glands. Pistillode
large 2-3-fid. Female fl. Calyx same as in male fl. Disc annular,
toothed. Ovary 1-3-celled ; styles 3, bifid. Fruit more or less
succulent, separating into 2-valved cocci.

Unarmed shrub	...	...	... *F. microcarpa.*
Spinous shrub	...	...	... *F. Leucopyrus.*

F. microcarpa, Blume Bijd. 580 ; Fl. Br. I. 5. 328 ; *F. Leucopyrus,*
Dalz. & Gibs. Bomb. Fl. 236 (not of Willd.). *Securinega obovata,* Muell.
Arg.; Brandis For. Fl. 455 ; Bedd. Fl. Sylv. 197. *Pandhar-phali,*

kodarsi, M. Throughout the presidency in deciduous forests, common.
Fl. hot season. Fr. R.S.

F. Leucopyrus, Willd. Sp. Pl. IV. 757; Fl. Br. I. 5. 328; *F. virosa*,
Dalz. & Gibs. Bomb. Fl. 236; *Securinega Leucopyrus*, Muell. Arg.;
Brandis For. Fl. 456; Bedd. Fl. Sylv. 197. *Vorepuvan*, M.; *kiran*, Sind.
Throughout the presidency and Sind, usually in open situations, very
common. Fl. H.S. Fr. R. S. The fruit of both the above species is a
white globose berry.

BREYNIA, Forst.

Glabrous shrubs. Leaves small, stipulate. Flowers small, monoicous, axillary. Calyx turbinate, 6-lobed. Stamens central, 3,
filaments united into a column. Disc 0. Styles 3, bifid. Fruit
dehiscent or indehiscent.

Calyx much enlarged in fruit ... ... *B. patens.*
Calyx not or little enlarged in fruit ... ... *B. rhamnoides.*

B. patens, Benth. in Gen. Plant. III. 277; Fl. Br. I. 5. 329; *Melantheopsis patens*, Muell. Arg.; Bedd. Fl. Sylv. 196; Brandis For. Fl. 455.
M. turbinata, Dalz. & Gibs. Bomb. Fl. 234; *P. turbinatus*, Roxb.; Grah.
Cat. Bomb. Pl. 180. Throughout the Konkan and North Kánara ghát
forests. Fl. May, June. Fr. Aug., Sept. Seeds with an arillus. Fruit
red-coloured.

B. rhamnoides, Muell. Arg.; Fl. Br. I. 5.330; Brandis For. Fl. 455;
Bedd. Fl. Sylv. 1961 (by error named *Melantheopsis patens*.) Throughout
the Konkan and North Kánara, often in hedges. Fl. H.S. Fr. R.S.

SAUROPUS, Blume.

Shrubs. Leaves alternate, distichous, membranous, entire. Flowers minute, monoecious, axillary. Male fl. Calyx 6-lobed; lobes
spurred inwards at the base. Stamens 3 on a triangular column;
anthers extrorse. Female fl. Calyx 6-cleft, persistent, accrescent;
styles 3, sessile, spreading with 3 recurved, incurved arms. Fruit
globose, depressed, albuminous.

S. quadrangularis, Muell. Arg. in Linnæa XXXII. 72; Fl. Br.
I. 5. 335; *Ceratogynum rhamnoides*, Dalz. & Gibs. Bom. Fl. 234.
Chikli, M. Konkan and North Kánara, common near the Arbail ghát,
in evergreen forests. The spurred petals of the minute male flower are
remarkable. Fl. June, Aug. Fr. July-Sept.

PUTRANJIVA, Wall.

Trees. Leaves alternate, stipulate, entire or serrulate. Flowers
dioicous, axillary, apetalous. Disc 0. Male fl. Calyx 2-5-partite.
Stamens 2-3, central, filaments free or connate. Female fl. Ovary
2-3-celled. Fruit indehiscent with a bony endocarp.

P. Roxburghii, Wall. Tent. Fl. Nep. 61. Cat. 6814; Fl. Br. I.
5. 336; Brandis For. Fl. 451; Bedd. Fl. Sylv. t. 275; Dalz. & Gibs.
Bomb. Fl. 236. *Putranjiva*, Vern. In the evergreen forests of the Konkan
and North Kánara, nowhere common. Fl. Apl., May. Fr. ripe next
Feb., Mch.

HEMICYCLIA, W. & A.

Trees or shrubs. Leaves alternate, petioled, entire. Flowers small, diœcious, apetalous. Males clustered at the nodes. Female flowers subsolitary. Male fl. Sepals 4-5, imbricate, inner larger. Stamens 4-23, round an orbicular disc, filaments free. Female fl. Disc flat, annular. Ovary obliquely ovoid, 1-celled ; ovules 2. Fruit with a hard endocarp. Seeds grooved, arillate.

> Fruit globose, small pisiform. Stamens 8-10 ... *H. sepiaria.*
> Fruit large, crowned with the stigma, ellipsoid. Sta-
> mens 5-8 ... ... *H. venusta.*

H. **sepiaria**, W. & A. in Edinb. New Phil. Jour. XIV. 297 ; Fl. Br. I. 5. 337 ; Dalz. & Gibs. Bom. Fl. 229 ; Bedd. Fl. Sylv. 198. From the Konkan southwards. Drupe red-coloured.

H. **venusta**, Thw. in Hook. Jour. Bot. VIII (1855) 272 ; Fl. Br. I. 5. 339; Dalz. & Gibs. Bom. Fl. 229 ; Bedd. Fl. Sylv. 198. "Hills in the Dhárwár district," Dalzell ; not seen by me.

CYCLOSTEMON, Blume.

Trees. Leaves alternate, entire or crenulate. Flowers diœcious, apetalous, axillary. Male fl. Sepals 4-6, broad, imbricate. Stamens 4-40, inserted round the margin of the disc, filaments free, erect. Female fl. Ovary 2-4-celled on the small disc, cells 2-ovuled Fruit indehiscent. Seeds with a caruncle, albumen fleshy.

C. **confertiflorus**, Hook f. Fl. Br. I. 5. 341. In the evergreen forests near the Devimone ghát of North Kánara. Fl. C.S.

BISCHOFIA, Blume.

A large tree. Leaves alternate, trifoliate, leaflets crenate. Flowers in axillary panicles, dioicous, apetalous. Calyx of 5-segments. Male fl. Sepals 5, concave, imbricating. Disc 0. Stamens 5. Pistillode short. Female fl. Ovary 3-celled ; cells 2-ovuled ; styles linear, entire. Fruit a globose drupe.

B. **javanica**, Blume Bijd. 1168 ; Brandis For. Fl. 446 ; Bedd. Fl. Sylv. t. 259 ; *Stylodiscus trifoliatus*, Dalz. & Gibs. Bomb. Fl. 235. *Boke,* Vern. In the evergreen forests of North Kánara, common on the Supa gháts. Fl. Mch.-Apl. Fr. ripe next Apl.

APOROSA, Blume.

Trees. Leaves alternate entire. Flowers diœcious, apetalous ; males in short dense spikes ; females sessile or in short few-flowered spikes. Male fl. Sepals 4. Stamens 2-5. Female fl. Ovary 2-celled ; stigmas small, plumose ; ovules 2 in each cell. Fruit dehiscent ; epicarp fleshy. Seeds plano-convex, albumen fleshy.

A. **Lindleyana**, Baill. Etudes Gen. Euphorb. 645 ; Fl. Br. I. 5. 349 ; Bedd. Fl. Sylv. t. 286 ; *Scepa Lindleyana,* Wight; Dalz. & Gibs. Bomb. Fl. 236. *Sali,* K. Common in the evergreen forests of the Konkan and North Kánara. Fl. C.S. Fr. ripe, June.

ANTIDESMA, Linn.

Shrubs or small trees. Leaves alternate, entire, penni-nerved.
Flowers small, dioicous; males in spikes; females in spikes or
racemes. Calyx 3-5-lobed, lobes imbricate. Petals 0. Stamens 2-5,
anthers didymous. Disc entire or lobed. Ovary 1-celled; ovules 2;
styles 3-4. Fruit an indehiscent, 1-seeded drupe.

Small tree. Calyx 5-7-partite. Stamens 4-7. Ovary
  tomentose ... ... *A. Ghæsembilla.*
Small tree. Calyx cupular, shortly 4-lobed. Stamens 3.
  Ovary glabrous ... ... *A. Bunius.*
A shrub. Calyx spreading, cup-shaped, 5-dentate.
  Stamens often 2. Ovary glabrous ... ... *A. diandrum.*
Small tree. Calyx 3-4 partite. Stamens 3-4. Ovary
  glabrous ... ... *A. Menasu.*

A. Ghæsembilla, Gærtn. Fruct. 1. 189, t. 39; Fl. Br. I. 5. 357;
Brandis For. Fl. 446; *A. pubescens*, Dalz. & Gibs. Bom. Fl. 236.
A. paniculatum, Dalz. & Gibs. Bom. Fl. 237. Joudhri, M. Throughout
the presidency, in North Kánara, a small tree in deciduous forests. Fl.
Apl.-June. Fr. Sept.-Oct. Leaves ovate, obtuse, sometimes slightly
cordate at the base. Spikes panicled.

A. Bunius, Spreng Syst. Veg. 1. 826; Fl. Br. I. 5. 358; Bedd. Fl.
Sylv. 200. *A. Alexiteria*, Willd. Grah. Cat. Bom. Pl. 186. *Amati*, M.
Throughout the western ghâts; Konkan and North Kánara in moist
forests, not mentioned by Dalzell; rare on the Khandála ghât. Fl.
May, June. Fr. Aug.-Sept. Fruit bright red.

A. diandrum, Roth. Nov. Sp. 369; Fl. Br. I. 5. 361; Brandis For.
Fl. 447; Dalz. & Gibs. Bomb. Fl. 237; Bedd. Fl. Sylv. 201; *A. lanceolatum*, Bedd. Fl. Sylv. 201. Throughout the presidency, common in
the Konkan and North Kánara in moist forests from the coast inland.
Fl. May-July. Fr. Aug.-Sept. Stamens usually 2. Fruit edible.

A. Menasu, Miq. Pl. Exsicc. Hohen. No. 104; Fl. Br. I. 5. 364;
A. lanceolatum, Dalz. & Gibs. Bomb. Fl. 237. In the evergreen forests
of the Konkan and North Kánara; also along nálás in deciduous moist
forests. Fl. Mch.-May. Fr. Sept. Oct. Leaves sometimes 10 in. long;
var. *linearifolia* is very common in the Siddápur táluka of North
Kánara.

BACCAUREA, Lour.

Trees. Leaves alternate, entire. Flowers diœcious or monoicous, often from the trunk or old branches, in racemes or spikes.
Calyx 4-5-cleft. Petals 0. Disc 0 or present. Stamens 4-10,
free. Ovary 2-5-celled, cells 2-ovuled; stigmas nearly sessile,
lobed. Fruit tardily loculicidally dehiscent. Seeds with a large
white arillus.

B. courtallensis, Muell. Arg. in DC. Prod. XV. II. 459; Fl. Br.
I. 5. 367; *B. sapida*, Bedd. Fl. Sylv. t. 280. North Kánara in moist
forests near Sungsal in the Ankola táluka. Fl. C. S.

JATROPHA, Linn.

Herbs, shrubs or trees, often glandular.　Leaves entire or lobed.
Flowers in terminal cymes, monoicous; female often petaliferous.
Sepals 5, imbricate.　Petals 5, free or connate.　Disc entire or of 5
glands.　Stamens many, filaments of the inner connate.　Ovary
2-4-celled; styles connate below, 2-fid; ovule 1 in each cell.　Fruit
a capsule with a bony endocarp.

> Petals free.
> Leaves 3-5-lobed, cordate.　Petioles and flowers with
> 　many long, stipitate viscous glands ... 　　... *J. glandulifera.*
> Eglandular.　Leaves 3-lobed, not cordate 　　... *J. nana.*
> Leaves peltate, pinnatifid, segments many.　Cymes
> 　scarlet 　　...　　...　　..　　... *J. multifida.*
> Petals connate.　Leaves entire or lobed ... 　　... *J. Curcas.*

J. glandulifera, Roxb. Fl. Ind. III. 688; Fl. Br. I. 5. 382; Dalz.
& Gibs. Bomb. Fl. 229.　*Jangli-erandi,* M.　Throughout the presidency,
common near the sea-coast at Kárwár, but not observed by me elsewhere.　Dalzell says "abundant at Punderpore."　Fl. throughout the
year.

J. nana, Dalz. & Gibs. Bomb. Fl. 229; Fl. Br. I. 5. 382.　*Kirkundi,* M.
" Rare, in stony places near Poona," Dalz.

J. multifida, Linn. Sp. Pl. 1006; Fl. Br. I. 5. 383.　*Chini-erandi,*
Vern.　Cultivated and naturalized throughout the presidency.

J. Curcas, Linn. Sp. Pl. 1006; Fl. Br. I. 5. 383; Brandis For. Fl.
442; Dalz. & Gibs. Bomb. Fl. Suppl. 77.　*Irundi, jaiphal,* Vern.;
kadandla, (Dhárwár).　Throughout the presidency, common in hedges
cultivated and naturalized.　Fl. Apl., May.　Fr. R.S.

ALEURITES, Forst.

Trees.　Leaves simple or lobed, 2-glandular at the base.　Flowers
monoicous, in large terminal panicles.　Calyx 2-3-partite.　Petals 5,
imbricate in bud.　Disc urceolate or glandular.　Stamens numerous,
on a conical torus.　Ovary 2-5-celled; cells 1-ovuled; styles 2-5,
2-cleft.　Fruit a large drupe; putamen 1-5-celled.　Albumen oily.

A. moluccana, Willd. Sp. Pl. IV. 590; Fl. Br. I. 5. 384; Bedd. Fl.
Sylv. t. 276; *A. triloba,* Dalz. & Gibs. Bomb. Fl. Suppl. 76.　*Akrod,*
jaiphal, Vern.　Planted in gardens throughout the Bombay Presidency;
it is often called, by Europeans, Belgaum walnut.　Fl. June-July.　Fr.
ripe next cold season.

CROTON, Linn.

Trees or shrubs, rarely herbs.　Leaves alternate, simple.　Flowers
small, in racemes or spikes, usually monoicous.　Calyx usually
4-partite, slightly imbricate in bud.　Petals as many as sepals,
developed in the males, rudimentary or absent in the females;
glands of the disc alternating with the petals.　Stamens 5 or nume-

rous, usually 10-20, filaments inflexed in bud. Ovary 2-4-celled ; cells 1-ovuled ; styles slender, 2-4-cleft. Capsule usually tricoccous, cocci 2-valved, 1-seeded. Seeds smooth with a caruncle.

Ovary lepidote or stellately tomentose. Trees or shrubs.

Shrub. Inflorescence and leaves densely silvery-lepidote. Leaves entire, 3-nerved at the base. Capsule ¼ inch ... *C. reticulatus.*

Small tree. Young leaves and inflorescence lepidote. Leaves glabrous when old, toothed, penninerved. Capsule ⅓ inch .. .. *C. oblongifolius.*

Small tree. Inflorescence stellately tomentose or glabrous, not lepidote. Leaves 3-5-nerved at the base, denticulate. Capsule scabrid, ⅓ inch ... ... *C. aromaticus.*

Shrub. Leaves penninerved ; mature glabrous, 3-nerved at the base. Capsule small, stellately tomentose, crustaceous, ¼ inch ... ... *C. Gibsonianus.*

Small tree. Leaves strongly triple-nerved, long petioled. Capsule large, obovoid, white, 1-inch, glabrous ... *C. Tiglium.*

Ovary glabrous. Young leaves softly stellately hairy, strongly 3-nerved. Capsule large, rugose, glabrous. A small tree... ... ... *C. Lawianus.*

C. reticulatus, Heyne in Wall. Cat. 7724 B. in part ; Fl. Br. I. 5. 386 ; *C. hypoleucus,* Dalz. & Gibs. Bomb. Fl. 231. *Panduray,* Vern. "In shady jungles on the Konkan hills." Dalzell. From the Konkan southwards, Fl. Br. I.

C. oblongifolius, Roxb. Fl. Ind. III. 685 ; Fl. Br. I. 5. 387 ; Brandis For. Fl. 440 ; Dalz. & Gibs. Bomb. Fl. 231 ; Beddome Fl. Sylv. 204. *Gunsur,* Vern. "Southern Konkan, rare," Dalzell. "A rare tree in our western forests," Beddome.

C. aromaticus, Linn. Sp. Pl. 1005 ; Fl. Br. I. 5. 388 ; Bedd. Fl. Sylv. 204. From the Konkan southwards, Fl. Br. I. ; on the Bababuden hills of Mysore, above 4,000 feet. Fl. Fr. R. & C.S.

C. Gibsonianus, Nimmo. in Grah. Cat. Bomb. Pl. 251 ; Fl. Br. I. 5. 392 ; Dalz. & Gibs. Bomb. Fl. 232. On the southern ghâts of North Kánara, in evergreen forests, common near the Falls of Gairsoppah ; near Násik, Dalzell. Fl. Fr. C.S. I am doubtful whether there are not two species here. Glands orange-coloured cordate. Capsule red when ripe, size of walnut, Grah. Glands small, orange-coloured, not cordate. Capsule small grey tomentose when ripe in specimens from North Kánara.

C. Tiglium, Linn. Sp. Pl. p. 1004 ; Fl. Br. I. 5. 393 ; Grah. Cat. Bomb. Pl. 181 ; Brandis For. Fl. 440. Purging croton. *Jamalgota, jaipa,* Vern. Naturalized in the South Konkan.

C. Lawianus, Nimmo in Grah. Cat. Bomb. Pl. 251 ; Fl. Br. I. 5. 394 ; Dalz. & Gibs. Bomb. Fl. 232. Bababuden Hills, Mysore, probably may be found in the southern parts of North Kánara, but not yet seen in the Bombay Presidency. Fruit with the calyx enlarged like in *Dimorphocalyx glabellus,* Thw.

GIVOTIA, Griff.

A small tree. Leaves alternate, cordate, sinuate toothed, 5-9-nerved at the base. Flowers diœcious, in axillary or terminal cymes. Disc entire or lobed. Sepals 5. Petals 5, connate Stamens 13-25 on a woolly receptacle, filaments connate. Ovary 2-3-celled; styles short, 2-fid; ovules 1 in each cell. Fruit a subglobose, 1-seeded drupe.

G. rottleriformis, Griff. in Calc. Jour. Nat. Hist. IV. 388; Fl. Br. I. 5. 395; Brandis For. Fl. 442; Dalz. & Gibs. Bomb. Fl. 223. Bedd. Fl. Sylv. t. 285. Throughout the dry districts of the Deccan, abundant in the Belgaum district near Gokák in dry, deciduous, open jungles. Fl. H.S.

BLACHIA, Baill.

Small trees or shrubs. Leaves alternate or opposite. Flowers monœcious; males subumbellate; females solitary or fascicled. Sepals 4-5, accrescent in the female. Petals 4-5, small, hyaline. Disc of 4-5 scales, alternating with the petals. Stamens 10-20, on a convex torus, filaments free. Petals 0 in the female flower. Ovary 3-4-celled; styles filiform, 2-partite, recurved; ovules 1 in each cell. Capsule 3-celled; cells 2-valved. Seeds ecarunculate, smooth, shining.

B. denudata, Benth. in Jour. Linn. Soc. XVII. 226; Fl. Br. I. 5 403; *Croton umbellatum,* Dalz. & Gibs. Bomb. Fl. 231. In the evergreen forests of the Konkan and North Kánara, common on the Supa ghâts. Fl. Oct., Dec. Fr. Dec., Feb.

DIMORPHOCALYX, Thw.

Trees. Leaves entire, coriaceous, alternate. Flowers in axillary or terminal few flowered spikes or the female subsolitary, diœcious. Calyx cupular, 5-toothed or lobed, much accrescent in fruit. Petals 5. Disc of 5 glands, alternating with the petals. Stamens 10-20, on a short torus; filaments stout, free or the inner connate. Ovary 3-celled; styles erect, 2-fid; ovule 1 in each cell. Fruit a 3-celled capsule, cells 2-valved.

D. Lawianus, Hook. f. Fl. Br. I. 5. 404; *Trigonostemon Lawianus,* Muell. Arg.: Bedd. Fl. Sylv. t. 273. In the evergreen forests of the Konkan and North Kánara, common on the southern ghâts; between Nilkund and Gairsoppah. Fl. Sept., Oct. Fr. Nov., Jany. Calyx segments of female flower unequal.

AGROSTISTACHYS, Dalzell.

Shrubs. Leaves long, entire or serrate, subsessile. Flowers diœcious, in bracteate spikes or racemes, males few within each bract; females solitary, long pedicelled. Male fl. Calyx of 2-5-valvate lobes. Petals 8, shorter. Disc-glands large, alternating with the petals. Stamens 8-13 on a convex receptacle; filaments free; anthers versatile, cells pendulous from the thickened connective. Female fl. 5-6-fid. Petals longer, caducous. Ovary 3-celled; styles thick, short, spreading, 2-fid. Fruit a 3-celled capsule or subfleshy. Seeds globose, albumen fleshy.

> Bracts of male 1-3-flowered; densely imbricate   ... *A. indica.*
> Flowers many under each bract, bracts remote   ... *A. longifolia.*

A. indica, Dalz. & Gibs. Bomb. Fl. 232; Fl. Br. I. 5. 406; Bedd. Fl. Sylv. 205. Throughout the Konkan and North Kánara in evergreen forests, on the banks of streams. Fl. R.S. Fr. Feb.

A. longifolia, Benth. in Gen. Pl. III. 303; Fl. Br. I. 5. 407; Bedd. Fl. Sylv. 205. Evergreen forests of North Kánara, on the Supa gháts, along streams. Fl. R.S. Fr. ripe Feb., Apl.

ADENOCHLÆNA, Baill.

Trees or shrubs. Leaves alternate, entire, 3 or penninerved. Flowers monœcious, terminal or axillary, in interrupted spikes; males many, females few. Disc 0. Sepals 4-6, valvate. Stamens 4-6, filaments free, exserted; anther-cells adnate to the thick connective, parallel. Female fl. Sepals 5-8, narrow, unequal, accrescent. Ovary 3-celled; styles long, connate at the base, bifid above, recurved, plumose; ovule 1 in each cell. Fruit a 3-celled capsule. Seeds albuminous, with a crustaceous testa.

A. indica, Bedd. MSS. Fl. Br. I. 5. 418; *Cephalocroton indicum*, Bedd. Fl. Sylv. t. 261. Evergreen forests of North Kánara, near the Falls of Gairsoppah. Fl. Oct., Dec.

TREWIA Linn.

Trees. Leaves simple, opposite. Flowers small, in axillary racemes, diœcious. Calyx 3-4-partite in the male flowers, valvate in the females, imbricate in bud. Petals and disc 0. Stamens numerous, free, on a central depressed torus. Ovary 3-4-celled; cells 1-ovuled; styles connate at the base, long, undivided. Fruit a drupe. Seeds albuminous.

> Female flowers solitary. Fruit globose, 1½ in.   ...   ... *T. nudiflora.*
> Female flowers in short racemes. Fruit ¼ in.   ...   ... *T. polycarpa.*

T. nudiflora, Linn. Sp. Pl. Ed. 3, Append. 1661; Fl. Br. I. 5. 423 Brandis For. Fl. 443; Dalz. & Gibs. Bomb. Fl. 231. *Petari*, Vern. Throughout the Konkan and North Kánara in moist forests, often along the banks of rivers and nálás. Fl. Dec., Feb. Fr. R.S.

T. polycarpa. Benth. in Gen. Plant. III. 319; Fl. Br. I. 5. 424; *T. nudiflora*, Bedd. Fl. Sylv. t. 281. The Konkan and North Kánara common, scarcely differs from *T. nudiflora*, Linn. Fl. C.S.

MALLOTUS, Lour.

Trees or shrubs. Leaves alternate or opposite, simple or lobed, sometimes peltate. Flowers usually diœcious, small, in terminal racemes or spikes. Calyx 3-5-partite; lobes valvate in the males, in the females spathaceous or valvately 3-6-lobed. Petals and disc 0. Stamens numerous, free or cohering at the base, on a central torus. Ovary 2-5-called; cells 1-ovuled; styles as many as the ovary cells, simple or connate at the base. Fruit a dry capsule of 2-5-cocci. Seeds albuminous; cotyledons broad.

Erect trees. Leaves opposite or alternate.
Leaves alternate, rusty white tomentose beneath, narrowly peltate. Capsule 3-4-coccous, muricate with soft white tomentose processes ... ... *M. albus.*
Leaves opposite, nearly glabrous, minutely glandular beneath. Capsule 3-dymous with soft, scattered tubercles ... ... *M. stenanthus.*
Leaves opposite, golden glandular beneath. Capsule with long soft, villous filaments ... ... *M. Lawii.*
Scandent shrub. Leaves alternate with soft tawny tomentum, 3-nerved. Capsule didymous, stellately tomentose ... ... *M. repandus.*
Erect tree. Leaves 3-nerved, glabrous above, puberulous beneath. Capsule tridymous, covered with crimson glandular powder, unarmed ... ... *M. philippinensis.*

M. albus, Muell. Arg. in Linnæa XXXIV. 188; Fl. Br. I. 5. 429; Brandis For. Fl. 444; Beddome Fl. Sylv. 208; *Rottlera mappoides,* Dalz. & Gibs. Bomb. Fl. 230. (Var. *occidentalis.*) Throughout the moist forests of the Konkan and North Kanara. Fl. Sept., Oct. Fr. ripe Nov., Dec. Leaves subpeltate, rusty tomentose or white beneath.

M. stenanthus, Muell. Arg. in Linnæa XXXIV. 191; Fl. Br. I. 5. 437. In the evergreen forests of North Kánara, on the gháts from Yellápur southwards. Fl. Sept., Nov. Fr. ripe C.S. Leaves sinuate, toothed, glandular at the base.

M. Lawii, Muell. Arg. in Linnæa XXXIV. 192; Fl. Br. I. 5. 438; Bedd. Fl. Sylv. 209; *Rottlera aureopunctata,* Dalz. & Gibs. Bomb. Fl. 230. Throughout the evergreen forests of the Konkan and North Kánara, common in the forests near the Devimone and Nilkund gháts. Fl. Oct., Dec. Fr. ripe Feb., Mch. Leaves penninerved, repand dentate.

M. repandus, Muell. Arg. in Linnæa XXXIV. 197; Fl. Br. I. 5. 442; Brandis For. Fl. 444; Bedd. Fl. Sylv. 210; *Rottlera tricocca,* Roxb.; Dalz. & Gibs. Bom. Fl. 230. Throughout the Konkan and North Kánara in moist forests, common in the South Konkan, also in the evergreen forests near Yellápur. Fl. Sept. Oct. Fr. ripe C.S.

M. philippinensis, Muell. Arg. in Linnæa XXXIV. 196; Brandis For. Fl. 444; Fl. Br. I. 5. 442; Bedd. Fl. Sylv. t. 289; *Rottlera tinctoria,* Dalz. & Gibs. Bomb. Fl. 230. *Kunkuma,* K.; *roen, kapila, shendri,* Vern. Throughout the dry forests of the presidency and in Sind. Fl. Nov., Jan. Fr. ripe Feb., May. The powder from the fruit yields the kamela dye used for colouring silk. c

CLEIDION, Blume.

Trees. Leaves simple, alternate. Flowers diœcious, males in racemes, females solitary, axillary. Calyx 3-5-partite, valvate in the males, imbricate in the females. Stamens numerous, on a conical, central receptacle ; anthers peltately attached, 4-celled ; connective produced. Ovary 2-3-celled ; cells 1-ovuled ; styles filiform, 2-cleft. Capsule 2-3-coccous. Seeds with fleshy albumen.

C. javanicum, Blume Bijd. 613 ; Fl. Br. I. 5. 444 ; Bedd. Fl. Sylv. t. 272 ; *Rottlera uranda*, Dalz. & Gibs. Bomb. Fl. 230. Evergreen forests of the Konkan and North Kanara, common in the Ainshi ghát forests. Fl. Oct., Dec. Fr. Jan., Feb.

MACARANGA, Thouars.

Trees or shrubs. Leaves alternate, large, peltate, entire or lobed. Flowers diœcious, in panicles, racemes or spikes. Calyx in males valvate, in females imbricate. Petals and disc wanting. Stamens 1 or more, central, on a convex receptacle ; anthers peltately attached, 3-4-celled. Ovary 2-6-celled, cells 1-ovuled. Fruit a small 1-5-celled, often glandular capsule. Seeds globose ; albumen fleshy.

Male bracts with a glandular appendage ; female bracts without glands.
Stamens 6-8......　...　...　...　...　...　...　... *M. indica.*
Bracts broad, toothed, tomentose ; smaller in the female.
Stamens 2-3　...　...　...　...　...　—　...　... *M. Roxburghii.*

M. indica, Wight, Ic. t. 1883 and 1949, f. 2 ; Fl. Br. I. 5. 446 ; Bedd. Fl. Sylv. t. 287. From the Konkan southwards. Fl. Br. I. ; " very common in the western forests of the Madras Presidency," Beddome ; not seen in North Kánara by me.

M. Roxburghii, Wight Ic. t. 1949, f. 4 ; Fl. Br. I. 5. 448 ; Dalz. & Gibs. Bomb. Fl. 228 ; *M. tomentosa*, Beddome. Fl. Sylv. t. 287. *Chanda*, Vern. Throughout the moist forests of the Konkan and North Kánara, very common. Fl. Feb., March. Fr. April, May.

HOMONOIA. Lour.

Shrubs. Leaves alternate, stipules deciduous. Flowers diœcious, in spikes or racemes or the females solitary. Calyx of males 3-partite and valvate, of the females 5-partite and imbricate. Petals and disc 0. Stamens numerous in a dense globose head of branched filaments. Ovary 3-4-celled, cells 1-ovuled ; styles 3-4, connate at the base, simple. Fruit a 3-4-celled capsule ; albumen fleshy.

Leaves lanceolate. Spikes long, slender ...　...　...　... *H. riparia.*
Leaves obovate. Spikes short, stout　...　...　...　... *H. retusa.*

H. riparia, Lour. Fl. Coch. 637 ; Fl. Br. I. 5. 455 ; Brandis For. Fl. 401 ; Bedd. Fl. Sylv. t. 212 ; *Adelia neriifolia*, Roth., Dalz. & Gibs Bomb. Fl. 231. In the beds of rivers and streams throughout the presidency, very common in Konkan and North Kánara rivers. Fl. Nov., March.

H. retusa, Muell. Arg. in Linnæa XXXIV. 200 ; Fl. Br. I. 5. 456 ; Brandis For. Fl. 445 ; *Adelia retusa,* Wight. Dalz. & Gibs. Bom. Fl. 231 ; Bedd. Fl. Sylv. 212. Throughout the presidency in river beds, often associated with *H. riparia,* common in North Kánara rivers. Fl. Nov., April.

RICINUS, Linn.

Trees or shrubs. Leaves alternate, peltate. Flowers monœcious, apetalous, in terminal subpanicled racemes. Calyx 5-divided, valvate in bud. Petals and disc wanting. Stamens numerous ; filaments variously connate on a plano-convex torus. Ovary 3-celled. Capsule dry, 3-coccous. Albumen oily.

R. communis, Linn. Sp. Pl. 1007 ; Fl. Br. I. 5. 457 ; Brandis For. Fl. 445 ; Dalz. & Gibs. Bomb. Fl. Suppl. 78. *Erandi,* Vern. ; *tirki,* Guj. ; *haralu,* K. Castor oil plant. Cultivated and naturalized near villages, throughout the presidency. Fl. Fr. Feb., March.

SAPIUM, P. Br.

Trees or shrubs. Leaves alternate, entire or toothed ; petiole 2-glandular at top. Flowers in terminal, simple or panicled spikes, monœcious, apetalous ; males several in each bract ; females in the lower part of the spike. Petals and disc 0. Calyx 2-3-lobed, toothed or split to the base into 2-3-valvate sepals. Stamens 2-3 ; anther-cells distinct. Ovary 2-3-celled ; styles spreading and recurved, connate at the base ; cells 1-ovuled. Fruit a tardily dehiscent 3-valved capsule. Seeds with fleshy albumen, usually long-persistent on the columella.

> Columella winged, persistent. Fruit capsular, size of a pea. ... *S. sebiferum.*
> Columella not persistent. Capsule large, woody...   ...   ... *S. indicum.*
> Fruit drupaceous, obscurely lobed, ½ inch ...   . *S. insigne.*

S. sebiferum Roxb. Fl. Ind. III. 693 ; Fl. Br. I. 5. 470 ; Dalz. & Gibs. Bomb. Fl. Suppl. 77 ; *Excœcaria sebifera,* Brandis For. Fl. 441. Chinese tallow tree. *Pipalyank,* Vern. Cultivated near Bombay. Fl. June. Fr. ripe Oct.

S. indicum, Willd. Sp. Pl. IV. 572 ; Fl. Br. I. 5. 471 ; Fl. Br. I. 5. 471 ; Grah. Cat. Bomb. Pl. 181. *Hurna,* M. Various parts of the South Konkan, Graham.

S. insigne, Benth. in Gen. Plant. III. 335 ; Fl. Br. I. 5. 471 ; Brandis For. Fl. 442 ; var. *malabarica, Excœcaria insignis,* Bedd. Fl. Sylv. 214 ; *Falconeria malabarica,* Dalz. & Gibs. Bomb. Fl. 227. *Ura, dudla,* M. Common near the coast of the Konkan and North Kánara, on dry rocky soil, usually on laterite, also in moist forests on the ghâts. Fl. C.S. Fr. Mch.

EXCŒCARIA, Linn.

Trees or shrubs. Leaves simple, usually alternate ; with paired stipules. Flowers small in terminal or axillary, androgynous spikes,

usually monœcious. Calyx 2-3-divided. Petals and disc 0. Stamens
2-3, free or connate. Ovary 2-4-celled; cells 1-ovuled; styles
2-4, simple, connate at the base. Fruit a 3-celled capsule; cocci
separating from the columella. Seeds globose; albumen fleshy.

Leaves alternate. Sepals minute, unequal ...   ... *E. Agallocha.*
Leaves opposite. Sepals orbicular, irregularly
    toothed ...   . *E. robusta.*

E. Agallocha, Linn. Sp. Pl. 1451; Fl. Br. I. 5. 472; Brandis
For. Fl. 442; Dalz. & Gibs. Bomb. Fl. 227; Bedd. Fl. Sylv. 255.
Geva, surúnd, phungali, M. In tidal marshes, along the coast of the
presidency, very common. Fl. July, Aug. Fr. Sept., Oct.

E. robusta, Hook, f. Fl. Br. I. 474. Konkan, Stocks.

ORDER 78. **URTICACEÆ**.

Herbs, shrubs or trees, often with milky sap. Leaves simple,
usually alternate. Stipules present, often deciduous. Flowers small,
in heads or cymes, usually monœcious or diœcious. Perianth simple
of 3-5 segments. Stamens as many as perianth-segments and
opposite to them. Ovary free, 1-celled, 1-ovuled. Micropyle superior.
Fruit various. Seeds with or without albumen.

Flowers bisexual or polygamous. Fruit a samara.
    Anthers erect in bud. Trees   ...   . HOLOPTELEA.
Fruit a drupe.
Cotyledons broad   ...   . CELTIS.
Cotyledons narrow   ...   . TREMA.
Filaments inflexed in bud. Ovule pendulous.
Male flowers capitate. Females solitary   ...   ... STREBLUS.
Male and female flowers in globose heads   ...   ... PLECOSPERMUM.
Male and female flowers spikate   ..   .. MORUS.
Flowers unisexual, males or all in globose heads or
    open or closed receptacles. Anthers erect.
Flowers inside closed receptacles   ...   . FICUS.
Male flowers on disciform receptacles. Females
    solitary .... ANTIARIS.
Flowers in globose, oblong or cylindric heads.
    Males monandrous   ...   . ARTOCARPUS.
Climbing shrubs. Male and female heads cymose ... CONOCEPHALUS.
Shrubs with stinging hairs. Stamens inflexed in bud.
    Ovule erect, orthotrope   ..   .. LAPORTEA.
Trees or shrubs without stinging hairs. Fruiting
    perianth dry or fleshy.
Fruiting perianth dry. · Stigma filiform   ...   ... BŒHMERIA.
Fruiting perianth fleshy.
Stigma sessile, subpeltate   ...   .   :.. VILLEBRUNEA.
Stigma penicillate. Leaves ashy beneath   ...   ... DEBREGEASIA.

HOLOPTELEA, Planch.

Trees. Leaves simple, alternate, unequal-sided; stipules caducous. Flowers bisexual, in lateral fascicles. Perianth campanulate, 4-8-lobed; lobes imbricate in bud. Stamens as many as perianth-lobes. Ovary free, 1-2-celled; styles 2; cells 1-ovuled. Fruit
a samara, surrounded by reticulate, obcordate wing; pedicel
articulate. Albumen 0.

H. integrifolia, Planch. in Ann. Soc. Nat. Scr. 3, X. 269 ; Fl. Br. I.
5. 481 ; Dalz. & Gibs. Bomb. Fl. 238 ; *Ulmus integrifolia*, Brandis For.
Fl. 431 ; Bedd. Fl. Sylv. t. 310. *Wawuli, papara,* M. Throughout the
presidency in deciduous forests. Fl. Feb., Mch. Fr. June-Aug. Leaves
entire, cotyledons plicate.

CELTIS, Linn.

Trees or shrubs. Leaves alternate, stipulate, entire or serrate.
Flowers polygamous, in axillary or lateral cymes. Perianth 4-5-
divided ; segments deciduous, imbricate in bud. Stamens 4-5, short,
surrounding a woolly torus. Ovary on a hairy disc ; stigmas 2,
sessile, deciduous. Fruit a globose drupe with a hard endocarp.
Albumen 0 or scanty.

C. tetrandra, Roxb. Fl. Ind. II. 63 ; Fl. Br. I. 5. 482 ; *C. Roxburghii*, Planch ; Dalz. & Gibs. Bomb. Fl. 237 ; *C. serotina*, Bedd. Fl. Sylv.
218. *Brumaj*, M. Throughout the Konkan and North Kánara, usually
in evergreen forests on the ghats, locally abundant. Fl. June, Sept.
Fr. Mch., May.

TREMA, Lour.

Trees. Leaves alternate, 3-7-nerved at the base ; stipules caducuous. Flowers monoicous, in axillary cymes. Sepals 4-5, subimbricate in bud. Stamens 5, longer than the sepals. Ovary sessile ;
style arms 2, linear, ovule pendulous. Fruit a small drupe with a
hard endocarp. Albumen fleshy.

T. orientalis, Blume Mus. Bot. II. 62 ; Fl. Br. I. 5. 484 ; *Sponia
orientalis*, Brandis For. Fl. 430 ; Bedd. Fl. Sylv. 219 ; *S. Wightii*,
Dalz. & Gibs. Bomb. Fl. 238. *Ranambada, kapashi, kargol*, M ; *bendakarke,
gol*, Vern. Charcoal tree. Throughout the Konkan and North Kánara
in moist forests, common. Fl. Mch.-Jany. Fr. Dec. Jany.

STREBLUS, Lour.

Tree or shrub. Leaves alternate, scabrid ; stipules small. Flowers dioicous, axillary, males in shortly pedunculate clusters, females
peduncled, solitary. Sepals 4, imbricate. Stamens 4, inflexed in
bud. Female fl. bracteate. Ovary 1-celled ; style arms long ; ovule
1, pendulous. Fruit globose, enclosed in the perianth. Albumen 0.

S. asper, Lour. Fl. Cochin II. 615 ; Fl. Br. I. 5. 489 ; Bedd.
Fl. Sylv. 220 ; Brandis For. Fl. 410 ; *Epicarpurus orientalis*, Blume,
Dalz. & Gibs. Bomb. Fl. 240. *Punje*, K.; *poi, kharota*, M.; *karvati*,
Vern. Throughout the presidency in dry open forests, common ; it is also
found in the evergreen forests of the North Kánara. Fl. Jany., Mch.
Fr. Apl., May.

PLECOSPERMUM, Trecul.

Trees or shrubs. Leaves alternate, entire. Flowers dioicous, in
globular heads. Male flowers bracteolate ; perianth 4-fid, segments
imbricate. Stamens free, 4, inflexed in bud. Female fl. Perianth

gamophyllous, 4-toothed.   Ovary free, 1-celled ; ovule 1, pendulous
style liliform, exserted.   Fruit an irregular syncarpium of akenes,
enclosed in the connate perianth lobes.   Albumen 0.

P. spinosum, Trecul. in Ann. Sc. Nat. Ser. 3, VIII. 124; Fl. Br. I.
5. 491; Brandis For. Fl. 401 ; Bedd. Fl. Sylr. t. 220.   In the dry
districts of the presidency, common in hedges in the Dhárwár district.
Fl. Jany., April.   Fr. May, June.

MORUS, Linn.

Trees or shrubs.   Leaves alternate, simple; stipules deciduous.
Flowers unisexual, spicate.   Male fl.   Sepals 4.   Stamens 4, oppo-
site to and longer than the   calyx-segments; anther-cells introrse.
Pistillode rudimentary.   Female fl.   Sepals 4, accrescent in fruit,
Ovary 1-celled ; ovule 1 pendulous.   Fruit a syncarpium of akenes
included in the succulent perianths.   Albumen fleshy.

Female spikes short, ovoid. Fruit black when ripe. *M. indica,*
Female spikes long, cylindric. Fruit cylindrical,
   yellowish-white   ...   . *M. lœvigata.*

M. indica, Linn. Sp. Pl. 986 ; Fl. Br. I. 5. 492 ; Brandis For. Fl.
408.   *Tut, ambat,* M.   Cultivated and run wild near villages in North
Kánara and elsewhere throughout the presidency.   Fl. Mch., June. Fr.
June, Aug.

M. lævigata, Wall. Cat. 4649 ; Fl. Br. I. 5. 492 ; Brandis For. Fl.
409.   Cultivated in gardens at Dhárwár and probably elsewhere in the
presidency.   Fl. C.S.   Fr. H. S.

FICUS, Linn.

Trees or shrubs.   Leaves alternate or rarely opposite, entire or
lobed.   Flowers unisexual on the inner surface of a globose or ovoid
receptacle, the mouth of which is closed by imbricate bracts.   Receptacles bracteate, unisexual, but usually androgynous with the
males near the mouth.   Flowers of 4 kinds, male, female, gall-flowers
and neuters. Male fl.   Perianth thin, 2-6-fid or partite.   Stamens
1-2 ; anthers of 2, distinct cells. Female fl. Perianth of the male or 0.
Ovary 1-celled ; style excentric, ovule 1, pendulous.   Akenes crustaceous or fleshy.

Leaves alternate, hispid.   Receptacles small, yellow or purple, ¼ in. peduncled, fascicled or in
  pairs; basal bracts 0.   ...   . *F. gibbosa.*
Leaves alternate, smooth or tomentose ; petiole
  not jointed to the blade.   Receptacles axillary,
  sessile or peduncled, tribracteate at the base.
Leaves more or less tomentose.   Receptacles
  sessile, usually in pairs.
Fruit pubescent, red. ⅓ in. in diameter ...   ... *F. bengalensis.*
Fruit pubescent, orange-yellow, 1 in. in diameter . *F. mysorensis.*
Fruit tomentose, grey, ⅓ in. in diameter ...   ... *F. tomentosa.*
Leaves glabrous,   Receptacles sessile, in pairs.

Fruit blood-red, ¾ in. in diameter ... ... *F. Benjamina.*
Fruit small, ½ in. yellow or reddish ... ... *F. retusa.*
Fruit small, ¼ in. greenish-yellow, dotted ... *F. Talboti.*
Leaves alternate. Receptacles in peduncled
axillary pairs. Bracts at the base 0, but 3,
small, free bracts low down on the peduncle.
Fruit globose, ¾ in. in diameter ... ... *F. nervosa.*
Leaves coriaceous or membranous; petiole long,
jointed to the blade. Receptacles sessile, in
pairs; basal-bracts 3, small.
Fruit when young white with dark spots, black
when ripe ... ... *F. Rumphii.*
Fruit dark purple; basal bracts broad, coriaceous. *F. religiosa.*
Receptacles in pairs from tubercles, purple black
with greenish dots. Basal-bracts brown, membranous ... ... *F. Arnottiana.*
Receptacles in clusters, small, dotted and whitish
yellow; basal-bracts 3, bifid ... ... *F. Tjakela.*
Receptacles crowded at the ends of the branches,
purple black; basal-bracts 3, minute, scarious ... *F. Tsiela.*
Receptacles globose ¼ in. white, flushed with red
and dotted; basal-bracts minute... ... ... *F. infectoria.*
Leaves rigid, coriaceous. Fruit peduncled, solitary, 1¼ in. in diameter, green, scabrid ... *F. callosa.*
Leaves coarsely-toothed or repand, scabrid.
Receptacles solitary, axillary, peduncled, small,
scabrid, green. A shrub ... ... *F. heterophylla.*
Leaves scabrous, hispid, toothed. Receptacles
peduncled, globose, 1 in. in diameter, yellow or
purple ... ... *F. asperrima.*
Leaves opposite. Receptacles peduncled, fascicled on the old wood or on leafy branches,
hispid, yellowish, sometimes hypogeal ... ... *F. hispida.*
Leaves alternate, membranous, glabrous. Receptacles large on short, axillary branches, scaly
and leafless, from the trunk, red ... ... *F. glomerata.*

F. gibbosa, Blume Bijd. 466; Fl. Br. I. 5. 496; *F. parasitica,* Koen.;
Bedd. Fl. Sylv. 224; Brandis For. Fl. 420; *Urostigma ampelos,* Dalz.
& Gibs. Bomb. Fl. 315; *U. volubile,* Dalz. & Gibs. Bomb. Fl. 242; *F.
tuberculata,* Bedd. Fl. Sylv. 224. *Datir,* Vern. Throughout the presidency, common on old walls or on the sides of wells, also epiphytic;
throughout the moist and dry forests of the western gháts. Fruit ripe
Apl., May.

F. bengalensis, Linn. Hort. Cliff. 471, n. 4; Fl. Br. I. 5. 499; Bedd.
Fl. Sylv. 222; Brandis For. Fl. 412; *F. indica,* Roxb.; *Urostigma
bengalense,* Dalz. & Gibs. Bomb. Fl. 240. Banyan tree. *Wad, alada, vadi,*
Vern. Throughout the presidency, wild or planted; self-sown throughout
the deciduous and evergreen forests of the western gháts. Dr. King in
his *Species of Ficus* says "really wild only in the sub-Himalayan forests
and on the lower slopes of the hill ranges of Southern India." Fruit
ripe, April-June.

F. mysorensis, Heyne in Roth. Nov. Sp. 390; Fl. Br. I. 5. 500; Brandis
For. Fl. 414; King. Sp. Ficus 19; Bedd. Fl. Sylv. 222; *F. pubescens,* Roth.
Urostigma dasycarpum, Dalz. & Gibs. Bomb. Fl. 242. *Bhurvar,* M.
Throughout the Konkan and North Kánara, in moist forests along the

gháts, also near villages and in open situations, self-sown or planted. Fr. ripe Apl.-May. Fruit when young flocculent tomentose, ripe glabrous, yellow, sometimes tinged with red.

F. tomentosa, Roxb. Fl. Ind. III. 550; Fl. Br. I. 5. 501; Brandis For. Fl. 414; Bedd. Fl. Sylv. 223. Donkey's banyan. *Kulgolu,* K. Throughout the presidency, common near the sea-coast of the Konkan and North Kánara, on sandstone rocks near Bádámi, Bijápur Collectorate. Fr. ripe Mch., Apl.

F. Benjamina, Linn. Mantiss. 129 (Excl. Syn. Rheede); Fl. Br. I. 5. 508; Bedd. Fl. Sylv. 223; *F. comosa,* Roxb.; Bedd. Fl. Sylv. 223; *Urostigma Benjamina,* Miq.; Dalz. & Gibs. Bomb. Fl. 242; var. *comosa.* Kz. For. Fl. 11. 446. In the moist forests of North Kánara, the variety *comosa* is found in the Southern Marátha Country, Dhárwár district, in deciduous forests, rare. Fr. ripe, H.S.

F. retusa, Linn. Mant. 129; Fl. Br. I. 5. 511; Brandis For. Fl. 417; Bedd. Fl. Sylv. 223; *Urostigma retusum* and *nitidum,* Dalz. & Gibs. Bomb. Fl. 241, 242. *Nandruk, pilala,* Vern. Throughout the presidency, commonly planted along road sides.

F. Talboti, King. Sp. Fic. 51, t. 63; Fl. Br. I. 5. 512. On the southern gháts of North Kánara, common in moist forests. Fr. ripe cold season. Leaves caudate acuminate.

F. nervosa, Roth. Nov. Sp. 338; Fl. Br. I. 5. 512; Bedd. Fl. Sylv. 223. In the evergreen forests of North Kánara from Supa southwards, a very large tree, without aereal roots. Fruit hard; walls thick, ripe Feb., Mch.

F. Rumphii, Blume Bijd. 437; Fl. Br. I. 5. 512; *F. cordifolia,* Roxb.; Brandis For. Fl. 416. *Pair,* Vern. Western gháts near Bombay, Brandis, at Khandála and probably throughout the Konkan gháts.

F. religiosa, Linn. Hort. Cliff. 471; Fl. Br. I. 5. 513; Bedd. Fl. Sylv. t. 314; Brandis For. Fl. 415; *Urostigma religiosum,* Dalz. & Gibs. Bomb. Fl. 241. *Arle, basri, pipal,* Vern.; *pipro,* Panch Maháls; *ashvatha,* M. Planted near temples and villages throughout the presidency. Fr. ripe May-July.

F. Arnottiana, Miq. Ann. Mus. III. 287; Fl. Br. I. 5. 513; *Urostigma cordifolium,* Dalz. & Gibs. Bomb. 242. *Paeer,* Vern. Throughout the Konkan and North Kánara, common on rocks near the coast, also in moist forests, Kumta, near the Yena rocks. Fr. ripe Feb.-Apl.

F. Tjakola, Burm. Fl. Ind. 227; Fl. Br. I. 5. 514. Throughout the Konkan and North Kánara, common in moist forests, abundant in the forests near Yellápur and generally on the gháts of North Kánara. Fr. ripe Mch., May. Stipules large, membranous red, deciduous. Young leaves appear in Feb. A very distinct species.

F. Tsiela, Roxb. Fl. Ind. III. 519; Fl. Br. I. 5. 515; Bedd. Fl. Sylv. 314; Brandis For. Fl. 415; *Urostigma pseudo-Tjiela,* Miq.; Dalz. & Gibs. Bomb. Fl. 241. *Pipri,* Vern.; *bili-basri,* K. Throughout the presidency; often planted along road sides. Fruit purple black when ripe, Apl.-Oct.

F. infectoria, Roxb. Fl. Ind. III. 550 ; Fl. Br. I. 5. 515 ; Brandis For. Fl. 414 ; Bedd. Fl. Sylv. 222; *Urostigma infectoria,* Dalz. & Gibs. Bomb. Fl. 241 ; *Ficus Lambertiana,* Miq. *Urostigma Lambertianum,* Dalz. & Gibs. Bomb. Fl. 241 ; *Ficus Wightiana,* Bedd. Fl. Sylv. 222. *Bassari, pakuri, lendra,* M. ; *hari basri,* Vern. Common throughout the presidency, usually in dry forests. Var. *Lambertiana* is common in North Kánara in moist forests near Yellápur and elsewhere. Var. *Wightiana* is found in the southern parts of North Kánara ; it is not so common as the former variety. Var. *infectoria* proper is often planted along road sides and is a very common tree throughout the presidency. The white dotted receptacles are sometimes more or less peduncled. I have a tree from North Kánara, formerly named at Calcutta *Ficus urophylla,* Miq.. but which has nothing to do with that species; it is closely allied to *Ficus infectoria,* var. *Lambertiana ;* or it may be a new species ; it has the peduncles ½ in long.

F. callosa, Willd. in Act. Acad. Berol. 1798, 102, t. 4 ; Fl. Br. I. 5. 516; *F. cinerascens,* Thw. ; Bedd. Fl. Sylv. 224. In the evergreen forests of the Konkan and North Kánara, a very large tree with pearly juice, without aëreal roots. Receptacles large, peduncled, solitary, ripe June-July.

F. heterophylla, Linn. fil. Suppl. 442 ; Fl. Br. I. 5. 518 ; Dalz. & Gibs. Bomb. Fl. 243 ; Brandis For. Fl. 424. *Datir,* M. Throughout the presidency, common in North Kánara, along the banks of streams and rivers. Fr. ripe May-June.

F. asperrima, Roxb. Fl. Ind. III. 554 ; Fl. Br. I. 5. 522 ; Dalz. & Gibs. Bomb. Fl. 243 ; Bedd. Fl. Sylv. 224. *Khargas,* K. ; *kharwat,* M. Throughout the moist forests of North Kánara and the Konkan, very common. Fr. ripe Mch., Apl.

F. hispida, Linn. f. Suppl. 442 ; Fl. Br. I. 5. 552 ; Brandis For. Fl. 423 ; Bedd. Fl. Sylv. 224 ; *Covellia oppositifolia,* Gasp.; Dalz. & Gibs. Bomb. Fl. 243 ; *C. dæmonum,* Dalz. & Gibs. Bom. Fl. 244. *Kurwat,* Vern. ; *dher-umber, kala-umber, kharoti, bokria,* M. ; *dhédu mera,* Panch Mahals. Throughout the Konkan and North Kánara, often along the banks of rivers and in moist situations, common in the moist forests near Kárwár. Receptacles hispid, sometimes hypogeal, ridged at the top. Ripe fruit R.S.

F. glomerata, Roxb. Fl. Ind. III. 558 ; Fl. Br. I. 5. 535 ; Brandis For. Fl. 422 ; Bedd. Fl. Sylv. 224; *Covellia glomerata,* Dalz. & Gibs. Bomb. Fl. 243. *Umbur,* M. ; *rumadi, atti,* K. Common throughout the presidency, near villages along road sides and near streams and rivers. Fr. ripe throughout the year at different times. One of the principal shade trees in the Mysore coffee plantations.

ANTIARIS, Lesch.

Trees. Leaves alternate, stipulate. Flowers monoicous, axillary. Male flowers on the surface of a fleshy disc, surrounded by imbricate

bracts. Sepals 4, imbricate. Stamens 3-8; filaments short or 0;
anthers 2-celled, extrorse. Female involucre 1-flowered, urceolate,
many cleft at the apex. Perianth 0. Ovary 1-celled; ovule 1, pendulous from the apex of the cell; style terminal, bifid. Fruit
fleshy. Albumen 0.

A, toxicaria, Leschen. in Ann. Mus. Paris XVI. 478, t. 22; Fl. Br. I.
5. 537; *A. innoxia*, Brandis For. Fl. 427; *A. saccidora*, Dalz. & Gibs.
Bomb. Fl. 244; Bedd. Fl. Sylv. t. 307. *Ajjanpatte*, K.; *karrat*, *chaudkura*,
M.; *jassoond*, *chandul*, Vern. In the evergreen forests of the Konkan and
North Kanara, common near Yellápur. Fl. Sept., Oct. Fr. C.S. A very
large tree.

ARTOCARPUS, Forst.

Trees with milky juice. Leaves alternate, entire or divided.
Flowers monoicous, minute, on the outside of globose or oblong
receptacles, the males and females on separate heads. Male fl. Perianth of 2-4-segments, imbricate in bud. Stamen 1. Female fl. Perianth
tubular, entire. Style simple, usually exserted. Ovary 1-celled,
ovule 1, pendulous. Fruit a syncarpium consisting of the enlarged
perianths, each enclosing a small nut. Albumen 0.

> Fruit tubercled, 1-2-feet long, oblong or cylindric   *A. integrifolia.*
> Fruit spinous, size of a lemon, spines hispid   ...   *A. hirsuta.*
> Fruit smooth, globose, 2-3-inches in diameter   ...   *A. Lakoocha.*

A. hirsuta, Lamk. Encycl. III. 201; Fl. Br. I. 5. 541; Brandis For.
Fl. 426; Bedd. Fl. Sylv. t. 308; Dalz. & Gibs. Bomb. Fl. 244. *Hebbal-
sina*, K.; *ran* or *patphunnas*, M. Anjeli wood. In the evergreen forests
of the Konkan and North Kanara, yields a valuable timber. Fl. Jan., Feb.
Fr. ripe. May.

A. integrifolia. Linn. f. Suppl. 412; Fl. Br. 1. 5. 541; Brandis For. Fl.
425; Dalz. & Gibs. Bomb. Fl. 244; Bedd. Fl. Sylv. 219. *Halsina*, K.;
phunnas, M. Jack fruit tree. Cultivated near villages throughout the
Konkan and North Kanara, often planted along road sides, said to be indigenous in the forests of the western gháts. Fl. C.S. Fr. ripe R.S.
(June-Aug.)

A. Lakoocha, Roxb. Fl. Ind. III. 524; Fl. Br. I. 5. 543; Brandis.
For. Fl. 426. Dalz. & Gibs. Bomb. Fl. 244; Bedd. Fl. Sylv. 219-
Wotomba, *badhar*, M.; *wonte*, K.; *lowi*, Vern. Evergreen forests of the
Konkan and North Kánara, common in the forests near Yellápur. Fl.
Mch., Apl. Fr. July, Aug.

A. incisa, L.; Fl. Br. I. 5. 589. *Vilayti phunnas*, M. Bread fruit tree,
cultivated near the coasts of the Konkan and North Kánara.

CONOCEPHALUS, Blume.

Climbing shrubs. Leaves alternate, simple, sometimes 3-nerved.
Stipules connate. Flowers in axillary, cymose heads. Male fl. Perianth tubular, 4-lobed, valvate. Stamens 2-4, erect. Female fl. Perianth oblong or clavate, 4-lobed. Ovary included; style undivided;
ovule erect. Albumen scanty or 0.

C. concolor, Dalz. & Gibs. Bomb. Fl. 239 ; Fl. Br. I. 5. 546. Konkan at the Phoonda ghát. Dalz. This may not be a *Conocephalus* ; it is given as a doubtful species in the Fl. Br. I. Dalzell does not say it is a climber.

LAPORTEA, Gaud.

Shrubs or trees with stinging hairs. Leaves alternate, entire or toothed ; stipules opposite, free or connate. Flowers mon- or diœcious, in axillary panicles, flowers and fruit often reflexed. Male fl. Sepals 4-5. Stamens 4-5, inflexed in bud. Pistillode clavate. Female fl. Sepals 4 or 0. Ovary oblique ; style linear, ovule erect. Achene flattened. Albumen 0.

L. crenulata, Gaud. in Freyc. Voy. Bot. 498 ; Fl. Br. I. 5. 550 ; Brandis For. Fl. 404 ; Bedd. Fl. Sylv. t. 306. Konkan, Stocks, does not seem to have been met with by any subsequent collector.

BOEHMERIA, Jacq.

Shrubs or small trees. Leaves opposite and alternate, toothed, 3-nerved. Flowers dioicous or monoicous in axillary, sessile or panicled clusters, usually unisexual. Male fl. Perianth 4-lobed, segments valvate. **Stamens 4. Pistillode clavate. Female fl.** Perianth tubular, compressed or ventricose, with a narrow, 2-4-toothed mouth. Ovary 1-celled ; ovule 1, erect. Akene dry. Seeds albuminous.

Leaves rugose. Flowers in axillary clusters　...　*B. malabarica.*
Leaves long petioled. Flower-clusters in axillary panicles　...　...　...　...　...　*B. platyphylla.*

B. malabarica, Wedd. Monogr. 355 ; Fl. Br. I. 5. 575 ; *B. travancorica,* Bedd. Fl. Sylv. 225 (*B. ramiflora,* t. 27, f. 2). Throughout the Konkan and North Kánara in evergreen forests, common on the Siddápur ghats. F. Nov., Feb. Fr. H. S.

B. platyphylla, Don Prod. 60 ; Fl. Br. I. 5. 578 ; Brandis For. Fl. 403 ; *Splitgerbera scabrella,* Dalz. & Gibs. Bomb. Fl. 239. Throughout the presidency, common in the moist forests near the sea-coast of the Konkan and North Kánara. Fl. Apl., Sept. Fr. R.S.

VILLEBRUNEA, Gaud.

Trees. Leaves alternate 3-5-nerved. Flowers mono- or dioicous in spicate or panicled clusters. Male fl. Perianth 4-fid, valvate. Stamens 4. Pistillode woolly. Female fl. Perianth tubular, ventricose, adnate to the ovary, mouth toothed. Ovary 1-celled ; ovule erect, stigma linear. Fruit a free akene. Albumen scanty.

V. integrifolia, Gaud. Bot. Bonite Voy. t. 91 ; Fl. Br. I. 5. 589. *Oreocnide acuminata,* Kz. For. Fl. II. 427. *O. sylvatica,* Bedd. Fl. Sylv. 225. t. 26. f. 4. Higher ghats of the Konkan, in moist forests. Stocks.

DEBREGEASIA, Gaud.

Shrubs or trees. Leaves alternate, trinerved, often white tomentose beneath. Flowers mono- or dioicous, males clustered, the females in small heads. Male fl. Perianth 3-4-partite, segments valvate in bud. Stamens 4. Pistillode glabrous or woolly. Female fl. Perianth tubular dilated below, mouth 4-toothed, adnate to the 1-celled ovary. Ovule erect; stigma tufted. Fruit a collection of small akenes on a fleshy torus, often yellow. Perianth accrescent. Albumen copious.

D. volutina, Gaud. Bot. Bonite Voy. t. 90 ; Fl. Br. I. 5. 590 ; Brandis For. Fl. 405 ; *Morocarpus longifolius*, Bedd. Fl. Sylv. 226, t. 26, f. 5. *Conocephalus nivens*, Wight. Dalz. & Gibs. Bomb. Fl. 239. *Capsi, kurgul*, Vern. Konkan and North Kánara, common in evergreen forests. Fl. Nov., Dec. Fr. Dec. Feb.

ORDER 79. CASUARINEÆ.

Leafless trees or shrubs, branchlets cylindric, grooved and jointed, internodes with a ring of small scales (leaves). Flowers unisexual, males in terminal spikes ; female in ovoid heads, bracteate and 2-bracteolate. Male fl. Sepals 1-2, circumciss at the base. Stamen 1, inflexed in bud. Female fl. Ovary minute, 1-celled; style 2-fid, arms filiform, stigmatose to the base. Ovules 2, collateral. Fruit a small cone, formed of the hardened bracts and bracteoles, enclosing the winged akenes.

CASUARINA, Forst.

Characters of the order.

C. equisetifolia, Forst. Char. Gen. 103, f. 53 ; Fl. Br. I. 5. 598 ; Brandis For. Fl. 453 ; Bedd. Fl. Sylv. t. 226. Beef wood of Australia. *Sura*, M. Planted throughout the presidency : there are large plantations of this tree in North Kánara, near the sea-coast. Fl. Sept.-Oct. Fruit ripe June. Flowers dioicous.

ORDER 80. SALICINEÆ.

Trees or shrubs. Leaves alternate, stipulate. Flowers dioicous in lateral catkins similar in both sexes, scales spirally arranged, each bearing 1 flower in its axil. Perianth 0. Disc cup-shaped or irregularly formed. Stamens 2 or more. Ovary 1-celled, style short or 0. Ovules few or many on 2-4 placentas, erect, anatropous. Capsule 2-4-valved. Seeds few or many, each with a pencil of long silky hairs, growing from the funicle. Albumen 0.

SALIX, L.

Characters as above. Leaves short petioled. Stamens generally 2, long exserted. Disc of 1-2, separate glands. Capsule 2-valved, the valves usually rolling back, placentas near the base of the valves.

S. tetrasperma, Roxb. Fl. Ind. III. 573 ; Fl. Br. I. 5. 626 ; Brandis For. Fl. 462 ; Bedd. Fl. Sylv. t. 302 ; Dalz. & Gibs. Bomb. Fl. 220. *Bacha, b'lasa*, M. ; *wallunj*, Vern. Throughout the presidency, along streams and river banks. Fl. Oct. Nov. Leaves glaucous beneath.

S. babylonica, Linn.; Fl. Br. I. 5. 629 ; Brandis For. Fl. 465.   Cultivated at Belgaum and elsewhere throughout the presidency.

ORDER 81.   GNETACEÆ.

Climbing shrubs with jointed stems. Leaves opposite, broad or scale like. Flowers monœcious or diœcious, in axillary or terminal bracteate spikes. Male flowers. Sepals 2-4 or tubular. Anthers 2-3 sessile or on a column of the connate filaments. Female flower. An erect ovule terminating above in a long tubular prolongation of its coat, resembling a style, and enclosed in an undivided perianth. Seed dry or drupaceous.

> Nearly leafless shrubs ...   .   *Ephedra.*
> Shrubs with broad leaves   ...   .   *Gnetum.*

EPHEDRA Linn.

Shrubs or undershrubs with nodose stems, branches articulate. Leaves reduced to a membranous sheath with 2 opposite lobes. Flowers diœcious in sessile or pedunculate spikes, opposite, in pairs or whorled, bracteate. Male flower; perianth of 2 membranous, opposite sepals. Anthers 2-10 sessile or stipitate. Female fl., a naked ovule with the outer coat produced into a styliform tube.

E. vulgaris, Rich. Conif. 26 ; Fl. Br. I. 5. 640. Upper Sind !

E. peduncularis. Boiss. Fl. Orient. V. 717 ; Fl. Br. I. 5. 641 ; *E. Altc,* Brandis For. Fl. 501.   Plains of Sind.   Fl. Mch., Apl.   Fr. ripe May.

GNETUM, Linn.

Climbing shrubs with jointed branches. Leaves elliptic, petiolate, penninerved. Flowers monœcious, in the axils of cup-shaped bracts and mixed with articulate hairs. Male fl. monandrous, protruding from a thick clavate sheath, slits terminal. Female fl. Ovule ovoid, inner integument with a toothed or fimbriate mouth. Fruit an oblong drupe.

G. scandens, Roxb. Fl. Ind. III. 518 ; Fl. Br. I. 5. 642 ; Brandis For. Fl. 502 ; Dalz. & Gibs. Bomb. Fl. 246. *Kumbal, umbli.* Throughout the moist forests of the Konkan and North Kánara ; very common in the evergreen forests of the Supa gháts. Fl. Mch.-Apl. Fr. ripe June.

ORDER 82.   PALMÆ.

Trees or shrubs, erect or climbing. Leaves pinnately or palmately divided, segments linear or lanceolate, folded longitudinally with numerous parallel nerves. The segments of the palmatifid leaves are frequently bifid, those of the pinnate leaves entire or irregularly lobed. Petiole broad based, usually amplexicaul or sheathing. Inflorescence terminal or axillary of simple or panicled spikes, enclosed when young in usually more than one spathe. Each flower usually 3-bracteate. Flowers hermaphrodite, unisexual or polygamous. Perianth of 6 segments in two series, those of the fertile flower often persistent in fruit. Stamens 6, rarely 3 or more, anthers versatile. Ovary of 3 carpels, free or united ; stigmas 3, usually sessile, undivided ; ovules 1-2, erect in each cell. Fruit a 1-3-celled drupe or berry or 3 distinct drupes or berries, often 1-2 aborted ; pericarp smooth, rough or retrorsely scaled. Albumen ruminate or even, solid or hollow.

Erect shrubs or trees.   Fruit without scales.
Leaves pinnately divided.
Pinnæ linear lanceolate, finely acuminate, upper connato and
  2-cleft. Ripe fruit orange-yellow. Seed 1 inch   ...   ...   ARECA.
Pinnæ obliquely dentate at apex. Ripe fruit purple, seed 2 inches. ACTINORHYTIS.
Pinnæ linear, auricled at base unequally bilobed at apex ...   ...   ARENGA.
Leaves bipinnate ; pinnæ wedge-shaped, erose, toothed.   Fruit
  fleshy, 2-seeded   ...   .   CARYOTA.
Leaves pinnate ; pinnæ rigid, linear ; petiole spinous   ...   ...   PHŒNIX.
Leaves pinnate ; petioles with a fibrous base. Fruit very large.
  Albumen with a large cavity.   ...   .   COCOS.
Leaves fan-shaped.
Fruit globose.   Albumen horny.
Flowers bisexual. Embryo apical. A lofty palm   ..   ...   CORYPHA.
Flowers polygamous. Embryo dorsal. A branched tufted palm. NANNORHOPS.
Flowers dioicous. Fruit large, albumen cartilaginous   ...   ...   BORASSUS.
Climbing shrubs. Fruit with retrorse scales   ...   .   CALAMUS.

ARECA, Linn.

Stems simple, annulate. Leaves pinnate, unarmed. Spathes solitary. Spadices branched, lax, pendant. Flowers monoicous, on the same inflorescence, sessile. Male flowers many, minute. Sepals 3. Petals 3, valvate. Stamens 3 or 6, filaments short, anthers linear. Female fl., much larger than the males. Calyx of 3 sepals, imbricate. Petals 3, imbricate, tips valvate ; perianth **accrescent. Ovary 1-celled ; ovule erect. Fruit ovoid ; albumen ruminate, embryo basilar.**

A. Catechu, Linn. Sp. Pl. 1189; Fl. Br. I. 6. 405 ; Brandis For. Fl. 551; Dalz. & Gibs. Bomb. Fl. Suppl. 95. Supári or betel-nut palm. *Pung*, M. ; *adiki*, K. Cultivated throughout the presidency, but nowhere so abundantly as in the southern tálukas of North Kánara (Sirsi and Siddápur). Fl. R. S.   Fr. C.S.

ACTINORHYTIS, H. Wendl.

Stems stout, annulate, unarmed. Leaves pinnate. Spathes 2 caducous. Spadix shortly peduncled with pendulous, flexuous branches. Flowers monoicous on the same spadix, near the base two males with a female between, above solitary or twin, bracteate males. Male flowers many, minute. Sepals 3, imbricate. Petals 3, valvate. Stamens 24-30, in fascicles ; anthers linear, versatile. Female flowers much larger than the males. Perianth accrescent. Sepals 3, imbricate. Petals 3, imbricate, valvate at the tips. Ovary

1-celled, ovule 1, pendulous. Fruit large, ellipsoid ; pericarp fleshy fibrous ; endocarp crustaceous, albumen ruminate.

A. capparia, Wendl. et Drude in Linnæa XXXIX. 184 ; *Areca cocoides,* Griff. Palms. Brit. India. 150. t. 230. B. *liám supári,* K. Planted throughout North Kánara in supári gardens of the Haiga Brahamins, probably introduced along with the supári, said to be indigenous in the islands of the Malay Archipelago. Fl. C.S. Fr. ripe next C.S.

ARENGA, Labill.

Stems simple, erect, covered with the remains of the fibrous leaf-sheaths. Leaves pinnate, white beneath ; pinnæ 1-2-auricled at the base. Spathes many, basilar. Spadices interfoliar, large, panicled. Males and females usually solitary and on separate spadices. Male fl. Sepals 3, imbricate in bud. Petals 3, valvate. Stamens indefinite. Female fl. Perianth of the male, accrescent. Staminodes many or 0. Ovary 3-celled, cells 1-ovuled. Fruit obovoid, 2-3-seeded ; stigmas terminal. Albumen not ruminate.

A. Wightii, Griff. in Calc. Jour. Nat. Hist. V. 475 ; Fl. Br. I. 6. 422. On the Ankola táluka gháts of North Kánara, common on the Mushki ghát at about 1,500 feet elevation, in moist forests, gregarious but very local ; very common on the gháts near the Falls of Gairsoppah in evergreen forests. Male flowers strongly scented. Fl. C.S. Fr. R.S.

CARYOTA, Linn

Tall palms with the trunk naked or sheathed. Leaves bipinnate, pinnules wedge-shaped, erose-toothed. Spathes 3-5, incomplete, tubular. Spadices interfoliar, peduncled with many pendulous, slender branches. Flowers monoicous, a female between two males. Male fl. Sepals 3, imbricate in bud. Corolla 3-partite, valvate. Stamens numerous, connate at the base ; anthers linear, basifixed. Female fl. Sepals as in the male. Petals imbricate in bud. Ovary 3-celled ; stigma 3-lobed, ovules erect. Fruit globose, 1-2-seeded. Albumen horny, ruminate.

C. urens, Linn. Fl. Zeyl. 187 ; Fl. Br. I. 6. 422 ; Dalz & Gibs. Bomb. Fl. 278 ; Brandis For. Fl. 550. *Mhár* palm. *Birly mhár,* Vern. ; *baini,* K. ; *birli,* M. Common, in evergreen forests throughout the Konkan and North Kánara. Fl. Fr. throughout the year.

PHŒNIX, Linn.

Shrubby or tall palms. Stems covered with the bases of the petioles, or rarely annulate. Leaves pinnate, pinnæ entire, lowest transformed into long spines, petiole with a fibrous amplexicaul sheath. Spathe complete, coriaceous. Spadices branched, erect or drooping. Flowers dioicous, small, sessile, coriaceous. Male fl. Calyx cupular, 3-toothed. Petals 3, valvate. Stamens 6 or 3, rarely 9. Female

fl. Calyx of the male, accrescent.   Petals 3, imbricate ; staminodes 6,
or a 6-toothed cup.   Ovaries 3 ; stigmas sessile.   Drupe 1-seeded.
Albumen with a longitudinal furrow, horny, not ruminate.

> Large trees. Leaves with alternate and opposite rigid pinnæ.
> Fruit small, yellow ...   .   *P. sylvestris*
> Stems slender, short or 0. Pinnæ fascicled.   Fruit black. Stem 0.
> Spadix peduncle very short   ...   .   *P. acaulis.*
> Stems short, 8-10 feet, rather slender. Fruiting spadix long
> peduncled   ...   .   *P. humilis.*

P. **sylvestris,** Roxb. Fl. Ind. III. 787 ; Fl. Br. I. 6. 425 ; Dalz. &
Gibs. Bomb. Fl. 278 ; Brandis For. Fl. 554.   Wild date palm. *Shindi,
khajur,* Vern. ; *ichil,* K. Very common in moist ground throughout the
dry districts of the presidency, usually along the banks and in the beds of
streams and water-courses, certainly indigenous. Fl. Jany.-Feb. Fr. ripe
June.

P. **acaulis,** Buch. ex. Roxb. Fl. Ind. III. 783 ; Fl. Br. I. 6. 426 ;
Dalz. & Gibs. Bomb. Fl. 278 ; Brandis For. Fl. 555.   Common on the
gháts, Dalz. There is a bulbiform stemmed palm which I cannot distinguish from the next species, common on the Sirsi táluka gháts. Fl. C.S.
Fr. ripe May-June.

P. **humilis,** Royle. Ill. 394, 397, 399 ; Fl. Br. I. 6. 426 ; var. *pedunculata,* Becc. Fl. Br. I. 6. 427. Common on the gháts of North Kínara.
Fl. C.S. Fr. ripe May, June. . Leaflets made into mats.

COCOS, Linn.

Tall palms with annulate stems. Leaves pinnate, petioles
amplexicaul with a fibrous base. Spathe simple, woody. Spadix
paniculate. Flowers monoicous, sessile, bracteate, male flowers
numerous on the upper branches. Male fl. Calyx of 3 sepals.
Petals 3, valvate in bud. Stamens 6 ; anthers erect. Female fl.
Sepals 3, imbricate, with 2 bracts at the base. Petals 3, smaller
than the sepals. Ovary 3-celled, surrounded with 6 staminodes.
Fruit 1-seeded, mesocarp woody-fibrous. Albumen fleshy, oily, with
a central cavity.

C. **nucifera,** Linn. ; Roxb. Fl. Ind. III. 614 ; Brandis For. Fl. 556 :
Dalz. & Gibs. Bomb. Fl. 279. Cocoanut palm. *Narel,* Vern. *Tengina,* K.
Planted throughout the presidency, and cultivated extensively along the
sea-coast. Fl. throughout the year. Fr. ripe 9-10 months after
flowering.

CORYPHA, Linn.

Tall, stout, annulate palms, dying after flowering and fruiting.
Leaves large, orbicular, flabellately multifid ; petiole spinous.
Spathes many, tubular. Spadix an immense terminal panicle.
Flowers small, clustered, bisexual. Calyx cupular, 3-fid. Petals 3,
valvate in bud. Stamens 6, equal, anthers dorsifixed. Ovary
3-lobed, 3-celled. Drupe usually solitary. Albumen horny.

C. **umbraculifora,** Linn. Sp. Pl. Ed. II. 1657; Fl. Br. I. 6. 428; Dalz. & Gibs. Bomb. Fl. Suppl. 94; Brandis For. Fl. 549. *Talipat* palm. *Tali, shri-tali,* K. In the moist forests of the Kumta and Honávar tálukas of North Kánara, covering extensive areas near Gairsoppah and Yena, also on the Yellápur ghats, sometimes planted in gardens near the sea-coast. Flowers at about the age of 40 years and dies down afterwards. The horny globose seeds (*bajarbet*) are made into necklaces and buttons and exported to the Persian Gulf ports from North Kánara. Segments of the leaves are used for writing on.

NANNORHOPS, H. Wendl.

A gregarious, tufted palm with prostrate branching rhizomes or stems. Leaves flabellate, rigid, plicate, segments 2-fid. Spathes tubular, sheathing. Spadix interfoliar, much branched. Flowers polygamous. Calyx tubular, 3-lobed. Corolla 3-partite, segments valvate. Stamens in herm. Fl. 6, in male about 9. Ovary trigonous. Fruit an ovoid or subglobose berry, varying in size from $\frac{1}{4}$ to $\frac{1}{2}$ in. in diameter. Albumen horny with a central cavity.

N. **Ritchieana,** H. Wendle in Bot. Zeit. 1879, 148; *Chamærops Ritchieana,* Griff.; Brandis For. Fl. 547. *Pharra,* Sind. Sind **on** dry arid hills, usually stemless, sometimes, however, forms a stem 6-8 feet high: Fr. ripe H. S.

BORASSUS, Linn.

Lofty, simple-stemmed palms. Leaves palmately fan-shaped; petiole spinously serrate. Spathes several, incomplete. Spadix with a few branches. Flowers dioicous, bracts large sheathing coriaceous. Male fl. in cylindrical catkins. Calyx and corolla 3-divided. Stamens 6. Female fl. solitary within the scales of the catkin. Calyx of 3 sepals. Petals 6, imbricate. Staminodes 6, connate in a ring round the ovary. Ovary usually 3-celled; stigmas sessile. Drupe large, containing 2-4, obcordate, fibrous, pyrenes. Albumen horny, turning hollow.

B. **flabellifer,** Linn.; *B. flabelliformis,* Roxb. Fl. Ind. III. 790; Brandis For. Fl. 544; Dalz. & Gibs. Bomb. Fl. 278. Palmyra tree. *Tad, tamar,* M. Planted throughout the presidency and Sind. Fl. Mch. Fr. May.

CALAMUS, Linn.

Scandent, rarely erect palms. Leaves alternate, pinnatisect, rachis often produced into an armed flagellum; sheath armed, produced into a ligula, with or without an armed flagellum. Spathes tubular or open, persistent and passing into bracts and bracteoles. Spadices axillary, usually elongate, sometimes produced into a flagellum. Flowers polygamo-dioicous, solitary within the spathules. Male fl. Calyx cupular, 3-toothed. Petals 3, valvate. Stamens 6. Female fl. Calyx of male. Corolla tubular below, 3-fid, valvate.

Ovary incompletely 3-celled, covered with retrorse scales; stigmas 3. Fruit globose or ellipsoid, clothed with deflexed, polished scales. Seed smooth or pitted. Albumen equable or ruminate.

C. Thwaitesii, Bcc. Fl. Br. I. 6. 441. Common in the evergreen forests at the foot of the Nilkund ghát of North Kánara. In flower and fruit Feb., Mch. Scandent. This species or a closely allied one is common near the sea-coast at Marmagoa. Sir J. Hooker remarks on specimens of this plant sent to Kew in the Fl. Br. 1. 6. 445, under *C. pseudo-tenuis*, Bcc. There is a species common in North Kánara on the Ainshi ghát, also in the ravines rear Kadra, which was referred to *C. flagellum*, Griff. at Kew. It has, however, an equable albumen, deeply foveolate, surrounded with a brown spongy covering. The outside of the fruit is as in the description in the Fl. Br. I. 6. 439. Fl. Fr. C. & H. seasons.

· **C. pseudo-tenuis,** Bcc. MSS. Fl. Br. I. 6. 445. Common on the Sapa ghâts of North Kánara. Fl. Mch., Apl. Fr. July. The minute male flowers in short, decurved spikelets and the small, beaked, brown fruit are characteristic.

C. Rotang, Linn. Sp. Pl. Ed. 1. 325 ; Ed. 2. 463 ; Fl. Br. I. 6. 447 ; *C. Roxburghii*, Griff. Palms Brit. Ind. 55, t. 195. Deccan peninsula and Ceylon, Fl. Br. I. That this species is indigenous in the ghâts or elsewhere in the Bombay Presidency is doubtful. Brandis says western ghâts and valleys of the Sátpudás, but it is doubtful whether he ever saw specimens of *C. Rotang*, L., from either the Bombay ghâts or the Bombay Sátpudás. Dalzell and Gibson and Graham give Linnæus' name of *C. Rotang* to what the natives call 'Bet.' which includes several species. There are several species undescribed from North Kánara, and further research will no doubt bring to light a number of others. The genus *Calamus* is difficult, as it is not always easy to match the male flowers and fruit of the same species. The cane brakes are generally in remote places, and so profusely armed that they are with difficulty entered. The flowers of some species appear at long intervals and during the rainy season or just at the beginning of the monsoon.

ORDER **83. GRAMINEÆ.**

Herbs, rarely shrubs or trees. Leaves· alternate, distichous, consisting of a tubular split sheath and a narrow linear blade joined to the sheath by a petiole (in bamboos) ; sheath terminating in a scarious or fringed ligule. Flowers hermaphrodite or unisexual, arranged in distichous 1 or many flowered spikelets, usually with 1-2 empty glumes (bracts) at the base. The flowering axis bears 1 or more distichous glumes (flowering glumes). Each flowering glume bears in its axil the palea, a transparent 2-nerved or keeled bract. The flowering glume embraces the palea with its incurved edges. Between the flowering glume and the palea is situated the flower, consisting of 2, 3 small scales (lodicules) 3 or more free stamens and the superior ovary, crowned with 2 plumose stigmas.

Ovary 1-celled.   Fruit a 1-seeded caryopsis, pericarp adherent to the testa, and sometimes to the palea and flowering glume.   Albumen farinaceous.   Embryo at base of albumen small or minute.

Stamens 6.
Filaments free.   Lodicules 3, 2.   Style deciduous.
  Embryo conspicuous on surface of caryopsis ...   ... BAMBUSA.
Filaments free.   Lodicules 0.   Style with a persistent
  base.   Embryo not apparent on surface of caryopsis... DENDROCALAMUS.
Filaments united in a slender tube.   Caryopsis linear,
  oblong   ...   . OXYTENANTHERA.

BAMBUSA, Schreber.

Large bamboos, usually in compact clumps.   Spikelets 5 to many-flowered, usually clustered and spiked, forming a gigantic leafless panicle.   Empty glumes 2 or more, lower shortest, the upper similar to the flowering glumes.   Palea 2-keeled, usually fimbriate.   Lodicules 3, ciliate, membranous.   Stamens 6.   Caryopsis with a longitudinal furrow, often adhering to the palea and flowering glume.

Branches spinescent.   Leaves with longitudinal nerves
  distinct, transverse not so ...   . *B. arundinacea.*
Unarmed.   Leaves with distinct, transverse veins   ... *B. vulgaris.*

B. **arundinacea,** Retz. Munro in Trans. Linn. Soc. 26, 103. Bedd. Fl. Sylv. 231 ; Brandis For. Fl. 564 ; Dalz. & Gibs. Bomb. Fl. 299 ; Gamble Ind. Timbers, 428. *Bans, Vern.* ; *dongi, bidrgala,* K. The large thorny bamboo, common throughout the presidency, usually in deciduous forests.

B. **vulgaris,** Wendl. ; Munro l.c. 106 ; Dalz. & Gibs. Bomb. Fl. 299 ; Bedd. Fl. Sylv. 232 ; Brandis For. Fl. 568 ; Gamble Ind. Timbers, 428. *Kulluk, bamboo,* Vern.   Cultivated in Kolhápur, Poona, Sátára, &c., but not indigenous.   Stems with green and yellow stripes.

OXYTENANTHERA, Munro.

Erect or scandent bamboos. •Spikelets often elongate and curved, verticellate, few flowered.   Flowers 1-3, the terminal one or the last but one fertile, rachilla obsolete.   Flowering glume many nerved, spinose-mucronate.   Palea in the fertile flowers convex on the back, keels obsolete, in the other flowers bicarinate.   Lodicules none.   Stamens 6, monadelphous, anthers with a mucro or bristle or with a few hairs.   Style slender, divided at the apex into 2-3 long and plicate stigmas.   Caryopsis linear-oblong with a longitudinal furrow.

2 flowers to spikelet.   Style hairy.   Anthers acute, not
  apiculate ...   . *O. Stocksii.*
Spikelets slender, 1-flowered.   Style glabrous.   Anthers
  apiculate ...   . *O. monostigma.*

O. **monostigma,** Bedd. Fl. Sylv. 233 ; Gamble Ind. Timbers, 429. *Bambusa Ritcheyi,* Munro. *Choua,* K.   Common throughout the Konkan and North Kánara ghâts, usually as undergrowth in deciduous forests. Culms often as large as those of *Dendrocalamus*, cavity small. A soft bamboo, young stems covered with deciduous brown tomentum. Flowers frequently. at least clumps or single stems are often found in flower.

O. **Stocksii,** Munro ; Trans. Linn. Soc. 26 to 130 ; Bedd. Fl. Sylv. 233. *Konda,* Vern. Commonly cultivated along the coast, rare in the ghât forests of North Kánara. A strong bamboo. used for punting poles and for making native umbrellas, Flowering culms are frequently found.

DENDROCALAMUS, Nees.

Unarmed bamboos. Inflorescence paniculate, spikelets congested in heads. Characters of *Bambusa.* Lodicules 0. Ovary hairy, style long filiform, undivided or 2-3-fid at the apex, base persistent. Caryopsis with a thick pericarp. Embryo not conspicuous on surface.

D. **strictus,** Nees, Munro Trans. Linn. Soc. 26 ; 147 ; Brandis For. Fl. 569 ; Bedd. Fl. Sylv. 235. ; Gamble. Ind. Timb., 430. Male bamboo. *Shib, busa, udha, medur, mace, mandgay,* Vern. Throughout the presidency, usually in deciduous forests—the common unarmed bamboo. I am unacquainted with any species of the genus *Arundinaria* from the Bombay Presidency, although there may be one or two species found on the higher Konkan ghâts. I have, however, a species of bamboo which Mr. Gamble says is *Teinostachyum Wightii.* It has thin culms ½ to ¾ in. in diameter, hollow, and forms large dense clumps along the margins of streams and nálás, common on the Supa ghâts, also in the Kárwár táluka and between Nilkund and Gairsoppah, in the evergreen forests of North Kánara. Flowers not seen. *Hooda,* M ; *woutenulgi,* K. This bamboo is much used in the construction of temporary bridges over the streams and nálás of the ghâts, during the south-west monsoon. Leaves large, 6-10 in. by 1½ to 2 in. ; ligula bearded. It is sometimes slender and semiscandent, the tips bending over like carriage whips, when without support.

NATIVE AND ENGLISH NAMES.

ww.ingramcontent.com/pod-product-compliance
ghtning Source LLC
ambersburg PA
HW020856270326
928CB00006B/736